T0211008

Linear Synchronous Motors

Transportation and Automation Systems

Second Edition

The ELECTRIC POWER ENGINEERING Series
Series Editor Leo L. Grigsby

Published Titles

Linear Synchronous Motors: Transportation and Automation Systems, Second Edition
Jacek Gieras, Zbigniew J. Piech, and Bronislaw Tomczuk

The Induction Machines Design Handbook, Second Edition
Ion Boldea and Syed Nasar

Computational Methods for Electric Power Systems, Second Edition
Mariesa L. Crow

Electric Energy Systems: Analysis and Operation
Antonio Gómez-Expósito, Antonio J. Conejo, and Claudio Cañizares

Distribution System Modeling and Analysis, Second Edition
William H. Kersting

Electric Machines
Charles A. Gross

Harmonics and Power Systems
Francisco C. De La Rosa

Electric Drives, Second Edition
Ion Boldea and Syed Nasar

Power System Operations and Electricity Markets
Fred I. Denny and David E. Dismukes

Power Quality
C. Sankaran

Electromechanical Systems, Electric Machines, and Applied Mechatronics
Sergey E. Lyshevski

Electrical Energy Systems, Second Edition
Mohamed E. El-Hawary

Electric Power Substations Engineering
John D. McDonald

Electric Power Transformer Engineering
James H. Harlow

Electric Power Distribution Handbook
Tom Short

Linear Synchronous Motors

Transportation and Automation Systems

Second Edition

Jacek F. Gieras
Zbigniew J. Piech
Bronisław Z. Tomczuk

CRC Press
Taylor & Francis Group
Boca Raton London New York

CRC Press is an imprint of the
Taylor & Francis Group, an **informa** business

CRC Press
Taylor & Francis Group
6000 Broken Sound Parkway NW, Suite 300
Boca Raton, FL 33487-2742

© 2012 by Taylor & Francis Group, LLC
CRC Press is an imprint of Taylor & Francis Group, an Informa business

No claim to original U.S. Government works

Printed in the United States of America on acid-free paper
Version Date: 20110621

International Standard Book Number: 978-1-13-807205-3 (paperback)

This book contains information obtained from authentic and highly regarded sources. Reasonable efforts have been made to publish reliable data and information, but the author and publisher cannot assume responsibility for the validity of all materials or the consequences of their use. The authors and publishers have attempted to trace the copyright holders of all material reproduced in this publication and apologize to copyright holders if permission to publish in this form has not been obtained. If any copyright material has not been acknowledged please write and let us know so we may rectify in any future reprint.

Except as permitted under U.S. Copyright Law, no part of this book may be reprinted, reproduced, transmitted, or utilized in any form by any electronic, mechanical, or other means, now known or hereafter invented, including photocopying, microfilming, and recording, or in any information storage or retrieval system, without written permission from the publishers.

For permission to photocopy or use material electronically from this work, please access www.copyright.com (http://www.copyright.com/) or contact the Copyright Clearance Center, Inc. (CCC), 222 Rosewood Drive, Danvers, MA 01923, 978-750-8400. CCC is a not-for-profit organization that provides licenses and registration for a variety of users. For organizations that have been granted a photocopy license by the CCC, a separate system of payment has been arranged.

Trademark Notice: Product or corporate names may be trademarks or registered trademarks, and are used only for identification and explanation without intent to infringe.

Visit the Taylor & Francis Web site at
http://www.taylorandfrancis.com

and the CRC Press Web site at
http://www.crcpress.com

Contents

Preface to the 2nd Edition

Twelve years have passed since the publication of the first edition of this book in 1999.

The growth in demand for linear motors is principally driven by the replacement of traditional mechanical (ball screws, gear trains, cams), hydraulic, or pneumatic linear motion systems in manufacturing processes, machining, material handling, and positioning with direct electromechanical drives. The linear motor market heavily depends on the semiconductor industry (applications) and permanent magnet industry (linear motors manufacture). A recent market study[1] shows that linear motors have impacted the linear motion market less than expected. The main obstacle that makes companies reluctant to replace mechanical, hydraulic, and pneumatic actuators with linear electric motors is the higher initial cost of installation of direct linear motor drives as compared to traditional mechanical drives.

The North American market for linear motors totaled only US$ 40 million in 1999 and grew by over 20% annually to reach the size of US$ 95 million in 2004. In the same year, the European linear motor market reached the size of US$ 114 million [67]. These numbers do not include special applications and large linear motors (rollercoasters, people movers, military applications). In 2007, the worldwide linear motor component market was estimated at about US$ 230 million, and the worldwide linear motor system market at about US$ 400 million [42]. The worldwide linear motor markets estimated growth between 2007 and 2009 was over 10% for components and over 15% for systems. Linear motors sales fell by close to 50% in 2009 due to global economic recession. The first signs of recovery were visible in the 2nd quarter of 2010. In the future, the main market players will be those manufacturers who can offer complete direct-drive systems.

---

[1] IMS Research, Wellingborough, UK

All the above numbers are very small in comparison with the world market for standard rotary electric motors. The worldwide market for electric motors grew from US\$ 12.5 billion in 2000 to US\$ 19.1 billion in 2005 and is predicted to reach US\$ 39.1 billion by the year 2015.

In comparison with the 1st edition of this book, the 2nd edition has been thoroughly revised and expanded, Chapter 4, Chapter 5, Appendix D, and List of Patents have been added, and at the end of each chapter, examples of calculations or mathematical models have been added.

The authors hope that the improved and updated new edition of *Linear Synchronous Motors* will find broad readerships including engineers, researchers, scientists, students, and all enthusiasts of linear motors and direct drives.

Prof. Jacek F. Gieras, FIEEE
e-mail: jgieras@ieee.org

Connecticut, June 2011

About the Authors

Jacek F. Gieras graduated in 1971 from the Technical University of Lodz, Poland, with distinction. He received his PhD degree in Electrical Engineering (Electrical Machines) in 1975, and the Dr habil. degree (corresponding to DSc), also in Electrical Engineering, in 1980 from the University of Technology, Poznan, Poland. His research area is Electrical Machines, Drives, Electromagnetics, Power Systems, and Railway Engineering. From 1971 to 1998 he pursued his academic career at several universities worldwide, including Poland (Poznan University of Technology and University of Technology and Life Sciences, Bydgoszcz), Canada (Queens University, Kingston, Ontario), Jordan (Jordan University of Sciences and Technology, Irbid) and South Africa (University of Cape Town). He was also a Central Japan Railway Company Visiting Professor at the University of Tokyo (Endowed Chair in Transportation Systems Engineering); Guest Professor at Chungbuk National University, Cheongju, South Korea; and Guest Professor at University of Rome La Sapienza, Italy. In 1987 he was promoted to the rank of Full Professor (life title given by the President of the Republic of Poland). Since 1998 he has been affiliated with United Technologies Corporation, USA, most recently with Hamilton Sundstrand[2] Applied Research. In 2007 he also became Faculty Member (Full Professor) of the University of Technology and Life Sciences in Bydgoszcz, Poland.

Prof. Gieras authored and coauthored 11 books, over 250 scientific and technical papers, and holds 30 patents. His most important books are *Linear Induction Motors*, Oxford University Press, 1994, UK, *Permanent Magnet Motors Technology: Design and Applications*, Marcel Dekker, New York, 1996, 2nd edition 2002, 3rd edition 2010 (Taylor & Francis), *Linear Synchronous Motors: Transportation and Automation Systems*, CRC Press, Boca Raton, Florida, 1999 (coauthor Z.J. Piech), *Axial Flux Permanent Magnet Brushless*

---

[2] United Technologies.

Machines, Springer-Kluwer Academic Publishers, Boston-Dordrecht-London, 2004, 2nd edition 2008 (coauthors R.Wang and M.J. Kamper), *Noise of Polyphase Electric Motors*, CRC Press – Taylor & Francis, 2005 (coauthors C. Wang and J.C. Lai), and *Advancements in Electric Machines*, Springer, Dordrecht-London-New York, 2008. He is a Fellow of IEEE, Hamilton Sundstrand Fellow (United Technologies Corporation, USA), Full Member of the International Academy of Electrical Engineering, and member of Steering Committees of numerous international conferences.

Zbigniew (Jerry) Piech graduated in 1975 from the Technical University of Wroclaw, Poland. He received his PhD degree in Electrical Engineering in 1981, also from the Technical University of Wroclaw. From 1975 to 1989 he was an academic staff member of the Department of Electrical Engineering at the University of Technology and Life Sciences, Bydgoszcz, Poland. In the 1990s, during employment at the United Technologies Research Center, East Hartford, CT, USA, his work has focused on linear motors for roped and ropeless elevators and magnetic bearings for aerospace applications. In 1997-1998 he served as Research and Development Director at Anorad Corporation, Hauppage, NY, one of the leading US companies in linear motor system technology. He returned to United Technologies Corporation in 1998 and joined Otis Engineering[3], Farmington, CT. His major contributions in the area of electric machines and power systems include development of a series of permanent magnet brushless motors for Otis elevators Gen2®with flat ropes, linear-motor-driven elevator door openers, electromagnetic brakes, and special actuators. Over the past 10 years he had a leading role in establishing permanent magnet motor technology as the standard for Otis Elevator hoist machines.

Dr Piech authored and coauthored 20 publications and holds 14 patents in the area of electric machine design and controls. He is Otis Fellow (United Technologies Corporation, USA) and member of Connecticut Academy of Sciences and Engineering (CASE).

Bronislaw Z. Tomczuk graduated in 1977 from the Opole University of Technology, Poland, with honors. He received his PhD and Dr habil. (corresponding to DSc) degrees from the Technical University of Lodz, Poland, in 1985 and 1995, respectively. Since 1978, he has been on the academic staff of the Technical University of Opole, Poland. In 2007 he was promoted to the rank of Full Professor (life title given by the President of the Republic of Poland). At present, he is a Full Professor in the Department of Electrical Engineering and Computer Science and the head of the Group of Industrial Electrical Engineering at the Opole University of Technology. His area of interests is 3D mathematical modeling of electromagnetic fields using numerical methods and its applications to CAD of transformers, linear motors, actuators, magnetic bearings, and other electromagnetic devices. Since the 1990s he

---

[3] United Technologies.

has been engaged in research collaboration with Polish industry. Since 2003 he has been a leader of three large-scale national research projects.

Prof. Tomczuk is author or coauthor of 5 books, over 170 scientific and technical articles, and holds 10 patents. For his research, he has obtained many professional distinctions and awards, including Gold Cross of Merit from the President of Poland, and Medal of National Committee of Education. He has been a member of Compumag Society (UK) since 2001, Association of Polish Electrical Engineers since 1977, and Polish Society of Theoretical and Applied Electrical Engineering since 1980.

1

Topologies and Selection

1.1 Definitions, Geometry, and Thrust Generation

Linear electric motors can drive a linear motion load without intermediate gears, screws, or crank shafts. A *linear synchronous motor* (LSM) is a linear motor in which the mechanical motion is in synchronism with the magnetic field, i.e., the mechanical speed is the same as the speed of the traveling magnetic field. The thrust (propulsion force) can be generated as an action of

- traveling magnetic field produced by a polyphase winding and an array of magnetic poles N, S,...,N, S or a variable reluctance ferromagnetic rail (LSMs with a.c. armature windings);
- magnetic field produced by electronically switched d.c. windings and an array of magnetic poles N, S,...,N, S or variable reluctance ferromagnetic rail (linear stepping or switched reluctance motors).

The part producing the traveling magnetic field is called the *armature* or *forcer*. The part that provides the d.c. magnetic flux or variable reluctance is called the *field excitation system* (if the excitation system exists) or *salient-pole rail, reaction rail*, or *variable reluctance platen*. The terms *primary* and *secondary* should rather be avoided, as they are only justified for linear induction motors (LIM) [63] or transformers. The operation of an LSM does not depend on, which part is movable and which one is stationary.

Traditionally, a.c. polyphase synchronous motors are motors with d.c. electromagnetic excitation, the propulsion force of which has two components: (1) due to the traveling magnetic field and d.c. current magnetic flux (synchronous component) and (2) due to the traveling magnetic field and variable reluctance in d- and q-axis (reluctance component). Replacement of d.c. electromagnets with permanent magnets (PMs) is common, except for LSMs for magnetically levitated vehicles. PM brushless LSMs can be divided into two groups:

- PM LSMs in which the input current waveforms are sinusoidal and produce a traveling magnetic field;

- PM d.c. linear brushless motors (LBMs) with position feedback, in which the input rectangular or trapezoidal current waveforms are precisely synchronized with the speed and position of the moving part.

Construction of magnetic and electric circuits of LSMs belonging to both groups is the same. LSMs can be designed as flat motors (Fig. 1.1) or tubular motors (Fig. 1.2). In d.c. brushless motors, the information about the position of the moving part is usually provided by an absolute position sensor. This control scheme corresponds to an *electronic commutation*, functionally equivalent to the mechanical commutation in d.c. commutator motors. Therefore, motors with square (trapezoidal) current waveforms are called *d.c. brushless motors*.

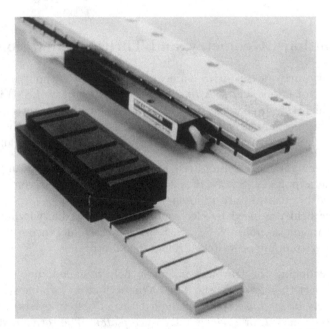

Fig. 1.1. Flat three-phase PM linear motors. Photo courtesy of Kollmorgen, Radford, VA, USA.

Instead of d.c. or PM excitation, the difference between the d- and q-axis reluctances and the traveling magnetic field can generate the reluctance component of the thrust. Such a motor is called the a.c. *variable reluctance* LSM. Different reluctances in the d- and q-axis can be created by making salient ferromagnetic poles using ferromagnetic and nonferromagnetic materials or using anisotropic ferromagnetic materials. The operation of LBMs can be regarded as a special case of the operation of LSMs.

In the case of LSMs operating on the principle of the traveling magnetic field, the speed v of the moving part

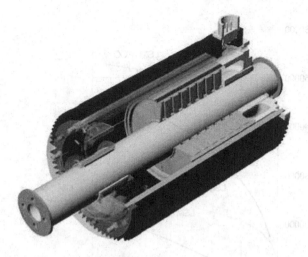

Fig. 1.2. Tubular PM LSM. Moving rod (reaction rail) contains circular PMs [192]

$$v = v_s = 2f\tau = \frac{\omega}{\pi}\tau \qquad (1.1)$$

is equal to the *synchronous speed* v_s of the traveling magnetic field and depends only on the input frequency f (angular input frequency $\omega = 2\pi f$) and pole pitch τ. It does not depend on the number of poles $2p$.

As for any other linear-motion electric machine, the useful force (thrust) F_x is directly proportional to the output power P_{out} and inversely proportional to the speed $v = v_s$, i.e.,

$$F_x = \frac{P_{out}}{v_s} \qquad (1.2)$$

Direct electromechanical drives with LSMs for factory automation systems can achieve speeds exceeding 600 m/min = 36 km/h and acceleration of up to 360 m/s^2 [67]. The thrust density, i.e., thrust per active surface $2p\tau L_i$

$$f_x = \frac{F_x}{2p\tau L_i} \text{ N/m}^2 \qquad (1.3)$$

of LSMs is higher than that of LIMs (Fig. 1.3).

The polyphase (usually three-phase) armature winding can be distributed in slots, made in the form of concentrated-parameter coils or made as a coreless (air cored) winding layer. PMs are the most popular field excitation systems for short traveling distances (less than 10 m), e.g., factory transportation or automation systems. A long PM rail would be expensive. Electromagnetic excitation is used in high-speed passenger transportation systems operating

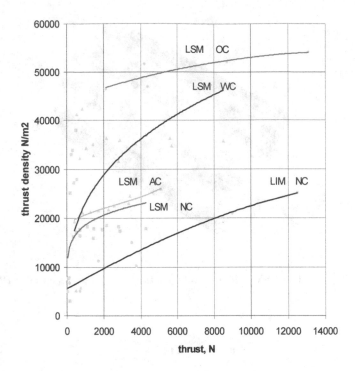

Fig. 1.3. Comparison of thrust density for single-sided LIMs and LSMs: AC — air cooling, NC — natural cooling, OC — oil cooling, WC — water cooling [67].

on the principle of magnetic levitation (maglev). The German system, *Transrapid*, uses vehicle-mounted steel core excitation electromagnets and stationary slotted armatures. Japanese MLX001 test train sets use onboard superconducting (SC) air-cored electromagnets and a stationary three-phase air-cored armature winding distributed along the guideway (Yamanashi Maglev Test Line).

A *linear stepping motor* has a concentrated armature winding wound on salient poles and PM excitation rail or variable reluctance platen (Fig. 1.4). The thrust is generated as an action of the armature magnetic flux and PM flux (active platen), or the armature magnetic flux and salient ferromagnetic poles (variable reluctance platen). Stepping motors have no position feedback.

The topology of a linear switched reluctance motor is similar to that of a stepping motor with variable reluctance platen. In addition, it is equipped with position sensors. The *turn-on* and *turn-off* instant of the input current is synchronized with the position of the moving part. The thrust is very sensitive to the turn-on and turn-off instant.

Fig. 1.4. PM linear stepping motors. Photo courtesy of Tokyo Aircraft Instrument, Co., Ltd., Japan.

In the case of a linear stepping or linear switched reluctance motor, the speed v of the moving part is

$$v = v_s = f_{sw}\tau \tag{1.4}$$

where f_{sw} is the fundamental switching frequency in one armature phase winding, and τ is the pole pitch of the reaction rail. For a rotary stepping or switched reluctance motor $f_{sw} = 2p_r n$, where $2p_r$ is the number of rotor poles and n is rotational speed in rev/s.

1.2 Linear Synchronous Motor Topologies

LSMs can be classified according to whether they are

- flat (planar) or tubular (cylindrical);
- single-sided or double-sided;
- slotted or slotless;
- iron cored or air cored;
- transverse flux or longitudinal flux.

The above topologies are possible for nearly all types of excitation systems. LSMs operating on the principle of the traveling magnetic field can have the following excitation systems:

- PMs in the reaction rail
- PMs in the armature (passive reaction rail)
- Electromagnetic excitation system (with winding)
- SC excitation system
- Passive reaction rail with saliency and neither PMs nor windings (variable reluctance motors)

LSMs with electronically switched d.c. armature windings are designed either as linear stepping motors or linear switched reluctance motors.

1.2.1 Permanent Magnet Motors with Active Reaction Rail

Fig. 1.5a shows a single-sided flat LSM with the armature winding located in slots and surface PMs. Fig. 1.5b shows a similar motor with buried-type PMs. In surface arrangement of PMs, the yoke (back iron) of the reaction rail is ferromagnetic, and PMs are magnetized in the normal direction (perpendicular to the active surface). Buried PMs are magnetized in the direction of the traveling magnetic field, and the yoke is nonferromagnetic, e.g., made of aluminum. Otherwise, the bottom leakage flux would be greater than the linkage flux, as shown in Fig. 1.6. The same effect occurs in buried type PM rotors of rotary machines in which the shaft must also be nonferromagnetic [70].

The so-called *Halbach array* of PMs also does not require any ferromagnetic yoke and excites stronger magnetic flux density and closer to the sinusoids than a conventional PM array [79]. The key concept of the Halbach array is that the magnetization vector should rotate as a function of distance along the array (Fig. 1.7).

It is recommended that be furnished a PM LSM with a *damper*. A rotary synchronous motor has a cage damper winding embedded in pole shoe slots. When the speed is different from the synchronous speed, electric currents are induced in damper circuits. The action of the armature magnetic field and damper currents allows for asynchronous starting, damps the oscillations,

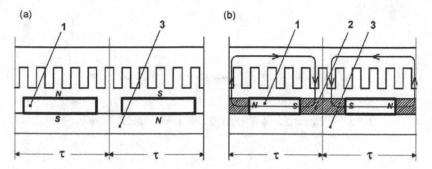

Fig. 1.5. Single sided flat PM LSMs with slotted armature core and (a) surface PMs, (b) buried PMs. 1 — PM, 2 — mild steel pole, 3 — yoke.

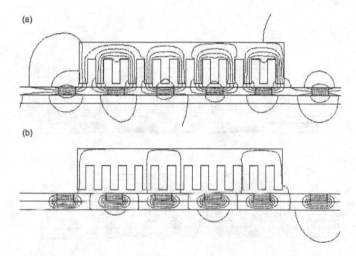

Fig. 1.6. Magnetic flux distribution in the longitudinal sections of buried-type PM LSMs: (a) nonferromagnetic yoke, (b) ferromagnetic yoke (back iron).

and helps to return to synchronous operation when the speed decreases or increases. Also, a damper circuit reduces the backward-traveling magnetic field. It would be rather difficult to furnish PMs with a cage winding so that the damper of PM LSMs has the form of an aluminum cover (Fig. 1.8a) or solid steel pole shoes (Fig. 1.8b). In addition, steel pole shoes or aluminum cover (shield) can protect brittle PMs against mechanical damage.

The *detent force*, i.e., attractive force between PMs and the armature ferromagnetic teeth, force ripple and some higher-space harmonics, can be reduced with the aid of skewed assembly of PMs. Skewed PMs can be arranged in one row (Fig. 1.9a), two rows (Fig. 1.9b), or even more rows.

Specification data of flat, single-sided PM LBMs manufactured by Anorad are shown in Table 1.1 [12], and motors manufactured by Kollmorgen are

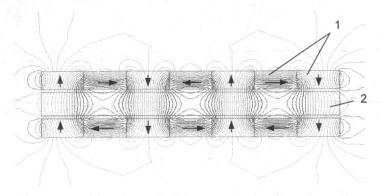

Fig. 1.7. Double-sided LSM with Halbach array of PMs. 1 — PMs, 2 — coreless armature winding.

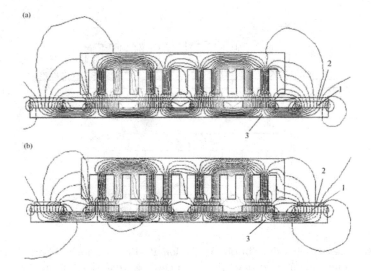

Fig. 1.8. Dampers of surface-type PM LSMs: (a) aluminum cover (shield), (b) solid steel pole shoes. 1 — PM, 2 — damper, 3 — yoke.

shown in Table 1.2 [112]. The temperature 25°C, 125°C, or 130°C for the thrust, current, resistance, and power loss is the temperature of the armature winding.

The EMF constant k_E in Tables 1.1 and 1.2 for sinusoidal operation is defined according to the equation expressing the EMF (induced voltage) excited by PMs without the armature reaction, i.e.,

$$E_f = c_E \Phi_f v_s = k_E v_s \tag{1.5}$$

Table 1.1. Flat three-phase, single-sided PM LBMs with natural cooling systems manufactured by Anorad, Hauppage, NY, USA

Parameter	LCD-T-1	LCD-T-2-P	LCD-T-3-P	LCD-T-4-P
Continuous thrust at 25°C, N	163	245	327	490
Continuous current at 25°C, A	4.2	6.3	8.5	12.7
Continuous thrust at 125°C, N	139	208	277	416
Continuous current at 125°C, A	3.6	5.4	7.2	10.8
Peak thrust (0.25 s), N	303	455	606	909
Peak current (0.25 s), A	9.2	13.8	18.4	27.6
Peak force (1.0 s), N	248	373	497	745
Peak current (1.0 s), A	7.3	11.0	14.7	22.0
Continuous power losses at 125°C, W	58	87	115	173
Armature constant, k_E, Vs/m	12.9			
Thrust constant (three phases), k_F N/A	38.6			
Resistance per phase at 25°C, Ω	3.2	2.2	1.6	1.1
Inductance, mH	14.3	9.5	7.1	4.8
PM pole pitch, mm	23.45			
Maximum winding temperature, °C	125			
Armature assembly mass, kg	1.8	2.4	3.6	4.8
PM assembly mass, kg/m	6.4			
Normal attractive force, N	1036	1555	2073	3109

Table 1.2. Flat three-phase, single-sided PM LBMs with natural cooling systems manufactured by Kollmorgen, Radford, VA, USA

Parameter	IC11-030	IC11-050	IC11-100	IC11-200
Continuous thrust at 130°C, N	150	275	600	1260
Continuous current at 130°C, A	4.0	4.4	4.8	5.0
Peak thrust, N	300	500	1000	2000
Peak current, A	7.9	7.9	7.9	7.9
Continuous power losses at 130°C, W	64	106	210	418
Armature constant, at 25°C, k_E, Vs/m	30.9	51.4	102.8	205.7
Thrust constant (three phases) at 25°C, k_F, N/A	37.8	62.9	125.9	251.9
Resistance, line-to-line, at 25°C, Ω	1.9	2.6	4.4	8.0
Inductance, line-to-line, mH	17.3	27.8	54.1	106.6
Electrical time constant, ms	8.9	10.5	12.3	13.4
Thermal resistance winding to external structure, °C/W	1.64	0.99	0.50	0.25
Maximum winding temperature, °C	130			
Armature assembly mass, kg	2.0	3.2	6.2	12.2
PM assembly mass, kg/m	5.5	7.6	12.8	26.9
Normal attractive force, N	1440	2430	4900	9850

Table 1.3. Slotted versus slotless LSMs

Quantity	Slotted LSM	Slotless LSM
Higher thrust density	x	
Higher efficiency in the lower speed range	x	
Higher efficiency in the higher speed range		x
Lower input current	x	
Less PM material	x	
Lower winding cost		x
Lower thrust pulsations		x
Lower acoustic noise		x

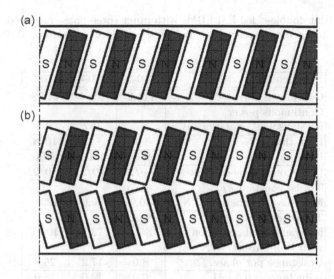

Fig. 1.9. Skewed PMs in flat LSMs: (a) one row, (b) two rows.

where Φ_f is the magnetic flux of the excitation system, and $k_E = c_E \Phi_f$. Thus, the armature constant k_E multiplied by the synchronous speed v_s gives the EMF E_f.

The thrust constant k_F in Tables 1.1 and 1.2 is defined according to the simplified equation for the electromagnetic thrust, i.e.,

$$F_{dx} = m_1 \frac{c_F}{2} \Phi_f I_a \cos \Psi = k_F I_a \cos \Psi \qquad (1.6)$$

for a sinusoidally excited LSM with equal reluctances in the d- and q-axis and for the angle between the armature current I_a and the q-axis $\Psi = 0^0$ ($\cos \Psi = 1$). Thus, the thrust constant $k_F = 0.5 m_1 c_F \Phi_f$ times the armature current I_a gives the thrust. Derivations of eqns (1.5) and (1.6) are given in Chapter 6.

In the case of six degrees of freedom (DOF) as, for example, in planar PM actuators [39, 99] eqn (1.6) takes the matrix form, i.e.,

$$
\begin{bmatrix}
F_{dx} \\
F_{dy} \\
F_{dz} \\
T_{dx} \\
T_{dy} \\
T_{dz}
\end{bmatrix}
=
\begin{bmatrix}
k_{Fx}(x,y,z,\phi) \\
k_{Fy}(x,y,z,\phi) \\
k_{Fz}(x,y,z,\phi) \\
k_{Tx}(x,y,z,\phi) \\
k_{Ty}(x,y,z,\phi) \\
k_{Tz}(x,y,z,\phi)
\end{bmatrix}
\times I_a \qquad (1.7)
$$

where k_{Fx}, k_{Fy}, k_{Fz} are the thrust constants, and k_{Tx}, k_{Ty}, k_{Tz} are the torque constants in the x, y, and z directions, repectively.

Double-sided, flat PM LSMs consist of two external armature systems and one internal excitation system (Fig. 1.10a), or one internal armature system

Table 1.4. Flat double-sided PM LBMs with inner three-phase air-cored series-coil armature winding manufactured by Trilogy Systems Corporation, Webster, TX, USA

Parameter	310-2	310-4	310-6
Continuous thrust, N	111.2	209.1	314.9
Continuous power for sinusoidal operation, W	87	152	230
Peak thrust, N	356	712	1068
Peak power, W	900	1800	2700
Peak/continuous current, A	10.0/2.8	10.0/2.6	10.0/2.6
Thrust constant k_F for sinusoidal operation, N/A	40.0	80.0	120.0
Thrust constant k_F for trapezoidal operation with Hall sensors, N/A	35.1	72.5	109.5
Resistance per phase, Ω	8.6	17.2	25.8
Inductance ± 0.5 mH	6.0	12.0	18.0
Heat dissipation constant for natural cooling, W/^{0}C	1.10	2.01	3.01
Heat dissipation constant for forced air cooling, W/$^{\circ}$C	1.30	2.40	3.55
Heat dissipation constant for liquid cooling, W/$^{\circ}$C	1.54	2.85	4.21
Number of poles	2	4	6
Coil length, mm	142.2	264.2	386.1
Coil mass, kg	0.55	1.03	1.53
Mass of PM excitation systems, kg/m	12.67 or 8.38		

and two external excitation systems (Fig. 1.10b). In the second case, a linear Gramme's armature winding can be used.

In *slotless motors* the primary winding is uniformly distributed on a smoooth armature core or does not have any armature core. Slotless PM LSMs are detent force free motors, provide lower torque ripple and, at high input frequency, can achieve higher efficiency than slotted LSMs. On the other hand, larger nonferromagnetic air gap requires more PM material, and the thrust density (thrust per mass or volume) is lower than that of slotted motors (Table 1.3). The input current is higher as synchronous reactances in the d- and q-axis can decrease to a low undesired value due to absence of teeth. Fig. 1.11a shows a single-sided flat slotless motor with armature core, and

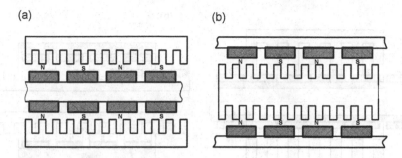

Fig. 1.10. Double-sided flat PM LSMs with: (a) two external armature systems, (b) one internal armature system.

Fig. 1.11b shows a double-sided slotless motor with inner air-cored armature winding (moving coil motor).

Table 1.4 contains performance specifications of double-sided PM LBMs with inner three-phase air-cored armature winding manufactured by Trilogy Systems Corporation, Webster, TX, USA [222]. Trilogy also manufactures motors with parallel wound coils as well as miniature motors and high-force motors (up to 9000 N continuous thrust).

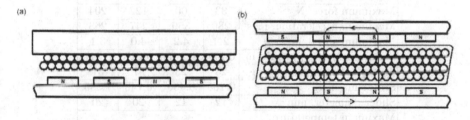

Fig. 1.11. Flat slotless PM LSMs: (a) single-sided with armature core, (b) double-sided with inner air-cored armature winding.

By rolling a flat LSM around the axis parallel to the direction of the traveling magnetic field, i.e., parallel to the direction of thrust, a tubular (cylindrical) LSM can be obtained (Fig. 1.12). A tubular PM LSM can also be designed as a double-sided motor or slotless motor.

Tubular single-sided LSMs LinMoT®[1] with movable internal PM excitation system (slider) and stationary external armature are manufactured by Sulzer Electronics AG, Zürich, Switzerland (Table 1.5). All active motor parts, bearings, position sensors, and electronics have been integrated into a rigid metal cylinder [125].

---

[1] LinMot®is a registered trademark of Sulzer Electronics AG, Zürich, Switzerland.

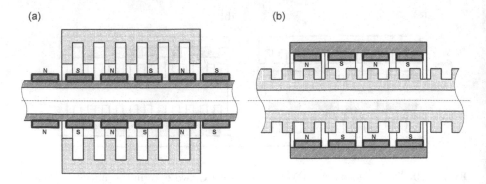

Fig. 1.12. Single-sided slotted tubular PM LSMs: (a) with external armature system, (b) with external excitation system.

Table 1.5. Data of tubular LSMs LinMot®manufactured by Sulzer Electronics AG, Zürich, Switzerland

Parameter	P01 23x80	P01 23X160	P01 37x120	P01 37x240
Number of phases	2			
Permanent magnets	NdFeB			
Maximum stroke, m	0.210	0.340	1.400	1.460
Maximum force, N	33	60	122	204
Maximum acceleration, m/s^2	280	350	247	268
Maximum speed, m/s	2.4	4.2	4.0	3.1
Stator (armature) length, m	0.177	0.257	0.227	0.347
Stator outer diameter, mm	23	23	37	37
Stator mass, kg	0.265	0.450	0.740	1.385
Slider diameter, mm	12	12	20	20
Maximum temperature of the armature winding, ^{0}C	90			

All the above-mentioned PM LSMs are motors with *longitudinal magnetic flux*, the lines of which lie in the plane parallel to the direction of the traveling magnetic field. LSMs can also be designed as *transverse magnetic flux* motors, in which the lines of magnetic flux are perpendicular to the direction of the traveling field. Fig. 1.13 shows a single-sided transverse flux LSM in which PMs are arranged in two rows. A pair of parallel PMs creates a two pole flux excitation system. A double-sided configuration of transverse flux motor is possible; however, it is complicated and expensive.

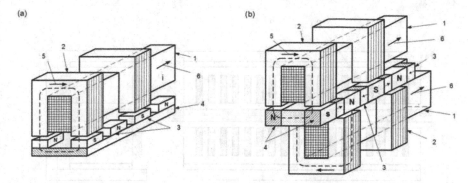

Fig. 1.13. Transverse flux PM LSM: (a) single-sided; (b) double-sided. 1 — armature winding, 2 — armature laminated core, 3 — PM, 4 — armature current, 5 — back ferromagnetic core, 6 — mild steel pole shoes, 7 — magnetic flux.

1.2.2 PM Motors with Passive Reaction Rail

The drawback of PM LSMs is the large amount of PM material that must be used to design the excitation system. Normally, expensive rare-earth PMs are requested. If a small PM LSM uses, say, 10 kg of NdFeB per 1 m of the reaction rail, and 1 kg of good-quality NdFeB costs US$ 130, the cost of the reaction rail without assembly amounts to US$ 1300 per 1 m. This price cannot be acceptable, e.g., in passenger transportation systems.

A cheaper solution is to apply the PM excitation system to the short armature that magnetizes the long reaction rail and creates magnetic poles in it. Such a linear motor is called the *homopolar* LSM.

The homopolar LSM as described in [59, 181] is a double sided a.c. linear motor that consists of two polyphase armature systems connected mechanically and magnetically by a ferromagnetic U-type yoke (Fig. 1.14). Each armature consists of a typical slotted linear motor stack with polyphase armature winding and PMs located between the stack and U-type yoke. Since the armature and excitation systems are combined together, the armature stack is oversized as compared with a conventional steel-cored LSM. The PMs can also be replaced by electromagnets [181, 186]. The variable reluctance reaction rail is passive. The saliency is created by using ferromagnetic (solid or laminated) cubes separated by a nonferromagnetic material. The reaction rail poles are magnetized by the armature PMs through the air gap. The traveling magnetic field of the polyphase armature winding and salient poles of the reaction rail produce the thrust. Such a homopolar LSM has been proposed for the propulsion of maglev trains of *Swissmetro* [181].

Further simplification of the double-sided configuration can be made to obtain a single-sided PM LSM shown in Fig. 1.15.

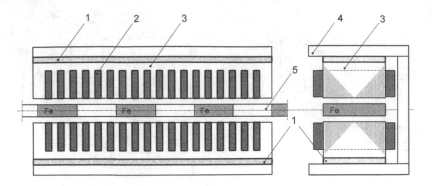

Fig. 1.14. Double-sided homopolar PM LSM with passive reaction rail. 1 — PM, 2 — armature winding, 3 — armature stack, 4 – yoke, 5 — reaction rail.

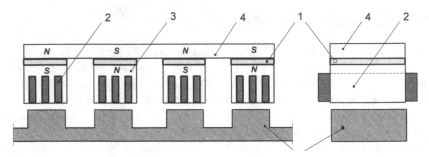

Fig. 1.15. Single-sided PM LSM with a passive reaction rail. 1 — PM, 2 — armature winding, 3 — armature stack, 4 — yoke, 5 — ferromagnetic reaction rail.

1.2.3 Motors with Electromagnetic Excitation

The electromagnetic excitation system of an LSM is similar to the salient pole rotor of a rotary synchronous motor. Fig. 1.16 shows a flat single-sided LSM with salient ferromagnetic poles and *d.c. field excitation winding*. The poles and pole shoes can be made of solid steel, laminated steel, or sintered powder. If the electromagnetic excitation system is integrated with the moving part, the d.c. current can be delivered with the aid of brushes and contact bars, inductive power transfer (IPT) systems [201], linear transformers, or linear brushless exciters.

1.2.4 Motors with Superconducting Excitation System

In large-power LSMs, the electromagnets with ferromagnetic core that produce the excitation flux can be replaced by coreless *superconducting* (SC) electromagnets. Since the magnetic flux density produced by the SC electromagnet is greater than the saturation magnetic flux density of the best

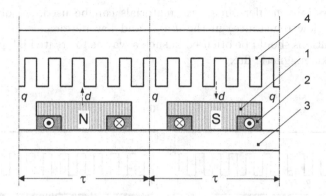

Fig. 1.16. Electromagnetic excitation system of a flat single-sided iron-cored LSM. 1 — salient pole, 2 — d.c. excitation winding, 3 — ferromagnetic rail (yoke), 4 — armature system.

laminated alloys ($B_{sat} \approx 2.4$ T for cobalt alloy), there is no need to use the armature ferromagnetic core. An LSM with SC field excitation system is a totally air-cored machine (Fig. 1.17).

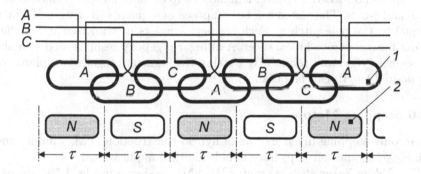

Fig. 1.17. Three-phase air-cored LSM with SC excitation system. 1 — armature coils, 2 — SC excitation coils.

1.2.5 Variable Reluctance Motors

The simplest construction of a *variable reluctance LSM* or *linear reluctance motor* (LRM) is that shown in Fig. 1.16 with d.c. excitation winding being removed. However, the thrust of such a motor would be low as the ratio of d-axis permeance to q-axis permeance is low. Better performance can be obtained when using *flux barriers* [183] or steel laminations [127]. To make flux barriers, any nonferromagnetic materials can be used. To obtain high permeance (low reluctance) in the d-axis and low permeance in the q-axis, steel laminations should be oriented in such a way as to create high permeance for the d-axis magnetic flux.

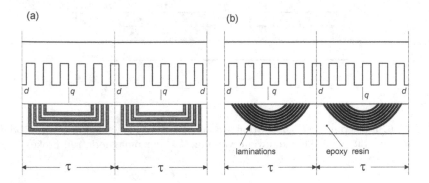

Fig. 1.18. Variable reluctance LSMs with (a) flux barriers, (b) steel laminations.

Fig. 1.18a shows a variable reluctance platen with flux barriers, and Fig. 1.18b shows how to arrange steel laminations to obtain different reluctances in the d- and q-axis. The platen can be composed of segments the length of which is equal to the pole pitch τ. Each segment consists of semicircular *lamellas* cut out from electrotechnical sheet. A filling, e.g., epoxy resin, is used to make the segment rigid and robust. By putting the segments together, a platen of any desired length can be obtained [81].

1.2.6 Stepping Motors

So far, only stepping linear motors of hybrid construction (PM, winding and variable reluctance air gap) have found practical applications.

The *hybrid linear stepping motor* (HLSM), as shown in Fig. 1.19, consists of two parts: the *forcer* (also called the *slider*) and the variable reluctance platen [40]. Both of them are evenly toothed and made of high-permeability steel. This is an early design of the HLSM, the so-called Sawyer linear motor [88]. The forcer is the moving part with two rare-earth magnets and two concentrated parameter windings. The tooth pitch of the forcer matches the tooth pitch on the platen. However, the tooth pitches on the forcer poles are spaced 1/4 or 1/2 pitch from one pole to the next. This arrangement allows for the PM flux to be controlled at any level between minimum and maximum by the winding so that the forcer and the platen line up at a maximum permeance position. The HLSM is fed with two-phase currents (90° out of phase), similarly as a rotary stepping motor. The forcer moves 1/4 tooth pitch per each full step.

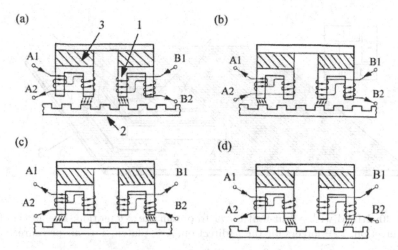

Fig. 1.19. Principle of operation of an HLSM: (a) initial position; (b) 1/4 tooth pitch displacement of the forcer; (c) 2/4 tooth pitch displacement; (d) 3/4 tooth pitch displacement. 1 — forcer, 2 — platen, 3 — PM.

There is a very small air gap between the two parts that is maintained by strong air flow produced by an air compressor [102]. The average air pressure is about 300 to 400 kpa and depends on how many phases are excited.

Table 1.6 shows specification data of HLSMs manufactured by Tokyo Aircraft Instrument Co., Ltd., Tokyo, Japan [122]. The *holding force* is the amount of external force required to break the forcer away from its rest position at rated current applied to the motor. The *step–to–step accuracy* is a measure of the maximum deviation from the true position in the length of each step. This value is different for full-step and microstepping drives. The *maximum start—stop speed* is the maximum speed that can be used when starting or stopping the motor without ramping that does not cause the motor to fall out of synchronism or lose steps. The *maximum speed* is the maximum linear speed that can be achieved without the motor stalling or falling out of synchronism. The *maximum load mass* is the maximum allowable mass applied to the forcer against the scale that does not result in mechanical damage. The *full-step resolution* is the position increment obtained when the currents are switched from one winding to the next winding. This is the typical resolution obtained with full-step drives and it is strictly a function of the motor construction. The *microstepping resolution* is the position increment obtained when the full-step resolution is divided electronically by proportioning the currents in the two windings (Chapter 6). This resolution is typically 10 to 250 times smaller than the full-step resolution [122].

HLSMs are regarded as an excellent solution to positioning systems that require a high accuracy and rapid acceleration. With a microprocessor controlled *microstepping mode* (Chapter 6), a smooth operation with standard

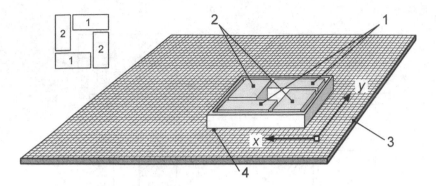

Fig. 1.20. HLSM with a four-unit forcer to obtain the x-y motion: 1 — forcers for the x-direction, 2 — forcers for the y–direction, 3 — platen, 4 — air pressure.

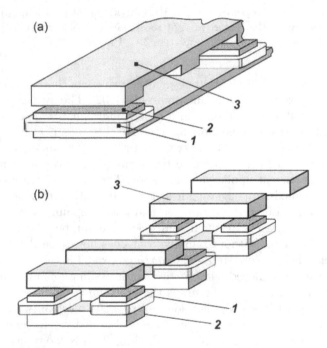

Fig. 1.21. Linear switched reluctance motor configurations: (a) longitudinal flux design, (b) transverse flux design. 1 — armature winding, 2 — armature stack, 3 — platen.

Table 1.6. Data of HLSMs manufactured by Tokyo Aircraft Instrument Co., Ltd., Tokyo, Japan.

Parameter	LP02-20A	LP04-20A	LP04-30A	LP60-20A
Driver	Bi-Polar Chopper			
Voltage, V	24 d.c.			
Resolution, mm	0.2	0.4	0.4	0.423
Holding Force, N	20	20	29.5	20
Step-to-step accuracy, mm	±0.03			
Comulative accuracy, mm	±0.2			
Maximum start-stop speed, mm/s	60	120	120	127
Maximum speed, mm/s	400	600	500	600
Maximum load mass, kg	3.0	3.0	5.0	3.0
Effective stroke, mm	330	300	360	310
Mass, kg	1.4	1.2	2.8	1.4

resolution of a few hundred steps/mm can be obtained. The advantages such as high efficiency, high throughput, mechanical simplicity, high reliability, precise open-loop operation, low inertia of the system, etc, have made these kind of motors more and more attractive in such applications as factory automation, high speed positioning, computer peripherals, facsimile machines, numerically controlled machine tools, automated medical equipment, automated laboratory equipment and welding robots. This motor is especially suitable for machine tools, printers, plotters and computer controlled material handling in which a high positioning accuracy and repeatability are the key problems.

When two or four forcers mounted at 90° and a special grooved platen ("waffle plate") are used, the x-y motion (two DOFs) in a single plane is obtained (Fig. 1.20). Specification data of the x-y HLSMs manufactured by Normag Northern Magnetics, Inc., Santa Clarita, CA, USA are given in Table 1.7 [161].

1.2.7 Switched Reluctance Motors

A *linear switched reluctance motor* has a doubly salient magnetic circuit with a polyphase winding on the armature. Longitudinal and transverse flux designs are shown in Fig. 1.21. A linear switched reluctance motor allows precise speed and position-controlled linear motion at low speeds and is not subject to design constraints (minimum speed limited by minimum feasible pole pitch) of linear a.c. motors [4].

Table 1.7. Data of $x - y$ HLSMs manufactured by Normag Northern Magnetics, Inc., Santa Clarita, CA, USA

Parameter	4XY0602-2-0	4XY2002-2-0	4XY2004-2-0	4XY2504-2-0
Number of forcer units per axis	1	1	2	2
Number of phases	2	2	2(4)	2(4)
Static thrust, N	13.3	40.0	98.0	133.0
Thrust at 1m/s, N	11.1	31.1	71.2	98.0
Normal attractive force, N	160.0	400.0	1440.0	1800.0
Resistance per phase, Ω	2.9	3.3	1.6	1.9
Inductance per phase, mH	1.5	4.0	2.0	2.3
Input phase current, A	2.0	2.0	4.0	4.0
Air gap, mm	0.02			
Maximum temperature, °C	110			
Mass, kg	3.2	0.72	2.0	1.5
Repeatability, mm	0.00254			
Resolution, mm	0.00254			
Bearing type	air			

1.3 Calculation of Forces

Neglecting the core loss and nonlinearity, the *energy stored in the magnetic field* is

$$W = \frac{1}{2}\Phi\mathcal{F} = \frac{1}{2}\Phi N i \qquad \text{J} \qquad (1.8)$$

where the magnetomotive force (MMF) $\mathcal{F} = Ni$. Introducing the reluctance $\Re = \mathcal{F}/\Phi$, the field energy is

$$W = \frac{1}{2}\Re\Phi^2 = \frac{1}{2}\frac{\mathcal{F}^2}{\Re} \qquad \text{J} \qquad (1.9)$$

The self-inductance

$$L = \frac{N\Phi}{i} \qquad (1.10)$$

is constant for $\Re = const.$ Hence,

$$W = \frac{1}{2}Li^2 \tag{1.11}$$

In a magnetic circuit with air gap $g > 0$, most of the MMF is expended on the air gap and most of the energy is stored in the air gap with its volume Ag where A is the cross section area of the air gap. Working in B and H units, the field energy per volume is

$$w = \frac{W}{Ag} = \frac{1}{2}B_g H = \frac{1}{2}\frac{B_g^2}{\mu_0} \quad \text{J/m}^3 \tag{1.12}$$

where B_g is the magnetic flux density in the air gap.

The magnetic quantities corresponding to electric quantities are listed in Table 1.8.

Table 1.8. Electric and corresponding magnetic quantities

Electric circuit	Unit	Magnetic circuit	Unit
Electric voltage, $V = \int_l E dl$	V	Magnetic voltage, $V_\mu = \int_l H dl$	A
EMF, E	V	MMF, $\mathcal{F}$	A
Current, I	A	Magnetic flux, Φ	Wb
Current density, J	A/m^2	Magnetic flux density, B	Wb/m^2=T
Resistance, R	Ω	Reluctance, $\Re$	1/H
Conductance, G	S	Permeance, G	H
Electric conductivity, ρ	S/m	Magnetic permeability, μ	H/m

The *force F_i associated with any linear motion* defined by a variable ξ_i of a device employing a magnetic field is given by

$$F_i = \frac{\partial W}{\partial \xi_i} \tag{1.13}$$

where W is the field energy in Joules according to eqn (1.8), F_i denotes the F_x, F_y, or F_z force component and ξ_i denotes the x, y, or z coordinate.

For a singly excited device

$$F_i = \frac{1}{2}\Phi^2\frac{d\Re}{d\xi_i} = \frac{1}{2}i^2\frac{dL}{d\xi_i} \tag{1.14}$$

where the magnetic flux $\Phi = const$, and electric current $i = const$.

Eqn (1.14) can be used to find the attractive force between two poles separated by an air gap $z = g$. Let us consider a linear electromagnetic actuator, electromagnet, or relay mechanism. The following assumptions are usually made: (a) leakage flux paths are neglected, (b) nonlinearities are neglected, and (c) all the field energy is stored in the air gap ($\mu_0\mu_r >> \mu_0$) where the magnetic permeability of free space $\mu_0 = 0.4\pi \times 10^{-6}$ H/m, and μ_r is the relative permeability. The volume of the air gap is Az, and the stored field energy is $W = 0.5(B_g^2/\mu_0)Az$. For a U-shaped electromagnet (two air gaps) the stored field energy is $W = 0.5(B_g^2/\mu_0)2Az = (B_g^2/\mu_0)Az$. With the displacement dz of one pole, the new air gap is $z + dz$, new stored energy is $W + dW = (B_g^2/\mu_0)A(z + dz)$, change in stored energy is $(B_g^2/\mu_0)Adz$, work done $F_z dz$ and the force

$$F_z = \frac{dW}{dz} = \frac{B_g^2}{\mu_0}A = \frac{1}{4}\frac{\mu_0(Ni)^2}{g^2}A \tag{1.15}$$

where $B_g = \mu_0 H = \mu_0[Ni/(2z)]$, $z = g$, i is the instantaneous electric current, and A is the cross section of the air gap (surface of a single pole shoe). Eqn (1.15) is used to find the normal (attractive) force between the armature core and reaction rail of linear motors. For a doubly excited device,

$$F_i = \frac{1}{2}i_1^2\frac{dL_{11}}{d\xi_i} + \frac{1}{2}i_2^2\frac{dL_{22}}{d\xi_i} + i_1 i_2\frac{dL_{12}}{d\xi_i} \tag{1.16}$$

where L_{11} is the self-inductance of the winding with current i_1, L_{22} is the self-inductance of the winding with current i_2, and L_{12} is the mutual inductance between coils 1 and 2. In simplified calculations, the first two terms are commonly zero.

1.4 Linear Motion

1.4.1 Speed-Time Curve

The *speed-time curve* is a graph that shows the variation of the linear speed versus time (Fig. 1.22a). In most cases, both for acceleration and deceleration periods, the speed is a linear function of time. Thus, the speed-time curve of a moving object is most often approximated by a trapezoidal function (Fig. 1.22a). The acceleration time is

$$t_1 = \frac{v_{const}}{a} \tag{1.17}$$

where v_{const} is the constant (steady state) speed. Similarly, the retardation time is

$$t_3 = \frac{v_{const}}{d} \tag{1.18}$$

where d is the deceleration. The time t_2 for a constant-speed running depends both on acceleration and deceleration, i.e.,

$$t_2 = t - t_1 - t_3 = t - v_{const}\left(\frac{1}{a} + \frac{1}{d}\right) \tag{1.19}$$

where the total time of run $t = t_1 + t_2 + t_3$. The total distance of run can be found on the basis of Fig. 1.22a, i.e.,

$$s = \frac{1}{2}v_{const}t_1 + v_{const}t_2 + \frac{1}{2}v_{const}t_3 = v_{const}t - \frac{v_{const}^2}{2}\left(\frac{1}{a} + \frac{1}{d}\right) \qquad (1.20)$$

or

$$kv_{const}^2 - tv_{const} + s = 0 \qquad (1.21)$$

where

$$k = \frac{1}{2}\left(\frac{1}{a} + \frac{1}{d}\right) \qquad (1.22)$$

The above quadratic eqn (1.21) allows one to find the constant speed as a function of the total time of run, acceleration, deceleration, and total distance of run, i.e.,

$$v_{const} = \frac{t}{2k} - \sqrt{\left(\frac{t}{2k}\right)^2 - \frac{s}{k}} \qquad (1.23)$$

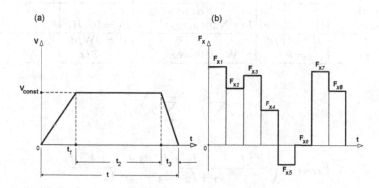

Fig. 1.22. Typical speed and thrust profiles: (a) speed-time curve, (b) thrust-time curve.

Table 1.9 compares basic formulae describing linear and rotational motions. There are two components of the linear acceleration: tangential $a = \alpha r$ and centripetal $a_r = \Omega^2 r$, where r is the radius.

1.4.2 Thrust-Time Curve

The *thrust-time curve* is a graph that shows the variation of the thrust versus time (Fig. 1.22b). The *rms* thrust (force in the x-direction) is based on the given duty cycle, i.e.,

Table 1.9. Basic formulae for linear and rotational motions

Linear motion			Rotational motion		
Quantity	Formula	Unit	Quantity	Formula	Unit
Linear displacement	$s = \theta r$	m	Angular displacement	θ	rad
Linear velocity	$v = ds/dt$ $v = \Omega r$	m/s	Angular velocity	$\Omega = d\theta/dt$	rad/s
Linear acceleration	$a = dv/dt$ $a_t = \alpha r$ $a_r = \Omega^2 r$	m/s^2	Angular acceleration	$\alpha = d\Omega/dt$	rad/s^2
Mass	m	kg	Moment of inertia	J	kgm^2
Force	$F = mdv/dt$ $= ma$	N	Torque	$T = Jd\Omega/dt$ $= J\alpha$	Nm
Momentum	$p = mv$	Ns	Angular momentum	$l = J\Omega$	kgm^{2}rad/s
Friction force	Dds/dt $= Dv$	N	Friction torque	$Dd\theta/dt$ $= D\Omega$	Nm
Spring force	Ks	N	Spring torque	$K\theta$	Nm
Work	$dW = Fds$	Nm	Work	$dW = Td\theta$	Nm
Kinetic energy	$E_k = 0.5mv^2$	J or Nm	Kinetic energy	$E_k = 0.5J\Omega^2$	J
Power	$P = dW/dt$ $= Fv$	W	Power	$P = dW/dt$ $= T\Omega$	W

$$F_{xrms}^2 \sum t_i = \sum F_{xi}^2 t_i$$

$$F_{xrms} = \sqrt{\frac{F_{x1}^2 t_1 + F_{x2}^2 t_2 + F_{x3}^2 t_3 + ... + F_{xn}^2 t_n}{t_1 + t_2 + t_3 + ... + t_n}} \qquad (1.24)$$

Similarly, in electric circuits, the rms or effective current is

$$I_{rms} = \sqrt{\frac{1}{T} \int_0^T i^2 dt}$$

since the average power delivered to the resistor R is

$$P = \frac{1}{T} \int_0^T i^2 R dt = R\frac{1}{T} \int_0^T i^2 dt = RI_{rms}^2$$

1.4.3 Dynamics

Fig. 1.23 shows the mass m sliding at velocity v with a viscous friction constant D_v on a surface. The applied instantaneous force is f_x, and the spring constant is k_s. According to d'Alembert's principle [2]

$$m\frac{dv}{dt} + D_v v + k_s \int v\, dt = F_x t \qquad (1.25)$$

The above equation can also be written as

$$m\ddot{x} + D_v \dot{x} + k_s x = F_x t \qquad (1.26)$$

where $\ddot{x} = d^2x/dt^2$ is the linear acceleration, $\dot{x} = dx/dt$ is the linear velocity, and x is the linear displacement. Eqns (1.25) and (1.26) are called *2nd order mass—spring—damper equations*. The inverse of stiffness (N/m) is the compliance (m/N) of an elastic element. The form of eqn (1.25) is similar to Kirchhoff's voltage equation for the RLC series circuit, i.e.,

$$L\frac{di}{dt} + Ri + \frac{1}{C}\int i\, dt = e \qquad (1.27)$$

where e is the instantaneous induced voltage (EMF), and i is the instantaneous current. Since

$$v = \frac{dx}{dt} \qquad \text{and} \qquad i = \frac{dq}{dt}$$

where q is the electrical charge, eqns (1.25) and (1.27) can be rewritten in the forms

$$m\frac{d^2x}{dt^2} + D_v \frac{dx}{dt} + k_s x = F_x \qquad (1.28)$$

$$L\frac{d^2q}{dt^2} + R\frac{dq}{dt} + \frac{1}{C}q = e \qquad (1.29)$$

Analogous systems are described by the same integro-differential equation or set of equations, e.g., mechanical and electrical systems.

In the mechanical system, energy stored in the mass is given by the kinetic energy $0.5mv^2$. Energy storage occurs in a spring from the displacement $x = \int v\, dt$, due to a force. This force is expressed in terms of the stiffness of spring k_s, as $f_x = k_s x = k_s \int v\, dt$.

Assuming a linear force-displacement relation, the work done is

$$W = \frac{1}{2}f_x x = \frac{1}{2}\frac{1}{k_s}f_x^2 \qquad \text{J} \qquad (1.30)$$

---

[2] d'Alembert's principle: The sum of forces acting on a body and forces of inertia is equal to zero.

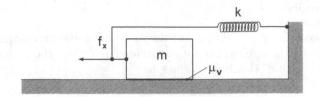

Fig. 1.23. A simple linear mechanical system.

A viscous friction element, such as a *dashpot*, is an energy-dissipating element. In the *mass–inductance analogy*, the inductance has stored electromagnetic energy $0.5Li^2$. Application of a voltage to a capacitance causes a proportional storage of charge, $q = \int i\,dt$, and the voltage component is $(1/C)\int i\,dt$.

Similarly, in the *mass–capacitance analogy*, the energy stored in a capacitance is $0.5Ce^2$. The energy storage of a spring, $0.5(1/k_s)f_x^2$, is analogous to that in the inductance, $0.5Li^2$, whence the inductance becomes the analogue of spring *compliance* $(K = 1/k_s)$.

1.4.4 Hamilton's Principle

The action integral

$$I = \int_{t_1}^{t_2} \mathcal{L}dt \tag{1.31}$$

has a stationary value for the correct path of motion, where $\mathcal{L}$ is the *Lagrangian*, so

$$\delta I = \delta \int_{t_1}^{t_2} \mathcal{L}dt = 0 \tag{1.32}$$

Mathematically, eqn (1.32) means that the variation of the action integral is equal to zero. Eqn (1.32) expresses the *principle of least action*, also called Hamilton's principle.[3] The Lagrangian of a mechanical system is defined as

$$\mathcal{L} = E_k - E_p \tag{1.33}$$

where E_k is the total kinetic energy, and E_p is the total potential energy.

Hamilton's principle can be extended to electromechanical systems. The Lagrangian of an electromechanical system is defined as

---

[3] The principle of least action was proposed by the French mathematician and astronomer Pierre-Louis Moreau de Maupertuis but rigorously stated only much later, especially by the Irish mathematician and scientist William Rowan Hamilton in 1835.

$$\mathcal{L} = \int_V (H^2 - E^2)dV \tag{1.34}$$

where H is the magnetic field intensity, E is the electric field intensity, and v is the volume. The principle of least action for electromechanical systems,

$$\delta I = \delta \int_{t_1}^{t_2} \left[\int_V (H^2 - E^2)dV \right] dt = 0 \tag{1.35}$$

was formulated by J. Larmor in 1890 and is called Larmor's principle.

1.4.5 Euler–Lagrange Equation

The Euler–Lagrange differential equation is the fundamental equation of variations in calculus. It states that if J is defined by an integral of the form

$$J = \int f[\xi(t), \dot{\xi}(t), t]dt \tag{1.36}$$

where ξ is the generalized coordinate and

$$\dot{\xi} = \frac{d\xi}{dt}$$

then J has a stationary value if the differential equation

$$\frac{\partial f}{\partial \xi} - \frac{d}{dt}\left(\frac{\partial f}{\partial \dot{\xi}}\right) = 0 \tag{1.37}$$

is satisfied. The Euler–Lagrange eqn (1.37) is expressed in time-derivative notation.

Hamilton's principle, also called the *principle of least action*, derived from d'Alambert's principle and the principle of *virtual work*, means that, for a real motion, the variation of action is equal to zero, i.e.,

$$\delta J = \delta \int \mathcal{L}(\xi, \dot{\xi}, t) = 0 \tag{1.38}$$

where $\mathcal{L} = E_k - E_p$ is called Lagrangian and is defined as a difference between kinetic and potential energy (1.33), or kinetic coenergy and potential energy.

Proof of Euler–Lagrange differential equation is given below.

$$\delta J = \delta \int_{t_1}^{t_2} \mathcal{L}(\xi, \dot{\xi}, t)dt = \int_{t_1}^{t_2} \left(\frac{\partial \mathcal{L}}{\partial \xi}\delta\xi + \frac{\partial \mathcal{L}}{\partial \dot{\xi}}\delta\dot{\xi} \right) dt$$

$$= \int_{t_1}^{t_2} \left[\frac{\partial \mathcal{L}}{\partial \xi}\delta\xi + \frac{\partial \mathcal{L}}{\partial \dot{\xi}}\frac{d}{dt}(\delta\xi) \right] dt \tag{1.39}$$

since $\delta\dot{\xi} = d(\delta\xi)/dt$. Now integrate the second term by parts using

$$u = \frac{\partial \mathcal{L}}{\partial \dot{\xi}} \qquad\qquad dv = d(\delta \xi)$$

$$du = \frac{d}{dt}\left(\frac{\partial \mathcal{L}}{\partial \dot{\xi}}\right) dt \qquad\qquad v = \delta \xi$$

because

$$d(uv) = u\,dv + v\,du \qquad\qquad \int_a^b u\,dv = uv\Big|_a^b - \int_a^b v\,du$$

Therefore,

$$\delta J = \frac{\partial \mathcal{L}}{\partial \dot{\xi}}\delta \xi\Big|_{t_1}^{t_2} + \int_{t_1}^{t_2}\left(\frac{\partial \mathcal{L}}{\partial \xi}\delta \xi - \frac{d}{dt}\frac{\partial \mathcal{L}}{\partial \dot{\xi}}\delta \xi\right) dt \qquad (1.40)$$

Only the path, not the endpoints, varies. So, $\delta \xi(t_1) = \delta \xi(t_2) = 0$, and (1.40) becomes

$$\delta J = \int_{t_1}^{t_2}\left(\frac{\partial \mathcal{L}}{\partial \xi} - \frac{d}{dt}\frac{\partial \mathcal{L}}{\partial \dot{\xi}}\right)\delta \xi\,dt \qquad (1.41)$$

Stationary values such that $\delta J = 0$ must be found. These must vanish for any small change δq, which gives from (1.41)

$$\frac{\partial \mathcal{L}}{\partial \xi} - \frac{d}{dt}\frac{\partial \mathcal{L}}{\partial \dot{\xi}} = 0$$

or

$$\frac{d}{dt}\frac{\partial \mathcal{L}}{\partial \dot{\xi}} - \frac{\partial \mathcal{L}}{\partial \xi} = 0 \qquad (1.42)$$

Eqn (1.42) is the Euler–Lagrange differential equation. Problems in the calculus of variations often can be solved by solution of the appropriate Euler–Lagrange equation.

The Euler–Lagrange equation for nonconservative systems, in which external forces and dissipative elements exist, takes the form

$$\frac{d}{dt}\left[\frac{\partial \mathcal{L}(\dot{\xi},\xi,t)}{\partial \dot{\xi}_k}\right] - \frac{\partial \mathcal{L}(\dot{\xi},\xi,t)}{\partial \xi_k} + \frac{\partial Ra(\dot{\xi},\xi,t)}{\partial \dot{\xi}_k} = Q_k \qquad (1.43)$$

in which the first term on the left-hand side represents forces of inertia of the system, the second term represents spring forces, the third term represents forces of dissipation, and Q_k on the right-hand side represents external forces. The Rayleigh dissipation function is defined as

$$Ra = \frac{1}{2}R\dot{\xi}^2 + \frac{1}{2}D_v\dot{\xi}^2 \qquad (1.44)$$

where R is the electric resistance and D_v is the mechanical dumping, e.g., viscous friction.

1.4.6 Traction

Let us consider a mechanism driven by a linear motor (Fig. 1.24a). The mechanism consists of a moving part, i.e., a linear–motor–driven car with the total mass m on a slope, a pulley, a cable, and a counterweight with its mass m_c. The efficiency of the system is η. The inertia of the pulley, and mass of the cable are neglected.

For the steady-state linear motion,

$$\eta(F_x + m_c g) = mg \sin \alpha + \mu mg \cos \alpha \qquad (1.45)$$

where $mg \sin \alpha$ is the force due to the gradient resistance, $\mu mg \cos \alpha$ is the force due to the friction resistance, and $g = 9.81$ m/s^2 is the gravitational acceleration. The coefficient of friction μ is approximately 0.2 for steel on dry steel, 0.06 for steel on oiled steel (viscous friction), 0.005 for linear bearings with rollers, and 0.002 to 0.004 for linear bearings with balls. Thus, the steady-state thrust (force produced by the linear motor) is

$$F_x = \frac{1}{\eta}(m \sin \alpha + \mu m \cos \alpha - m_c)g \qquad (1.46)$$

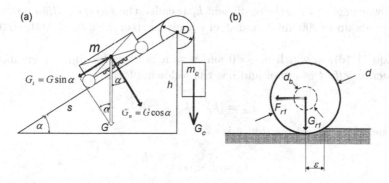

Fig. 1.24. Linear–motor–driven mechanism: (a) slope, (b) sketch for calculating the rolling resistance.

When the moving part runs up the gradient with an acceleration a the acceleration thrust is higher since the term $(m + m_c)a$ is added, i.e.,

$$F_{xpeak} = \frac{1}{\eta}[(m \sin \alpha + \mu m \cos \alpha - m_c)g + (m + m_c)a] \qquad (1.47)$$

The thrust according to eqn (1.47) is often called the peak thrust. Similarly, if the car runs up with a deceleration d, the braking force is

$$F_{xb} = \eta[(m \sin \alpha + \mu m \cos \alpha - m_c)g - (m + m_c)a] \qquad (1.48)$$

Note that, for the braking mode, the counterbalancing force is $1/\eta(F_{xb} + m_c g)$.

The linear–motor–driven car is furnished with wheels. The rolling force (Fig. 1.24b)

$$F_{r1} = \frac{\epsilon G \cos \alpha}{0.5d}$$

where $G \cos \alpha = G_n = mg \cos \alpha$. Including friction in wheel-axle bearings,

$$F_{r2} = F_{r1} + \frac{0.5d_b \mu G \cos \alpha}{0.5d} = \frac{\epsilon + 0.5\mu d_b}{0.5d} G \cos \alpha$$

where d_b is the diameter of the bearing journal. An additional resistance due to uneven track and hunting can be added by introducing a coefficient $\beta > 1$. Thus, the total rolling force

$$F_r = k_r G \qquad (1.49)$$

where

$$k_r = \beta \frac{(\epsilon + 0.5\mu d_b) \cos \alpha}{0.5d} \qquad (1.50)$$

In railway engineering, the coefficient k_r is called the *specific rolling resistance*. For speeds up to 200 km/h and steel wheels on steel rails, $k_r = 0.002$ to 0.012.

Eqn (1.46), in which $m_c = 0$ and $\eta \approx 1$, is known in railway engineering as *traction effort equation* and has the following form:

$$F_x = (k_r + k_g + k_a)G \qquad (1.51)$$

The *specific gradient resistance* is

$$k_g = \pm \sin \alpha = \pm \frac{h}{s} \qquad (1.52)$$

since the force due to gravity is $G_t = G \sin \alpha$. The "+" sign is for a car moving up the gradient, and the "−" sign is for a car moving down the gradient. Neglecting the inertia of rotating masses of the car, the *specific acceleration resistance* is

$$k_a = \frac{a}{g} \qquad (1.53)$$

where a is the linear acceleration or deceleration. For a curvelinear track, the *specific curve resistance* should be taken into account, i.e.,

$$k_c = \frac{0.153S + 0.1b}{R_c} \qquad (1.54)$$

where S is the circumference of wheel in meters, b is the mean value of all fixed wheel bases with $b < 3.3S$ in meters, and R_c is the track curve radius in meters. For example, if the wheel radius is $R = 0.46$ m, the wheel base is $b = 3$ m and the track curve radius is $R_c = 5$ km, the circumference of wheel $S = 2\pi R = 2\pi \times 0.46 = 2.89$ m, and the specific curve resistance is $k_c = (0.153 \times 2.89 + 0.1 \times 3.0)/5000 = 0.148 \times 10^{-3}$.

For high-speed trains the air resistance force

$$F_{air} = 0.5 C \rho v^2 A \quad \text{N} \tag{1.55}$$

should be added to eqn (1.51). The coefficient $C = 0.2$ for cone- or wedge-shaped nose, $C = 2.1$ for flat-front trains, and $C = 0.75$ for automobiles. The air density is ρ, the speed v is in meter/second and the front surface area A is in square meters. At 20°C and 1 atm, the air density is $\rho = 1.21$ kg/m^3. For example, if $C = 0.2$, $v = 200$ km/h $= 200/3.6 = 55.55$ m/s, and $A = 4 \times 3 = 12$ m^2, the air resistance force is $F_{air} = 0.5 \times 0.2 \times 1.21 \times (55.55)^2 \times 12 = 4480.6$ N. Eqn (1.55) gives too small values of the air resistance force for high-speed maglev trains with wedge shaped front cars.

1.5 Selection of Linear Motors

Given below are examples that show how to calculate the basic parameters of linear motion drives and how to choose a linear electric motor with appropriate ratings. This is a simplified selection of linear motors, and more detailed calculation of parameters, especially the thrust, is recommended (Chapter 3).

When designing a linear motor drive, it is always necessary to consider its benefits in comparison with traditional drives with rotary motors and mechanical gears, or ball screws transferring rotary motion into translatory motion [63]. *The authors take no responsibility for any financial losses resulting from wrong decisions and impractical designs.*

Examples

Example 1.1

A moving part of a machine is driven by a linear motor. The linear speed profile can be approximated by a trapezoidal curve (Fig. 1.22a). The total distance of run $s = 1.8$ m is achieved in $t = 0.5$ s with linear acceleration $a = 4g$ at starting, and linear deceleration $d = 3g$ at braking. Find the steady-state speed v_{const}, acceleration time t_1, acceleration distance s_1, constant speed time t_2, constant speed distance s_2, deceleration time t_3, and deceleration distance s_3.

Solution

According to eqn (1.22),

$$k = \frac{1}{2}\left(\frac{1}{4g} + \frac{1}{3g}\right) = 0.02973 \text{ s}^2/\text{m}$$

The constant speed according to eqn (1.23)

$$v_{const} = \frac{0.5}{2 \times 0.02973} - \sqrt{\left(\frac{0.5}{2 \times 0.02973}\right)^2 - \frac{1.8}{0.02973}} = 5.22 \text{ m/s}$$

$$= 18.8 \text{ km/h}$$

The time of acceleration

$$t_1 = \frac{5.22}{4g} = 0.133 \text{ s}$$

The distance corresponding to acceleration

$$s_1 = \frac{1}{2}5.22 \times 0.133 = 0.347 \text{ m}$$

The time of deceleration

$$t_3 = \frac{5.22}{3g} = 0.177 \text{ s}$$

The distance corresponding to deceleration

$$s_3 = \frac{1}{2}5.22 \times 0.177 = 0.462 \text{ m}$$

The time corresponding to the steady-state speed

$$t_2 = 0.5 - 0.133 - 0.177 = 0.19 \text{ s}$$

The distance corresponding to the steady-state speed

$$s_2 = 5.22 \times 0.19 = 0.991 \text{ m}$$

and

$$s_1 + s_2 + s_3 = 0.347 + 0.991 + 0.462 = 1.8 \text{ m}$$

Example 1.2

The specification data of a mechanism with linear motor shown in Fig. 1.24a are as follows: $m = 500$ kg, $m_c = 225$ kg, $\eta = 0.85$, $\alpha = 30^0$, $\epsilon = 0.00003$ m, $\mu = 0.005$, $d = 0.03$ m, $d_b = 0.01$ m, and $\beta = 1.3$. The car is moving up the slope. Find the thrust of the linear motor for (a) steady state, (b) starting with acceleration $a = 1$ m/s^2, (c) braking with deceleration $d = 0.75$ m/s^2. The mass of the cable, pulley, and gears is neglected.

Solution

The weight of the car

$$G = 500 \times 9.81 = 4905 \text{ N}$$

The specific rolling resistance according to eqn (1.50)

$$k_r = 1.3 \frac{0.00003 + 0.5 \times 0.005 \times 0.01}{0.5 \times 0.03} \cos 30^0 = 0.00412$$

The steady-state thrust according to eqn (1.46)

$$F_x = \frac{1}{\eta} \left(\sin \alpha + k_r - \frac{m_c}{m} \right) G = \frac{1}{0.85} \left(\sin 30^0 + 0.00412 - \frac{225}{500} \right) \times 4905$$

$$= 312.3 \text{ N}$$

At starting with acceleration $a = 1$ m/s^2 — eqn (1.47),

$$F_{xpeak} = \frac{1}{0.85} \left[\left(\sin 30^0 + 0.00412 - \frac{225}{500} \right) \times 4905 + (500 + 225) \times 1.0 \right]$$

$$= 1165.2 \text{ N}$$

The peak thrust of the linear motor should not be lower than the above value. At braking with deceleration $a = 0.75$ m/s^2 — eqn (1.48),

$$F_{xb} = 0.85 \left[\left(\sin 30^0 + 0.00412 - \frac{225}{500} \right) \times 4905 - (500 + 225) \times 0.75 \right]$$

$$= -236.6 \text{ N}$$

Example 1.3

A 1.5-kW, 1.5 m/s linear electric motor operates with almost constant speed and with the following thrust profile: 1600 N for $0 \leq t \leq 3$ s, 1200 N for $3 \leq t \leq 10$ s, 700 N for $10 \leq t \leq 26$ s, 500 N for $26 \leq t \leq 38$ s. The overload capacity factor $F_{xmax}/F_{xr} = 2$. Find the thermal utilization coefficient of the motor.

Solution

In accordance with eqn (1.2), the rated thrust produced by the linear motor is

$$F_{xr} = \frac{P_{out}}{v} = \frac{1500}{1.5} = 1000 \text{ N}$$

The linear motor has been properly selected since the maximum load for the given thrust profile 1600 N is less than the maximum thrust determined by the overload capacity factor, i.e.,

$$F_{xmax} = 2F_{xr} = 2 \times 1000 = 2000 \text{ N}$$

The *rms* thrust based on the given duty cycle

$$F_{xrms} = \sqrt{\frac{\sum F_{xi}^2 t_i}{\sum t_i}} = \sqrt{\frac{1600^2 \times 3 + 1200^2 \times 7 + 700^2 \times 16 + 500^2 \times 12}{3 + 7 + 16 + 12}}$$

$$= 867.54 \text{ Nm}$$

The coefficient of thermal utilization of the motor

$$\frac{F_{xrms}}{F_{xr}} \times 100\% = \frac{867.54}{1000.0} \times 100\% = 86.7\%$$

The linear motor, e.g., IC11-200 (Table 1.2) with continuous thrust 1260 N ($F_{xrms} = 867.54 < 1260$ N) and peak thrust 2000 N can be selected.

Example 1.4

In a factory transportation system, linear–motor–driven containers with steel wheels run on steel rails. Each container is driven by a set of 2 linear motors. The loaded container runs up the gradient and accelerates from $v = 0$ to $v_{const} = 18$ km/h in $t_1 = 5$ s, then it runs with constant speed 18 km/h, and finally it decelerates from $v = 18$ km/h to $v = 0$ in $t_3 = 5$ s. The total time of running is $t = 20$ s. Then containers are unloaded within minimum 10 s, and they run back to the initial position where they are loaded again. The time of loading is minimum 20 s. The speed and force curves of unloaded containers running down the gradient are the same as those of loaded containers, i.e., acceleration in 5 s to $v_{const} = 18$ k/h, run with constant speed and deceleration in 5 s to $v = 0$. The mass of each container, including the load and linear motor, is $m_c = 1200$ kg, without load $m'_c = 300$ kg. The rise in elevation is $h = 3$ m, and specific rolling resistance is $k_r = 0.0025$. The efficiency of the system can be assumed 100%. Find the length of the track, thrust curve, and *rms* thrust.

Solution

The movement of the container can be approximated by a trapezoidal speed-time curve (Fig. 1.22a). The linear acceleration

$$a = \frac{v_{const} - 0}{t_1 - 0} = \frac{18/3.6}{5} = 1 \text{ m/s}^2$$

The linear deceleration

$$d = \frac{0 - v_{const}}{(t_1 + t_2 + t_3) - (t_1 + t_2)} = \frac{-v_{const}}{t_3} = \frac{-18/3.6}{5} = -1 \text{ m/s}^2$$

where $t = t_1 + t_2 + t_3 = 20$ s is the total time of run. The time of running with constant speed

$$t_2 = t - t_1 - t_3 = 20 - 5 - 5 = 10 \text{ s}$$

The total distance of run is equal to the length of the track. According to eqn (1.20),

$$s = \frac{1}{2}vt_1 + vt_2 + \frac{1}{2}vt_3 = \frac{1}{2}\frac{18}{3.6}5 + \frac{18}{3.6}10 + \frac{1}{2}\frac{18}{3.6}5 = 75 \text{ m}$$

The specific gradient resistance

$$k_g = \pm\frac{h}{s} = \pm\frac{3}{75} = \pm0.04$$

The specific acceleration and deceleration resistances

$$k_a = \frac{a}{g} = \frac{1.0}{9.81} = 0.102 \qquad k'_a = \frac{d}{g} = \frac{-1.0}{9.81} = -0.102$$

The weight of loaded and empty container

$$G = m_c g = 1200 \times 9.81 = 11772 \text{ N}$$

$$G' = m'_c g = 300 \times 9.81 = 2943 \text{ N}$$

The thrust produced by linear motors when the loaded container moves up the gradient

- the loaded container accelerates ($t_1 = 5$ s)

$$F_{xpeak} = (k_r + k_g + k_a)G = (0.0025 + 0.04 + 0.102) \times 11,772 \approx 1701 \text{ N}$$

or $1701/2 = 850.5$ N per one linear motor;
- the loaded container runs with constant speed ($t_2 = 10$ s)

$$F_x = (k_r + k_g)G = (0.0025 + 0.04) \times 11,772 = 500.3 \text{ N}$$

or $500.3/2 \approx 250.2$ N per one linear motor;
- the loaded container decelerates ($t_3 = 5$ s)

$$F_{xb} = (0.0025 + 0.04 - 0.102) \times 11,772 = -700.43 \text{ N}$$

or each linear motor should be able to produce a braking force $-700.43/2 \approx -350.2$ N during the last 5 s of run.

The thrust produced by linear motors when the unloaded container moves down the gradient

- the unloaded container accelerates (5 s)

$$F'_x = (k_r - k_g + k_a)G' = (0.0025 - 0.04 + 0.102) \times 2943 = 189.8 \text{ N}$$

or $189.8/2 = 94.9$ N per one linear motor;
- the unloaded container runs with constant speed (10 s)

$$F''_x = (k_r - k_g)G' = (0.0025 - 0.04) \times 2943 = -110.4 \text{ N}$$

or one linear motor should produce $-110.4/3 = 55.2$ N braking force;
- the unloaded container decelerates (5 s)

$$F'''_x = (0.0025 - 0.04 - 0.102) \times 2943 = -410.55 \text{ N}$$

or each linear motor should produce a braking force of $-410.55/2 \approx 205.3$ N.

The *rms* thrust developed by two linear motors

$$F_{rms} = \frac{(1701^2 \times 5 + 500.3^2 \times 10 + 700.43^2 \times 5 + 0 + \dots}{(5 + 10 + 5 + 10 + 5 + 10 + 5 + 20)^{1/2}}$$

$$\frac{\dots + 189.8^2 \times 5 + 110.4^2 \times 10 + 410.55^2 \times 5 + 0)^{1/2}}{(5 + 10 + 5 + 10 + 5 + 10 + 5 + 20)^{1/2}} = 542.1 \text{ N}$$

The overload capacity factor

$$\frac{F_{xpeak}}{F_{rms}} = \frac{1701.0}{542.1} \approx 3.14$$

It will probably be difficult to find a linear motor with 3.14 peak-to-rms thrust ratio. If no such linear motor is available, the selected linear motor should develop the peak thrust minimum $1701/2 = 850.5$ N, and its rated thrust can be higher than $542.1/2 \approx 271.1$ N.

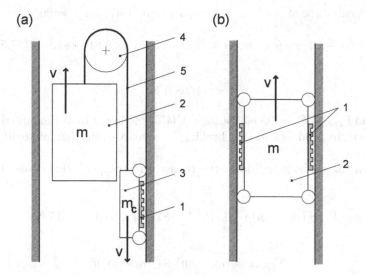

Fig. 1.25. Linear electric motor driven elevators: (a) with a rope; (b) ropeless.
1 — armature of a linear motor, 2 — car (load), 3 — counterweight, 4 — sheave,
5 — rope.

Example 1.5

A linear–motor–driven rope elevator is shown in Fig. 1.25a. The linear motor
is built in the counterweight. The mass of the car with load is $m = 819$
kg, the mass of counterweight is $m_c = 1187.5$ kg, the steady-state speed is
$v = 1.0$ m/s, the acceleration at starting is $a = 1.0$ m/s^2, and the linear motor
efficiency is $\eta = 0.6$. Neglecting the friction, rope mass, and sheave mass, find
the steady-state and peak thrust developed by the linear motor and its power
consumption at steady state.

Solution

The efficiency of the hoistway is assumed to be $\eta = 100\%$. The thrust at
steady state speed when the car is going up can be found on the basis of eqn
(1.46) in which $\alpha = 90°$,

$$F_x + m_c g = mg$$

$$F_x = (m - m_c)g = (819 - 1187.5) \times 9.81 = -3615 \text{ N}$$

The car is retarded when going up, and the linear motor should produce a
steady-state braking force -3615 N. In the case of drive failure, the elevator
car will be moving up, not down, because $m_c > m$.

The peak force at starting — compare eqn (1.47) for $\alpha = 90°$

$$F_{xpeak} = (m - m_c)g + (m + m_c)a = (819 - 1187.5) \times 9.81 + (819 + 1187.5) \times 1.0$$

$$= -1608.5 \text{ N}$$

The ratio $F_{xpeak}/F_x = 1608.5/3615 \approx 0.445$. The linear motor mounted in the counterweight produces smaller braking force at starting than at steady-state speed.

When the car is going down the thrust and the peak thrust are, respectively,

$$F_x = (m_c - m)g = (1187.5 - 819.0) \times 9.81 = 3615 \text{ N}$$

$$F_{xpeak} = (m_c - m)g + (m + m_c)a$$

$$= (1187.5 - 819.0) \times 9.81 + (819.0 + 1187.5) \times 1.0 = 5621.5 \text{ N}$$

The output power of the linear motor at steady-state speed

$$P_{out} = F_x v = 3615.0 \times 1.0 = 3615 \text{ W}$$

The electric power absorbed by the linear motor

$$P_{in} = \frac{P_{out}}{\eta} = \frac{3615}{0.6} = 6025 \text{ W}$$

Example 1.6

Consider a ropeless version of the elevator (Fig. 1.25b). The rope sheave and counterweight have been eliminated, and the linear motor built in the counterweight has been replaced by car-mounted linear motors. Assuming the mass of the loaded car $m = 4583$ kg, $a = 1.1$ m/s^2, $v = 10.0$ m/s, linear motor efficiency $\eta = 0.97$ and two linear motors per car, find the output and input power of linear motors.

Solution

The efficiency of hoistway is assumed to be $\eta = 100\%$. When the car is going up, the requested steady-state thrust is

$$F_x = mg = 4583.0 \times 9.81 = 44,960.0 \text{ N}$$

The requested peak thrust

$$F_{xpeak} = m(g + a) = 4583.0(9.81 + 1.1) \approx 50,000.0 \text{ N}$$

The ratio $F_{xpeak}/F_x = 50,000/44,960 = 1.112$. The steady-state output power of linear motors

$$P_{out} = F_x v = 44,960.0 \times 10.0 = 449,600.0 \text{W}$$

or $449,600.0/2 = 224,800.0$ W per one linear motor. It is recommended that two linear motors rated at minimum 225 kW each be chosen, and steady-state thrust $44.96/2 \approx 22.5$ kN at 10.0 m/s and peak thrust $50.0/2 = 25.0$ kN be developed.

When the car is going down, the steady-state breaking force is

$$F_{xb} = -mg = -4583.0 \times 9.81 = -44,960.0 \text{ N}$$

and the transient braking force is smaller,

$$F'_{xb} = -mg + ma = -4583.0 \times 9.81 = 4583.0 \times 1.1 = -39,518.8 \text{ N}$$

The following power can be recovered when regenerative braking is applied:

$$P_b = \eta F_{xb} v = 0.97 \times |44,960.0| \times 10.0 = 436,112.0 \text{ W}$$

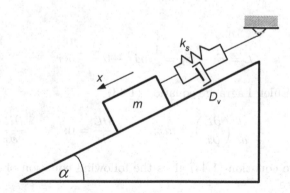

Fig. 1.26. Mass suspended from a linear spring.

Example 1.7

Mass m suspended from a linear spring with spring constant k_s slides on a plane with viscous friction D_v (Fig. 1.26). Find the equation of motion of the mass.

The generalized coordinate is $\xi = x$. The number of degrees of freedom DOF $= 3 - 2 = 1$. The constraint equation $y = z = 0$.

Solution

To solve this problem, Euler–Lagrange equation (1.43) for nonconservative systems will be used.

First method: $E_k \neq 0$, $E_p = 0$, $Q_k \neq 0$

Kinetic energy

$$E_k = \frac{1}{2}m\dot{x}^2$$

Rayleigh dissipation function according to eqn (1.44)

$$Ra = \frac{1}{2}D_v\dot{x}^2$$

External force

$$Q = mg\sin(\alpha) - k_s x$$

Lagrangian

$$\mathcal{L} = E_k - E_p = \frac{1}{2}m\dot{x}^2 - 0 = \frac{1}{2}m\dot{x}^2$$

Derivatives in Euler–Lagrange equation (1.43)

$$\frac{\partial \mathcal{L}}{\partial \dot{x}} = m\dot{x}; \qquad \frac{d}{dt}\left(\frac{\partial \mathcal{L}}{\partial \dot{x}}\right) = m\ddot{x}; \qquad \frac{\partial \mathcal{L}}{\partial x} = 0; \qquad \frac{\partial Ra}{\partial \dot{x}} = D_v\dot{x}$$

Euler–Lagrange equation (1.43) gives the following equation of motion (mechanical balance),

$$m\ddot{x} + D_v\dot{x} + k_s x = mg\sin(\alpha)$$

or

$$m\frac{d^2x}{dt^2} + D_v\frac{dx}{dt} + k_s x = mg\sin(\alpha)$$

Second method: $E_k \neq 0$, $E_p \neq 0$, $Q_k = 0$

Kinetic energy and Rayleigh dissipation function are the same as in the first case. Potential energy

$$E_p = \frac{1}{2}k_s x^2 - mgx\sin(\alpha)$$

Lagrangian

$$\mathcal{L} = E_k - E_p = \frac{1}{2}m\dot{x}^2 + mgx\sin(\alpha) - \frac{1}{2}k_s x^2$$

Derivatives

$$\frac{\partial \mathcal{L}}{\partial \dot{x}} = m\dot{x}; \qquad \frac{d}{dt}\left(\frac{\partial \mathcal{L}}{\partial \dot{x}}\right) = m\ddot{\xi}; \qquad \frac{\partial \mathcal{L}}{\partial x} = mg\sin(\alpha) - k_s x$$

Euler–Lagrange differential equation

$$m\ddot{x} + D_v\dot{x} + k_s x = mg\sin\alpha$$

Both methods give the same results.

2

Materials and Construction

2.1 Materials

All materials used in the construction of electrical machines can be divided into three groups:

1. Active materials, i.e., electric conductors (magnet wires), superconductors, electrical steels, sintered powders and PMs
2. Insulating materials
3. Construction materials

All current conducting materials (with high electric conductivity), magnetic flux conducting materials (with high magnetic permeability), and PMs are called *active materials*. They serve in the excitation of the EMF and MMF, concentrate the magnetic flux in the desired place or direction, and help to maximize the electromagnetic forces. Ferromagnetic materials are divided into *soft ferromagnetic materials*, i.e., with a narrow hysteresis loop, and *hard ferromagnetic materials* or *PMs*, i.e., with a wide hysteresis loop.

Insulating materials isolate electrically the current-carrying conductors from the other parts of electrical machines.

There are no insulating materials for the magnetic flux. Leakage fluxes can only be reduced by a proper shaping of the magnetic circuit or using electromagnetic or electrodynamic screens (shielding).

Construction materials are necessary for structural purposes intended for the transmission and withstanding of mechanical loads and stresses. In the electrical machine industry, mild carbon steel, alloyed steel, cast iron, wrought iron, non-ferromagnetic steel, nonferromagnetic metals and their alloys, and plastic materials are used as construction materials.

2.2 Laminated Ferromagnetic Cores

From the electromagnetic point of view, *laminated ferromagnetic cores* are used to improve the propagation of electromagnetic waves in conductive ferromagnetic materials. In thin ferromagnetic sheets, i.e., with their thicknesses below 1 mm, the *skin effect* at power frequencies 50 to 60 Hz practically does not exist. Since in laminated cores eddy currents are reduced, the damping effect of the electromagnetic field by eddy currents is reduced, too. The alternating magnetic flux occurs in the whole sheet cross section, and its distribution is practically uniform inside the laminated stack. Considering the skin effect, stacking factor, hysteresis losses, eddy-current losses, reactive power (magnetizing current), and easy stamping, the best thickness is 0.5 to 0.6 mm for 50-Hz electrical machines, and 0.2 to 0.35 mm for 400-Hz electrical machines [226].

The *main losses* in a ferromagnetic core with its mass m_{Fe} at any frequency f and given magnetic flux density B are calculated as

$$\Delta P_{Fe} = k_{Fe}\Delta p_{1/50}\left(\frac{f}{50}\right)^{4/3} B^2 m_{Fe} \tag{2.1}$$

where $k_{Fe} > 1$ is the coefffcient for including the difference in the distribution of the magnetic field in the core and in the sample in which the specific core losses have been measured, and for including the losses due to rotational magnetic reversal and the "work hardening" during stamping; $\Delta p_{1/50}$ is the specific core loss at $f = 50$ Hz and $B = 1$ T; f is the frequency of the magnetic field; and B is the magnetic flux density.

Better results are obtained if the losses are divided into *hysteresis lossess* and *eddy-current losses*, i.e.,

$$\Delta P_{Fe} = \Delta P_h + \Delta P_e = \left[k_h c_h\left(\frac{f}{50}\right)B^2 + k_e c_e\left(\frac{f}{50}\right)^2 B^2\right]m_{Fe} \tag{2.2}$$

where $k_h = 1$ to 2, $k_e = 2$ to 3, $c_h = 2$ to 5 Ws/(T^{2}kg) is the hysteresis constant, and $c_e = 0.5$ to 23 Ws^2/(T^{2}kg) is the eddy-current constant. The thicker the sheet, the higher the constants c_h and c_e. Eqn (2.2) can only be used if accurate values of c_h and c_e are known. In most cases the constants c_h and c_e are not specified.

The armature core losses ΔP_{Fe} can be calculated on the basis of the *specific core losses* and masses of teeth and yoke, i.e.,

$$\Delta P_{Fe} = \Delta p_{1/50}\left(\frac{f}{50}\right)^{4/3} [k_{adt}B_t^2 m_t + k_{adc}B_c^2 m_c] \tag{2.3}$$

where $k_{adt} > 1$ and $k_{ady} > 1$ are the factors accounting for the increase in losses due to metallurgical and manufacturing processes, $\Delta p_{1/50}$ is the specific core loss in W/kg at 1 T and 50 Hz, B_t is the magnetic flux density in a tooth, B_c is the magnetic flux density in the core (yoke), m_t is the mass of the teeth, and m_c is the mass of the core. For the teeth, $k_{adt} = 1.7$ to 2.0, and for the core, $k_{adc} = 2.4$ to 4.0 [113].

The external surfaces of electrotechnical steel sheets are covered with a thin layer of ceramic materials or oxides to electrically insulate the adjacent laminations in a stack. This insulation limits the eddy currents induced in the core due to a.c. magnetic fluxes. The thickness of the insulation is expressed with the aid of the *stacking* factor (insulation factor):

$$k_i = \frac{\sum_{i=1}^{n} d_i}{\sum_{i=1}^{n}(d_i + 2\Delta_i)} < 1 \qquad (2.4)$$

where d_i is the thickness of the ith lamination, and Δ_i is the thickness of the insulation layer of the ith lamination measured on one side. For the stack consisting of laminations of equal thickness d with the thickness of the insulation layer Δ (one side), the stacking factor is

$$k_i = \frac{d}{d + 2\Delta} \qquad (2.5)$$

For cold-rolled electrical steel sheets, the stacking factor is $k_i = 0.95$ to 0.98; for hot-rolled sheets this factor is smaller.

2.2.1 Electrical Sheet-Steels

Typical magnetic circuits of electrical machines and electromagnetic devices are laminated and are mainly made of *cold-rolled* electrotechnical steel sheets, i.e.,

- oriented (anisotropic) textured,
- nonoriented with silicon content,
- nonoriented without silicon.

Nowadays, *hot-rolled* electrotechnical steel sheets are almost never used. Electrotechnical steel sheets have crystal structure. *Oriented steel sheets* are used for the ferromagnetic cores of transformers, transducers, and large synchronous generators. *Nonoriented steel sheets* are used for construction of large, medium, and low-power rotary electrical machines, micromachines, small transformers and reactors, electromagnets, and magnetic amplifiers. Addition of 0.5% to 3.25% of silicon (Si) increases the maximum magnetic permeability corresponding to the critical magnetic field intensity, reduces the area of the hysteresis loop, increases the resistivity (reduces eddy current losses), and practically excludes *ageing* (increase in the steel losses with time). Thus, owing to the silicon content, the specific core losses are substantially reduced. On the other hand, silicon reduces somewhat permeability in strong fields (saturation magnetic flux density), increases hardness of laminations

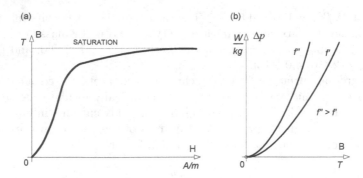

Fig. 2.1. Typical characteristics of electrotechnical steel sheets: (a) magnetization curve B-H, (b) specific core loss curves Δp-B at $f = const$.

and, as a consequence, shortens the life of stamping tooling (fast wear of the punching die).

The increase of power loss with time or *ageing* is caused by an excessive carbon content in the steel. Modern non-oriented fully processed electrical steels are free from magnetic ageing [203].

Fig. 2.1 shows a typical magnetization curve B-H and specific core loss curve Δp-B of electrotechnical steel sheets. The B-H curves are obtained by increasing the magnetic field intensity H from zero in a virgin sample (never magnetized before) as a set of top points of hysteresis loops. Specific core loss curves Δp-B are measured with the aid of Epstein's apparatus. The shape of the Δp-B curve such as that in Fig. 2.1b is only valid for steel sheets with crystal structure.

Silicon steels are generally specified and selected on the basis of allowable *specific core losses* (W/kg or W/lb). The most universally accepted grading of electrical steels by core losses is the American Iron and Steel Industry (AISI) system (Table 2.1), the so called "M-grading". The M number, e.g., M19, M27, M36, etc., indicates maximum specific core losses in W/lb at 1.5 T and 50 or 60 Hz, e.g., M19 grade specifies that losses shall be below 1.9 W/lb at 1.5 T and 60 Hz. Electrical steel M19 offers nearly the lowest core loss in this class of material and is probably the most common grade for motion control products (Fig. 2.2). The specific core loss curve of electrical steel M19 at 50 Hz is plotted in Fig. 2.3.

Nonoriented electrical steels are Fe-Si alloys with random orientation of crystal cubes and practically the same properties in any direction in the plane of the sheet or ribbon. Nonoriented electrical steels are available as both *fully processed* and *semiprocessed* products. Fully processed steels are annealed to optimum properties by the manufacturer and ready for use without any additional processing. Semiprocessed steels always require annealing after stamping to remove excess carbon and relieve stress. Better grades of silicon steel are always supplied fully processed, while semiprocessed silicon steel is avail-

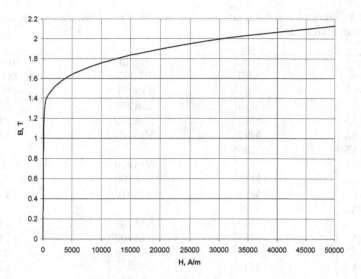

Fig. 2.2. Magnetization curve of fully processed Armco DI-MAX nonoriented electrical steel M19.

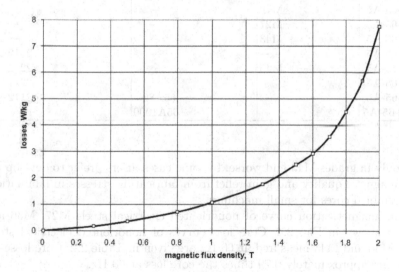

Fig. 2.3. Specific core loss curve of fully processed Armco DI-MAX nonoriented electrical steel M19 at 50 Hz.

Table 2.1. Silicon-steel designations specified by European, American, Japanese, and Russian standards

Europe IEC 404-8-4 (1986)	U.S. AISI	Japan JIS 2552 (1986)	Russia GOST 21427 0-75
250-35-A5	M15	35A250	2413
270-35-A5	M19	35A270	2412
300-35-A5	M22	35A300	2411
330-35-A5	M36	—	—
270-50-A5	—	50A270	—
290-50-A5	M15	50A290	2413
310-50-A5	M19	50A310	2412
330-50-A5	M27	—	—
350-50-A5	M36	50A350	2411
400-50-A5	M43	50A400	2312
470-50-A5	—	50A470	2311
530-50-A5	M45	—	2212
600-50-A5	—	50A600	2112
700-50-A5	M47	50A700	—
800-50-A5	—	50A800	2111
350-65-A5	M19	—	—
400-65-A5	M27	—	—
470-65-A5	M43	—	—
530-65-A5	—	—	2312
600-65-A5	M45	—	2212
700-65-A5	—	—	2211
800-65-A5	—	65A800	2112
1000-65-A5	—	65A1000	—

able only in grades M43 and worse. In some cases, users prefer to develop the final magnetic quality and gain relief from fabricating stresses in laminations or assembled cores for small machines.

The magnetization curve of nonoriented electrical steels M27, M36 and M43 is shown in Fig. 2.4. Core loss curves of nonoriented electrical steels M27, M36 and M43 measured at 60 Hz are given in Table 2.2. Core losses at 50 Hz are approximately 0.79 times the core loss at 60 Hz.

Table 2.3 contains magnetization curves B-H and specific core loss curves Δp-B of three types of cold-rolled, nonoriented electrotechnical steel sheets, i.e., Dk66, thickness $d = 0.5$ mm, $k_i = 0.96$, 7740 kg/m^3 (Surahammars Bruk AB, Sweden), H-9, $d = 0.35$ mm, $k_i = 0.96$, 7650 kg/m^3 (Nippon Steel Corporation, Japan) and DI-MAX EST20, $d = 0.2$ mm, $k_i = 0.94$, 7650 kg/m^3 (Terni–Armco, Italy).

For modern high-efficiency, high-performance applications, there is a need for operating a.c. devices at higher frequencies, i.e., 400 Hz to 10 kHz. Because

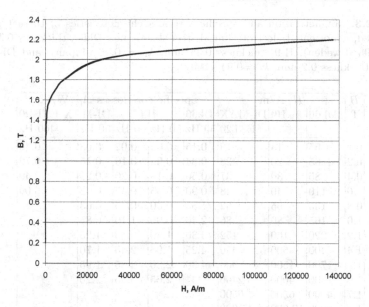

Fig. 2.4. Magnetization curve of fully processed Armco DI-MAX nonoriented electrical steels M27, M36, and M43. Magnetization curves for all these three grades are practically the same.

Table 2.2. Specific core losses of Armco DI-MAX nonoriented electrical steels M27, M36, and M43 at 60 Hz

| Magnetic flux density T | Specific core losses W/kg | | | | | | | |
| | 0.36 mm | | 0.47 mm | | | 0.64 mm | | |
	M27	M36	M27	M36	M43	M27	M36	M43
0.20	0.09	0.10	0.10	0.11	0.11	0.12	0.12	0.13
0.50	0.47	0.52	0.53	0.56	0.59	0.62	0.64	0.66
0.70	0.81	0.89	0.92	0.97	1.03	1.11	1.14	1.17
1.00	1.46	1.61	1.67	1.75	1.87	2.06	2.12	2.19
1.30	2.39	2.58	2.67	2.80	2.99	3.34	3.46	3.56
1.50	3.37	3.57	3.68	3.86	4.09	4.56	4.70	4.83
1.60	4.00	4.19	4.30	4.52	4.72	5.34	5.48	5.60
1.70	4.55	4.74	4.85	5.08	5.33	5.99	6.15	6.28
1.80	4.95	5.14	5.23	5.48	5.79	6.52	6.68	6.84

Table 2.3. Magnetization and specific core loss characteristics of three types of cold-rolled, nonoriented electrotechnical steel sheets, i.e., Dk66, thickness 0.5 mm, $k_i = 0.96$ (Sweden); H-9, thickness 0.35 mm, $k_i = 0.96$ (Japan); and DI-MAX EST20, thickness 0.2 mm, $k_i = 0.94$, (Italy).

B	H, A/m			Specific core losses Δp, W/kg				
T	Dk66	H9	DI-MAX EST20	Dk66 50 Hz	H9 50 Hz	60 Hz	DI-MAX 50 Hz	EST 20 400 Hz
0.1	55	13	19	0.15	0.02	0.02	0.08	0.30
0.2	65	20	28	0.24	0.06	0.10	0.15	0.70
0.4	85	30	37	0.50	0.15	0.20	0.25	2.40
0.6	110	40	48	0.90	0.35	0.45	0.42	6.00
0.8	135	55	62	1.55	0.60	0.75	0.63	
1.0	165	80	86	2.40	0.90	1.10	0.85	
1.2	220	160	152	3.30	1.30	1.65	1.25	
1.4	400	500	450	4.25	1.95	2.45	1.70	
1.5	700	1500	900	4.90	2.30	2.85	1.95	
1.6	1300	4000	2400		2.65	3.35	2.20	
1.7	4000	6500	6500					
1.8	8000	10,000	17,000					
1.9	15,000	16,000						
2.0	22,500	24,000						
2.1	35,000							

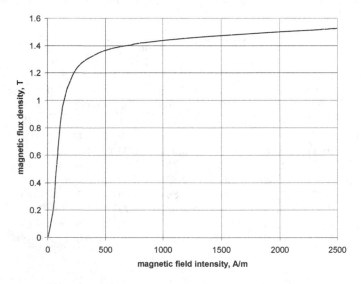

Fig. 2.5. Magnetization curve of Arnon^{TM} 5 nonoriented electrical steel.

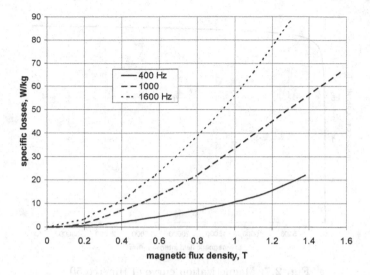

Fig. 2.6. Specific core loss curves of Arnon^{TM} 5 nonoriented electrical steel.

of the thickness of the standard silicon ferromagnetic steels 0.25 mm (0.010")
or more, core loss due to eddy currents is excessive. Nonoriented electrical
steels with thin gauges (down to 0.025 mm thick) for ferromagnetic cores of
high-frequency rotating machinery and other power devices are manufactured,
e.g., by Arnold Magnetic Technologies Corporation, Rochester, NY, USA.
Arnold has two standard nonoriented lamination products: Arnon^{TM} 5 (Figs
2.5 and 2.6) and Arnon^{TM} 7. At frequencies above 400 Hz, they typically
have less than half the core loss of standard-gauge nonoriented silicon steel
laminations.

2.2.2 High-Saturation Ferromagnetic Alloys

Cobalt—iron alloys with Co content ranging from 15% to 50% have the highest
known saturation magnetic flux density about 2.4 T at room temperature.
They are the natural choice for applications where mass and space saving are of
prime importance. Additionally, the iron—cobalt alloys have the highest Curie
temperatures of any alloy family and have found use in elevated temperature
applications. The nominal composition, e.g., for Hiperco 50 from Carpenter,
PA, USA is 49% Fe, 48.75% Co, 1.9% V, 0.05% Mn, 0.05% Nb, and 0.05% Si.

The specific mass density of Hiperco 50 is 8120 kg/m^3, modulus of elastic-
ity 207 GPa, electric conductivity 2.5×10^6 S/m, thermal conductivity 29.8
W/(m K), Curie temperature 940^{0}C, specific core loss about 76 W/kg at 2 T
and 400 Hz, and thickness from 0.15 to 0.36 mm. The magnetization curve of

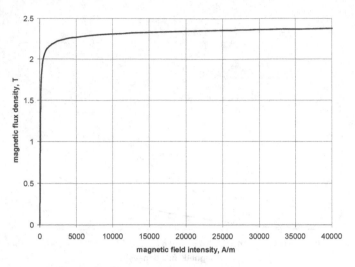

Fig. 2.7. Magnetization curve of Hiperco 50.

Table 2.4. Specific losses (W/kg) in 0.356 mm iron—cobalt alloy strips at 400 Hz.

	Magnetic flux density		
Alloy	1.0 T	1.5 T	2.0 T
Hiperco 15	30	65	110
Permendur 24	42	105	160
Hiperco 27	53	110	180
Rotelloy 5	40	130	200
Hiperco 50	25	44	76
Hiperco 50A	14	31	60
Hiperco 50HS	43	91	158
Rotelloy 3	22	55	78
Permendur 49	22	55	78
Rotelloy 8	49	122	204
HS 50	—	—	375

Hiperco 50 is shown in Fig. 2.7. Specific losses (W/kg) in 0.356 mm strip at 400 Hz of iron—cobalt alloys from Carpenter are given in Table 2.4.

2.2.3 Permalloys

Small electrical machines and micromachines working in humid or chemical-active atmospheres must have stainless ferromagnetic cores. The best corrosion-resistant ferromagnetic material is *permalloy* (NiFeMn), but on the other hand, its saturation magnetic flux density is lower than that of electrotechnical steel sheets. Permalloy is also a good ferromagnetic material for cores of small transformers used in electronic devices and in electromagnetic A/D converters, where a rectangular hysteresis loop is required.

2.2.4 Amorphous Materials

Core losses can be substantially reduced by replacing standard electrotechnical steels with *amorphous magnetic alloys*. Amorphous ferromagnetic sheets, in comparison with electrical sheets with crystal structure, do not have arranged–in–order, regular inner crystal structure (lattice). Table 2.5 shows physical properties, and Table 2.6 shows specific core loss characteristics of commercially available iron-based METGLAS®[1] amorphous ribbons 2605C0 and 2605SA2 [143].

Table 2.5. Physical properties of iron based METGLAS® amorphous alloy ribbons (AlliedSignals, Inc., Morristown, NJ, USA)

Quantity	2605CO	2605SA1
Saturation magnetic flux density, T	1.8	1.59 annealed 1.57 cast
Specific core losses at 50 Hz and 1 T, W/kg	less than 0.28	about 0.125
Specific density, kg/m^3	7560	7200 annealed 7190 cast
Electric conductivity, S/m	0.813×10^6 S/m	0.769×10^6 S/m
Hardness in Vicker's scale	810	900
Elastic modulus, GN/m^2	100 to 110	100 to 110
Stacking factor	less than 0.75	less than 0.79
Crystallization temperature, ^{0}C	430	507
Curie temperature, ^{0}C	415	392
Maximum service temperature, ^{0}C	125	150

METGLAS amorphous alloy ribbons are produced by rapid solidification of molten metals at cooling rates of about 10^6 ^{0}C/s. The alloys solidify before the atoms have a chance to segregate or crystallize. The result is a metal alloy with a glass-like structure, i.e., a noncrystalline frozen liquid. METGLAS alloys for electromagnetic applications are based on alloys of iron, nickel, and cobalt. Iron based alloys combine high saturation magnetic flux density with low core losses and economical price. Annealing can be used to alter *magnetostriction* to develop hysteresis loops ranging from flat to square.

---

[1] METGLAS® is a registered trademark of AlliedSignal, Inc.

Table 2.6. Specific core losses of iron based METGLAS® amorphous alloy ribbons (AlliedSignals, Inc., Morristown, NJ, USA).

| Magnetic flux density, B | Specific core losses, Δp, W/kg | | | |
| T | 2605CO | | 2605SA1 | |
	50 Hz	60 Hz	50 Hz	60 Hz
0.05	0.0024	0.003	0.0009	0.0012
0.10	0.0071	0.009	0.0027	0.0035
0.20	0.024	0.030	0.0063	0.008
0.40	0.063	0.080	0.016	0.02
0.60	0.125	0.16	0.032	0.04
0.80	0.196	0.25	0.063	0.08
1.00	0.274	0.35	0.125	0.16

Owing to very low specific core losses, amorphous alloys are ideal for power and distribution transformers, transducers, and high-frequency apparatus. Application to the mass production of motors is limited by hardness, up to 1100 in Vicker's scale. Standard cutting methods as a guillotine or blank die are not suitable. The mechanically stressed amorphous material cracks. Laser and EDM cutting methods melt the amorphous material and cause undesirable crystallization. In addition, these methods make electrical contacts between laminations, which contribute to the increased eddy-current and additional losses. *General Electric* cut amorphous materials in the early 1980s using chemical methods, but these methods were very slow and expensive [148]. Recently, the problem of cutting hard amorphous ribbons has been overcome by using a liquid jet [194]. This method makes it possible to cut amorphous materials in ambient temperature without cracking, melting, crystallization, and electric contacts between isolated ribbons. The face of the cut is very smooth. It is possible to cut amorphous materials on profiles that are suitable for manufacturing laminations for rotary machines, linear machines, chokes, and any other electromagnetic apparatus.

2.2.5 Solid Ferromagnetic Materials

Solid ferromagnetic materials, as cast steel and cast iron are used for salient poles, pole shoes, solid rotors of special induction motors, and reaction rails (platens) of linear motors. Table 2.7 shows magnetization characteristics B-H of a mild carbon steel (0.27% C) and cast iron. Fig. 2.8 shows B-H curves for three types of solid steels: Steel 35 (Poland), Steel 4340 (U.S.), and FeNiCo-MoTiAl alloy (U.S.).

Electrical conductivities of carbon steels are from 4.5×10^6 to 7.0×10^6 S/m at 20^0C.

Table 2.7. Magnetization curves of solid ferromagnetic materials: 1 — carbon steel (0.27%C), 2 — cast iron.

| Magnetic flux density, B | Magnetic field intensity, H | |
| | Mild carbon steel 0.27% C | Cast iron |
T	A/m	A/m
0.2	190	900
0.4	280	1600
0.6	320	3000
0.8	450	5150
1.0	900	9500
1.2	1500	18,000
1.4	3000	28,000
1.5	4500	
1.6	6600	
1.7	11,000	

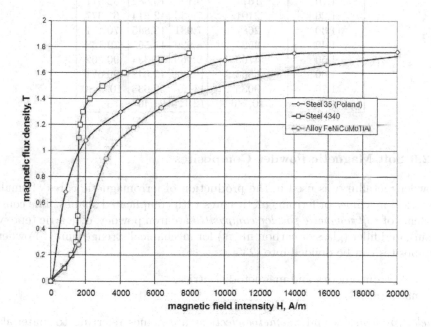

Fig. 2.8. Magnetization curves of solid steels.

Table 2.8. Magnetization and specific core loss characteristics of nonsintered *Accucore*, TSC Ferrite International, Wadsworth, IL, USA

Magnetization curve		Specific core loss curves		
Magnetic flux density, B	Magnetic field intensity, H	60 Hz	100 Hz	400 Hz
T	A/m	W/kg	W/kg	W/kg
0.10	152	0.132	0.242	1.058
0.20	233	0.419	0.683	3.263
0.30	312	0.772	1.323	6.217
0.40	400	1.212	2.072	9.811
0.50	498	1.742	2.976	14.088
0.60	613	2.315	3.968	18.850
0.70	749	2.954	5.071	24.295
0.80	909	3.660	6.305	30.490
0.90	1107	4.431	7.650	37.346
1.00	1357	5.247	9.039	44.489
1.10	1677	6.129	10.582	52.911
1.20	2101	7.033	12.214	61.377
1.30	2687	7.981	13.845	70.151
1.40	3525	8.929	15.565	79.168
1.50	4763	9.965	17.394	90.302
1.60	6563	10.869	19.048	99.671
1.70	9035	11.707	20.635	109.880
1.75	10,746	12.125	21.407	

2.2.6 Soft Magnetic Powder Composites

Powder metallurgy is used in the production of ferromagnetic cores of small electrical machines or ferromagnetic cores with complicated shapes. The components of *soft magnetic powder composites* are iron powder, dielectric (epoxy resin), and filler (glass or carbon fibers) for mechanical strengthening. Powder composites can be divided into [235]

- dielectromagnetics and magnetodielectrics,
- magnetic sinters.

Dielectromagnetics and *magnetodielectrics* are names referring to materials consisting of the same basic components: ferromagnetic (mostly iron powder) and dielectric (mostly epoxy resin) material [235]. The main tasks of the dielectric material are insulation and binding of ferromagnetic particles. In practice, composites containing up to 2% (of their mass) of dielectric materials are considered as *dielectromagnetics*. Those of higher content of dielectric material are considered as *magnetodielectrics* [235].

Magnetics International, Inc., Burns Harbor, IN, USA, has developed a new soft powder material, *Accucore*, which is competitive to traditional steel laminations [1]. The magnetization curve and specific core loss curves of the nonsintered *Accucore* are given in Table 2.8. When sintered, *Accucore* has higher saturation magnetic flux density than nonsintered material. The specific density is 7550 to 7700 kg/m^3.

2.3 Permanent Magnets

2.3.1 Demagnetization Curve

A *permanent magnet* (PM) can produce magnetic flux in an air gap with no exciting winding and no dissipation of electric power. As any other ferromagnetic material, a PM can be described by its *B-H* hysteresis loop. PMs are also called *hard magnetic materials*, which mean ferromagnetic materials with a wide hysteresis loop.

The basis for the evaluation of a PM is the portion of its hysteresis loop located in the upper left-hand quadrant, called the *demagnetization curve* (Fig. 2.9). If a reverse magnetic field intensity is applied to a previously magnetized, say, toroidal specimen, the magnetic flux density drops down to the magnitude determined by the point K. When the reverse magnetic flux density is removed, the flux density returns to the point L according to a minor hysteresis loop. Thus, the application of a reverse field has reduced the *remanence*, or *remanent magnetism*. Reapplying a magnetic field intensity will again reduce the flux density, completing the minor hysteresis loop by returning the core to approximately the same value of flux density at the point K as before. The minor hysteresis loop may usually be replaced with little error by a straight line called the *recoil line*. This line has a slope called the *recoil permeability* μ_{rec}.

As long as the negative value of applied magnetic field intensity does not exceed the maximum value corresponding to the point K, the PM may be regarded as being reasonably permanent. If, however, a greater negative field intensity H is applied, the magnetic flux density will be reduced to a value lower than that at point K. On the removal of H, a new and lower recoil line will be established.

The general relationship between the magnetic flux density B, intrinsic magnetization $B_{in} = \mu_0 M$ due to the presence of ferromagnetic material, and magnetic field intensity H may be expressed as [133, 166]

$$B = \mu_0 H + B_{in} = \mu_0(H + M) = \mu_0(1 + \chi)H = \mu_0 \mu_r H \qquad (2.6)$$

in which $\mathbf{B}$, $\mathbf{H}$, $\mathbf{B}_{in}$, and $\mathbf{M}$ are parallel or antiparallel vectors, so that eqn (2.6) can be written in a scalar form. The magnetic permeability of free space $\mu_0 = 0.4\pi \times 10^{-6}$ H/m. The relative magnetic permeability of ferromagnetic materials $\mu_r = 1 + \chi >> 1$. The magnetization vector $\mathbf{M} = \chi\mathbf{H}$ is proportional to the magnetic susceptibility χ of the material. The flux density $\mu_0 H$ would be present within, say, a toroid if the ferromagnetic core were not in place. The

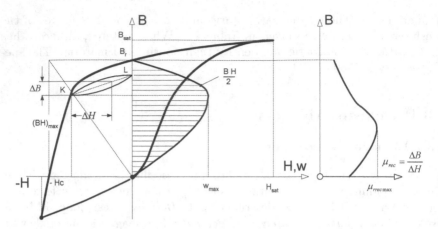

Fig. 2.9. Demagnetization curve, recoil loop, energy of a PM, and recoil magnetic permeability.

flux density B_{in} is the contribution of the ferromagnetic core. The intrinsic magnetization from eqn (2.6) is $B_{in} = B - \mu_0 H$.

A PM is inherently different from an electromagnet. If an external field H_a is applied to the PM, as was necessary to obtain the hysteresis loop of Fig. 2.9, the resultant magnetic field is

$$H = H_a + H_d \qquad (2.7)$$

where $-H_d$ is a potential existing between the poles, 180^0 opposed to B_{in}, proportional to the intrinsic magnetization B_{in}. In a closed magnetic circuit, e.g., toroidal circuit, the magnetic field intensity resulting from the intrinsic magnetization $H_d = 0$. If the PM is removed from the magnetic circuit,

$$H_d = -\frac{M_b B_{in}}{\mu_o} \qquad (2.8)$$

where M_b is the coefficient of demagnetization dependent on the geometry of a specimen. Usually $M_b < 1$ (see Appendix B).

2.3.2 Magnetic Parameters

PMs are characterized by the following parameters.

Remanent magnetic flux density B_r, or *remanence*, is the magnetic flux density corresponding to zero magnetic field intensity.

Coercive field strength H_c, or *coercivity*, is the value of demagnetizing field intensity necessary to bring the magnetic flux density to zero in a material previously magnetized.

Saturation magnetic flux density B_{sat} corresponds to high values of the magnetic field intensity when an increase in the applied magnetic field produces no further effect on the magnetic flux density. In the *saturation region* the alignment of all the *magnetic moments of domains* is in the direction of the external applied magnetic field.

Recoil magnetic permeability μ_{rec} is the ratio of the magnetic flux density to magnetic field intensity at any point on the demagnetization curve, i.e.,

$$\mu_{rec} = \mu_0 \mu_{rrec} = \frac{\Delta B}{\Delta H} \tag{2.9}$$

where the *relative recoil permeability* $\mu_{rrec} = 1$ to 3.5.

Maximum magnetic energy per unit produced by a PM in the external space is equal to the maximum magnetic energy density per volume, i.e.,

$$w_{max} = \frac{(BH)_{max}}{2} \quad \text{J/m}^3 \tag{2.10}$$

where the product $(BH)_{max}$ corresponds to the maximum energy density point on the demagnetization curve with coordinates B_{max} and H_{max} (Fig. 2.9).

Form factor of the demagnetization curve characterizes the concave shape of the demagnetization curve, i.e.,

$$\gamma = \frac{(BH)_{max}}{B_r H_c} = \frac{B_{max} H_{max}}{B_r H_c} \tag{2.11}$$

For a square demagnetization curve $\gamma = 1$ and for a straight line (rare-earth PM) $\gamma = 0.25$.

Owing to the leakage fluxes, PMs used in electrical machines are subject to nonuniform demagnetization. Therefore, the demagnetization curve is not the same for the whole volume of a PM. To simplify the calculation, in general, it is assumed that the whole volume of a PM is described by one demagnetization curve with B_r and H_c about 5% to 10% lower than those for uniform magnetization.

The leakage flux causes the magnetic flux to be distributed nonuniformly along the height $2h_M$ of a PM. As a result, the MMF produced by the PM is not constant. The magnetic flux is higher in the neutral cross section and lower at the ends, but the behavior of the MMF distribution is the opposite (Fig. 2.10).

The PM surface is not equipotential. The magnetic potential at each point on the surface is a function of the distance to the neutral zone. To simplify the calculation, the magnetic flux, which is a function of the MMF distribution along the height h_M per pole, is replaced by an equivalent flux. This equivalent flux goes through the whole height h_M and exits from the surface of the poles. To find the equivalent leakage flux and the whole flux of a PM, the equivalent magnetic field intensity needs to be found, i.e.,

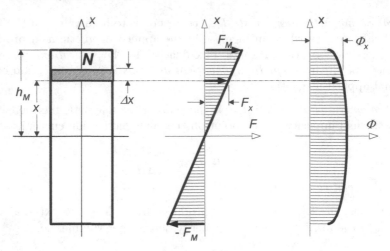

Fig. 2.10. Distribution of the MMF and magnetic flux along the height h_M of a rectangular PM.

$$H = \frac{1}{h_M} \int_0^{h_M} H_x dx = \frac{\mathcal{F}_M}{h_M} \qquad (2.12)$$

where H_x is the magnetic field intensity at a distance x from the neutral cross section, and $\mathcal{F}_M$ is the MMF of the PM per pole (MMF $= 2\mathcal{F}_M$ per pole pair).

The equivalent magnetic field intensity (2.12) allows the equivalent leakage flux of the PM to be found, i.e.,

$$\Phi_{lM} = \Phi_M - \Phi_g \qquad (2.13)$$

where Φ_M is the full equivalent flux of the PM, and Φ_g is the air gap magnetic flux. The *coefficient of leakage flux* of the PM,

$$\sigma_{lM} = \frac{\Phi_M}{\Phi_g} = 1 + \frac{\Phi_{lM}}{\Phi_g} > 1 \qquad (2.14)$$

simply allows the air gap magnetic flux to be expressed as $\Phi_g = \Phi_M/\sigma_{lM}$.

The following leakage permeance expressed in the flux Φ-MMF coordinate system corresponds to the equivalent leakage flux of the PM:

$$G_{lM} = \frac{\Phi_{lM}}{\mathcal{F}_M} \qquad (2.15)$$

An accurate estimation of the leakage permeance G_{lM} is the most difficult task in calculating magnetic circuits with PMs (Appendix A and Appendix B). This problem exists only in the circuital approach since using the field approach and, e.g., the finite element method (FEM), the leakage permeance can be found fairly accurately.

The average equivalent magnetic flux and equivalent MMF mean that the magnetic flux density and magnetic field intensity are assumed to be the same in the whole volume of a PM. The full energy produced by the magnet in the outer space is

$$W = \frac{BH}{2}V_M \qquad \text{J} \qquad (2.16)$$

where V_M is the volume of the PM or a system of PMs.

2.3.3 Magnetic Flux Density in the Air Gap

Let us consider a simple PM circuit with rectangular cross section consisting of a PM with height per pole h_M, width w_M, length l_M, two mild-steel yokes with average length $2l_{Fe}$ and an air gap of thickness g. From the Ampère's circuital law,

$$2H_M h_M = H_g g + 2H_{Fe} l_{Fe} = H_g g \left(1 + \frac{2H_{Fe} l_{Fe}}{H_g g}\right) = H_g k_{sat}$$

where H_g, H_{Fe}, and H_M are the magnetic field intensities in the air gap, mild steel yoke, and PM, respectively. The coefficient

$$k_{sat} = 1 + \frac{2H_{Fe} l_{Fe}}{H_g g} \qquad (2.17)$$

takes into account the magnetic voltage drops in the mild steel and is called *saturation factor* of the magnetic circuit.

Since $\Phi_g \sigma_{lM} = \Phi_M$ or $B_g S_g = B_M S_M / \sigma_{lM}$, where B_g is the air gap magnetic flux density, B_M is the PM magnetic flux density, S_g is the cross section area of the air gap, and $S_M = w_M l_M$ is the cross section area of the PM, the following equation can be written:

$$\frac{V_M}{2h_M} \frac{1}{\sigma_{lM}} B_M = \mu_0 H_g \frac{V_g}{g}$$

where $V_g = S_g g$ is the volume of the air gap, and $V_M = 2h_M S_M$ is the volume of the PM. The fringing flux in the air gap has been neglected. Multiplying through the equation for magnetic voltage drops and for magnetic flux, the air gap magnetic flux intensity is found as

$$B_g = \mu_0 H_g = \sqrt{\frac{\mu_0}{\sigma_{lM}} k_{sat}^{-1} \frac{V_M}{V_g} B_M H_M} \qquad (2.18)$$

Assuming $k_{sat} \approx 1$ and $\sigma_{lm} \approx 1$,

$$B_g \approx \sqrt{\mu_o \frac{V_M}{V_g} B_M H_M}$$ (2.19)

For a PM circuit, the magnetic flux density B_g in a given air gap volume V_g is directly proportional to the square root of the energy product $(B_M H_M)$ and the volume of magnet $V_M = 2 h_M w_M l_M$.

Following the trend toward smaller packaging, smaller mass, and higher efficiency, the material research in the field of PMs has focused on finding materials with high values of the maximum energy product $(BH)_{max}$.

The air gap magnetic flux density B_g can be estimated analytically on the basis of the demagnetization curve, air gap, and leakage permeance lines and recoil lines (Appendix A). Approximately, for an LSM with armature ferromagnetic stack and surface configuration of PMs, it can be found on the basis of the balance of magnetic voltage drops that

$$\frac{B_r}{\mu_0 \mu_{rrec}} h_M = \frac{B_g}{\mu_0 \mu_{rrec}} h_M + \frac{B_g}{\mu_0} g$$

where μ_{rrec} is the relative permeability of the PM (relative recoil permeability). Hence,

$$B_g = \frac{B_r h_M}{h_M + \mu_{rrec} g} \approx \frac{B_r}{1 + \mu_{rrec} g / h_M}$$ (2.20)

The air gap magnetic flux density is proportional to the remanent magnetic flux density B_r and decreases as the air gap g increases. Eqn (2.20) can only be used for preliminary calculations.

Demagnetization curves are sensitive to the temperature. Both B_r and H_c decrease as the magnet temperature increases, i.e.,

$$B_r = B_{r20}[1 + \frac{\alpha_B}{100}(\vartheta_{PM} - 20)]$$ (2.21)

$$H_c = H_{c20}[1 + \frac{\alpha_H}{100}(\vartheta_{PM} - 20)]$$ (2.22)

where ϑ_{PM} is the temperature of the PM, B_{r20} and H_{c20} are the remanent magnetic flux density and coercive force at 20°C, respectively, and $\alpha_B < 0$ and $\alpha_H < 0$ are temperature coefficients for B_r and H_c in %/^{0}C, respectively.

2.3.4 Properties of Permanent Magnets

In electric motors technology, the following PM materials are used:

- Alnico (Al, Ni, Co, Fe);
- Ferrites (ceramics), e.g., barium ferrite $BaO \times 6Fe_2O_3$ and strontium ferrite $SrO \times 6Fe_2O_3$;
- Rare-earth materials, i.e., samarium—cobalt SmCo and neodymium—iron—boron NdFeB.

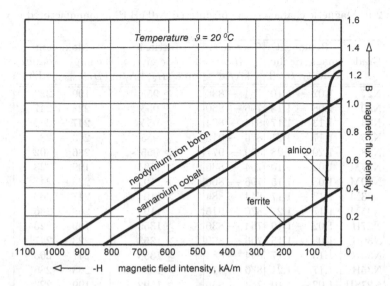

Fig. 2.11. Demagnetization curves of different permanent magnet materials.

Demagnetization curves of these PM materials are given in Fig. 2.11.

The main advantages of *Alnico* are its high magnetic remanent flux density and low temperature coefficients. The temperature coefficient of B_r is $-0.02\%/°C$, and maximum service temperature is 520°C. These advantages allow a quite high air gap flux density and high operating temperatures. Unfortunately, coercive force is very low, and the demagnetization curve is extremely nonlinear. Therefore, it is very easy not only to magnetize but also to demagnetize Alnico. Sometimes, Alnico PMs are protected from the armature flux and, consequently, from demagnetization, using additional soft-iron pole shoes. Alnico magnets dominated the PM machines industry from the mid 1940s to about 1970 when ferrites became the most widely used materials [166].

Barium and strontium *ferrites* were invented in the 1950s. A ferrite has a higher coercive force than that of Alnico, but at the same time has a lower remanent magnetic flux density. Temperature coefficients are relatively high, i.e., the coefficient of B_r is $-0.20\%/°C$ and the coefficient of H_c is $-0.27\%/°C$. The maximum service temperature is 400°C. The main advantages of ferrites are their low cost and very high electric resistance, which means no eddy-current losses in the PM volume. Barium ferrite PMs are commonly used in small d.c. commutator motors for automobiles (blowers, fans, windscreen wipers, pumps, etc.) and electric toys. Ferrites are produced by powder metallurgy. Their chemical formulation may be expressed as $MO\times6(Fe_2O_3)$, where M is Ba, Sr, or Pb. Strontium ferrite has a higher coercive force than barium ferrite. Lead ferrite has a production disadvantage from an environmental point of view. Ferrite magnets are available in *isotropic* and *anisotropic* grades.

Table 2.9. Magnetic characteristics of sintered NdFeB PMs manufactured in China

Grade	Remanent magnetic flux density B_r, T	Coercivity H_c, kA/m	Intrinsic coercive force iH_c, kA/m	Maximum energy product $(BH)_{max}$, kJ/m^3
N27	1.02 — 1.10	764 — 836	≥ 955	199 — 223
N30	1.08 — 1.15	796 — 860	≥ 955	223 — 247
N33	1.13 — 1.17	844 — 884	≥ 955	247 — 263
N35	1.17 — 1.21	876 — 915	≥ 955	263 — 286
N38	1.20 — 1.28	899 — 971	≥ 955	286 — 302
N27M	1.02 — 1.10	764 — 836	≥ 1194	199 — 223
N30M	1.08 — 1.15	796 — 860	≥ 1194	223 — 247
N33M	1.13 — 1.17	844 — 884	≥ 1194	247 — 263
N35M	1.17 — 1.21	876 — 915	≥ 1194	263 — 286
N27H	1.02 — 1.10	764 — 836	≥ 1353	199 — 223
N30H	1.08 — 1.15	796 — 860	≥ 1353	223 — 247
N33H	1.13 — 1.17	844 — 884	≥ 1353	247 — 263
N35H	1.17 — 1.21	876 — 915	≥ 1353	263 — 286
N27SH	1.02 — 1.10	764 — 836	≥ 1592	199 — 223
N30SH	1.08 — 1.15	796 — 860	≥ 1592	223 — 247
N33SH	1.13 — 1.17	844 — 884	≥ 1592	247 — 263
N35SH	1.16 — 1.22	876 — 915	≥ 1592	263 — 279
N25UH	0.97 — 1.05	748 — 812	≥ 1910	183 — 207
N27UH	1.02 — 1.10	764 — 836	≥ 1910	199 — 223

During the last three decades, great progress regarding available energy density $(BH)_{max}$ has been achieved with the development of *rare-earth PMs*. The rare-earth elements are in general not rare at all, but their natural minerals are widely mixed compounds. To produce one particular rare-earth metal, several others, for which no commercial application exists, have to be refined. This limits the availability of these metals. The first generation of these new alloys were invented in the 1960s and based on the composition $SmCo_5$, which has been commercially produced since the early 1970s. Today it is a well established hard magnetic material. $SmCo_5$ has the advantage of high remanent flux density, high coercive force, high energy product, linear demagnetization curve, and low temperature coefficient. The temperature coefficient of B_r is −0.03%/°C to −0.045%/°C, and the temperature coefficient of H_c is −0.14%/°C to −0.40%/°C. Maximum service temperature is 250° to 300°C. It is well suited to build motors with low volume and, consequently, high specific power and low moment of inertia. The cost is the only drawback. Both Sm and Co are relatively expensive due to their supply restrictions.

With the discovery of a second generation of rare-earth magnets on the basis of cost-effective neodymium (Nd) and iron (Fe), a remarkable progress with regard to lowering raw material costs has been achieved. This new generation of rare-earth PMs was announced by Sumitomo Special Metals, Japan,

Table 2.10. Physical properties of sintered NdFeB PMs manufactured in China

Grade	Operating temperature ^0C	Temperature coefficient for B_r %/^0C	Curie temp. ^0C	Specific mass density g/cm^3	Recoil permeability
N27	≤ 80	-0.11	310	7.4 — 7.5	1.1
N30	≤ 80	-0.11	310	7.4 — 7.5	1.1
N33	≤ 80	-0.11	310	7.4 — 7.5	1.1
N35	≤ 80	-0.11	310	7.4 — 7.5	1.1
N38	≤ 80	-0.11	310	7.4 — 7.5	1.1
N27M	≤ 100	-0.11	320	7.4 — 7.5	1.1
N30M	≤ 100	-0.11	320	7.4 — 7.5	1.1
N33M	≤ 100	-0.11	320	7.4 — 7.5	1.1
N35M	≤ 100	-0.11	320	7.4 — 7.5	1.1
N27H	≤ 120	-0.10	340	7.4 — 7.5	1.1
N30H	≤ 120	-0.10	340	7.4 — 7.5	1.1
N33H	≤ 120	-0.10	340	7.4 — 7.5	1.1
N35H	≤ 120	-0.10	340	7.4 — 7.5	1.1
N27SH	≤ 150	-0.10	340	7.4 — 7.5	1.1
N30SH	≤ 150	-0.10	340	7.4 — 7.5	1.1
N33SH	≤ 150	-0.10	340	7.4 — 7.5	1.1
N35SH	≤ 150	-0.10	340	7.4 — 7.5	1.1
N25UH	≤ 170	-0.10	340	7.4 — 7.5	1.1
N27UH	≤ 170	-0.10	340	7.4 — 7.5	1.1

Table 2.11. Magnetic characteristics of bonded NdFeB PMs manufactured in China

Grade	Remanent magnetic flux density B_r, T	Coercivity H_c, kA/m	Intrinsic coercive force iH_c, kA/m	Maximum energy product $(BH)_{max}$, kJ/m^3
N36G	≥ 0.70	≥ 170	≥ 210	32 — 40
N44Z	≥ 0.47	≥ 360	≥ 540	40 — 48
N52Z	≥ 0.55	≥ 360	≥ 500	48 — 56
N60Z	≥ 0.58	≥ 380	≥ 680	56 — 64
N68G	≥ 0.60	≥ 410	≥ 1120	64 — 72
N76Z	≥ 0.65	≥ 400	≥ 720	70 — 80
N84Z	≥ 0.70	≥ 450	≥ 850	80 — 88

Table 2.12. Physical properties of bonded NdFeB PMs manufactured in China

Grade	Maximum operating temperature ^{0}C	Temperature coefficient for B_r %/^{0}C	Curie temp. ^{0}C	Specific mass density g/cm^3
N36G	70	≤ -0.13	300	6.0
N44Z	110	≤ -0.13	350	6.0
N52Z	120	≤ -0.13	350	6.0
N60Z	120	≤ -0.13	350	6.0
N68G	150	≤ -0.13	305	6.0
N76G	150	≤ -0.13	360	6.0
N48Z	150	≤ -0.13	360	6.0

in 1983 at the 29th Annual Conference of Magnetism and Magnetic Materials held in Pittsburg. The Nd is a much more abundant rare-earth element than Sm. NdFeB magnets, which are now produced in increasing quantities, have better magnetic properties than those of SmCo, but unfortunately only at room temperature. The demagnetization curves, especially the coercive force, are strongly temperature dependent. The temperature coefficient of B_r is $-0.095\%/^\circ$C to $-0.15\%/^\circ$C and the temperature coefficient of H_c is $-0.40\%/^\circ$C to $-0.70\%/^\circ$C. The maximum service temperature is 170^0C, and Curie temperature is 300 to 360°C. The NdFeB is also susceptible to corrosion. NdFeB magnets have great potential for considerably improving the *performance–to–cost* ratio for many applications. For this reason they have a major impact on the development and application of PM apparatus.

According to the manufacturing processes, rare earth NdFeB PMs are clasified into *sintered* PMs (Tables 2.9 and 2.10) and *bonded* PMs (Table 2.11 and 2.12).

2.4 Conductors

Armature windings of electric motors are made of solid *copper conductor wires* with round or rectangular cross sections. When the cost or mass of the motor are paramount, e.g., long armature LSMs for transportation systems, magnetically levitated vehicles, hand tools, etc., *aluminum conductor wires* can be more suitable.

2.4.1 Magnet Wire

The *magnet wire* or *winding wire* is an insulated copper or aluminum conductor typically used to wind electromagnetic devices such as machines and transformers. *American Wire Gauge* is the standard used to represent successive diameters of wire. The system is based on the establishment of two arbitrary sizes: 4/0 defined exactly as 0.4600 inch diameter, and 36, defined as exactly 0.0050 inch diameter. The ratio of these two sizes is 92, and the sizes, between the two are based on the 39th root of 92, or approximately 1.123, so the nominal diameter of each gauge size increases approximately by this factor between AWG 36 and AWG 4/0, and decreases by this factor between AWG 36 and AWG 56, which is the smallest practical diameter for commercial magnet wire. Nominal wire diameters to AWG 44 are rounded to the fourth decimal place and are not necessarily rounded to the nearest digit.

There are a number of film insulation types ranging from temperature Class 105°C to Class 240°C. Each film type has its own unique set of characteristics to suit specific needs of the user. A very common wire used in many applications is single build polyurethane Class 155°C with a nylon topcoat. It is stocked in most sizes and is a good general-purpose insulation for undecided customers. Armored polyester insulation is another option when a higher temperature class is desired.

Bondable wire has a thermoplastic adhesive film superimposed over standard film insulation. When activated by heat or solvent, the bond coating cements the winding turn-to-turn to create a self-supporting coil. The use of bondable wire can eliminate the need for bobbins, tape, or varnishes.

2.4.2 Resistivity and Conductivity

Electric conductivity σ of materials used for windings of electrical machines is given in Table 2.13. The electric conductivity is temperature dependent.

The variation of resistance R, electric resistivity ρ and electric conductivity σ with temperature in temperature range from 0°C to 150°C is expressed by the following equations:

$$R(\vartheta) = R_{20}[1 + \alpha_{20}(\vartheta - 20)] \tag{2.23}$$

$$\rho(\vartheta) = \rho_{20}[1 + \alpha_{20}(\vartheta - 20)] \tag{2.24}$$

$$\sigma(\vartheta) = \frac{\sigma_{20}}{1 + \alpha_{20}(\vartheta - 20)} \tag{2.25}$$

where R_{20}, ρ_{20}, σ_{20}, α_{20} are the resistance, resistivity, conductivity, and temperature coefficient at 20°C, respectively, and ϑ is the given temperature. For copper wires, $\alpha = 0.00393$ 1/^{0}C, and for aluminum wires $\alpha = 0.00403$ 1/^{0}C. For $\vartheta > 150$°C, additional electric temperature coefficient β_{20} must be introduced, e.g., for resistance

$$R_\vartheta = R_{20}[1 + \alpha_{20}(\vartheta - 20) + \beta_{20}(\vartheta - 20)^2] \tag{2.26}$$

At temperatures higher than room temperature (up to 1000°C), the resistivity of copper, aluminum, and brass changes more or less linearly with temperature, while the resistivity of steels abruptly increases (Fig. 2.12).

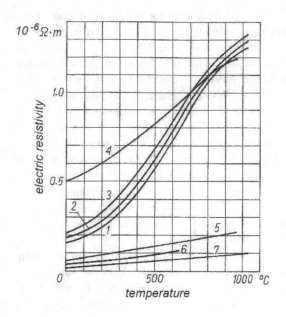

Fig. 2.12. Variation of resistivity ρ of metals with temperature: 1 — mild steel 0.11%C, 2 — mild steel 0.5%C, 3 — mild steel 1% C, 4 — stainless and acid resistant steels, 5 — brass 60%Cu, 6 — aluminum, 7 — copper [126].

Table 2.13. Electric resistivity, conductivity, and temperature coefficient at 20°C

Material	Electric resistivity $\times 10^{-6}$ Ωm	Electric conductivity $\times 10^6$ S/m	Electric temperature coefficient at 20°C 1/K or 1/°C
Aluminum	0.0278	36	+0.00390
Brass (58% Cu)	0.059	17	+0.00150
Brass (63% Cu)	0.071	14	+0.00150
Carbon	40	0.025	−0.00030
Cast iron	1	1	—
Constantan	0.48	2.08	−0.00003
Copper	0.0172	58	+0.00380
Gold	0.0222	45	—
Graphite	8.00	0.125	−0.00020
Iron (pure)	0.10	10	—
Mercury	0.941	1.063	+0.00090
Mild steel	0.13	7.7	+0.00660
Nickel	0.087	11.5	+0.00400
Platinum	0.111	9	+0.00390
Silver	0.016	62.5	+0.00377
Zinc	0.061	16.5	+0.00370

Normally, the resistivity of copper used for electric conductors is $\rho = 0.017241 \times 10^{-6}$ Ωm at 20°C, conductivity $\sigma = 58 \times 10^6$ S/m, the specific mass density 8900 kg/m^3, coefficient of linear expansion 1.68×10^{-5} 1/K, specific heat to 390 Ws/K, and unit heat conductivity 3.75×10^2 W/(K m). Various impurities degrade the electrical conductivity of copper.

The resitivity of aluminum, in normally refined form, is $\rho = 0.0282 \times 10^{-6}$ Ωm at 20°C, conductivity $\sigma = 35.5 \times 10^6$ S/m, specific mass density 2640 kg/m^3 for cast aluminum, 2700 kg/m^3 for drawn aluminum, coefficient of linear expansion 2.22×10^{-5} 1/K, specific heat 810 Ws/K, and unit heat conductivity 2.0×10^2 W/(K m).

Table 2.14. Maximum temperature rise $\Delta\vartheta$ for armature windings of electrical machines according to IEC and NEMA (based on 40° ambient temperature)

Rated power of machines, length of core and voltage	Insulation class				
	A °C	E °C	B °C	F °C	H °C
IEC a.c. machines < 5000 kVA (resistance method)	60	75	80	100	125
IEC a.c. machines ≥ 5000 kVA or length of core ≥ 1 m (embedded detector method)	60	70	80	100	125
NEMA a.c. machines ≤ 1500 hp (embedded detector method)	70	—	90	115	140
NEMA a.c. machines > 1500 hp and ≤ 7 kV (embedded detector method)	65	—	85	110	135

Table 2.15. Magnet wire insulation summary

Thermal class	Insulation type
105°C	Oleoresinous enamel (plain enamel) polyurethane
130°C	Polyurethane HT, polyurethane—nylon
155°C	Polyester
180°C	Polyester—imide, polyester—imide—nylon
200°C	Polyester—imide—amide, polytetrafluoroethylene (Teflon)
220°C	Polyamide, Kapton® tape
500°C	Aluminum oxided, ceramic coated

Table 2.16. Insulation specifications according to MWS Wire Industries, www.mwswire.com

Thermal class	Insulation type	MWS Product code	NEMA Standard (MW1000)	Fed specifi (JW11
105°C	Plain Enamel — Available 40–44 AWG	PE	None	No
	Formvar (RD)	F	MW 15	/
	Formvar (SQ and rect)	F	MW 18	/
	Polyurethane bondable	PB	MW 3	/
	Formvar bondable	FB	MW 19	/
	Polyurethane nylon bondable	PNB	MW 29	/
130°C	Polyurethane nylon[1]	PN	MW 28	/
155°C	Polyurethane 155[1]	P155	MW 79	/
	Polyurethane nylon 155[1]	PN155	MW 80	/
180°C	Polyurethane 180[1]	P180	MW 82	No
	Polyurethane nylon 180[1]	PN180	MW 83	No
	Polyester—imide	PT	MW 30	/
	Polyester nylon[1]	PTN	MW 76	/
	Solderable polyester[1]	SPT	MW 77	/
	Solderable polyester nylon[1]	SPTN	MW 78	/
	Polyester—imide bondable[1]	PTB	None	No
	Polyester—amide—imide bondable[1]	APTB	None	No
	Solderable polyester bondable[1]	SPTB	None	No
200°C	Glass fibers (RD)	Glass	MW 44	/
	Glass fibers (SQ and RECT)	Glass	MW 43	/
	Dacron glass (RD)	Dglas	MW 45	/
	Dacron glass (SQ and RECT)	Dglass	MW 46	/
	Polyester 200[1]	PT200	MW 74	/
	Polyester A/I Topcoat[1] (RD)	APT	MW 35	/
	Polyester A/I Topcoat[1] (SQ and RECT)	APT	MW 36	/
	Polyester A/I Polyamide—imide (RD)	APTIG	MW 35	No
	Polyester A/I Polyamide—imide (SQ and RECT)	APTIG	MW 73	No
	Polytetrafluoloethylene (Teflon[1])	Teflon	None	No
240°C	Polyimide–ML[2] (RD)	ML	MW 16	/1
	Polyimide–ML[2] (SQ and RECT)	ML	MW 20	/1

[1] UL–recognized insulations
[2] Registered trademark of DuPont Corp.
RD = round, RECT = rectangular, SQ = square

2.5 Insulation Materials

2.5.1 Classes of Insulation

The *maximum temperature rise* for the windings of electrical machines is determined by the temperature limits of insulating materials. The maximum temperature rise in Table 2.14 assumes that the temperature of the cooling medium $\vartheta_c \leq 40^0$C. The winding can reach the maximum temperature of

$$\vartheta_{max} = \vartheta_c + \Delta\vartheta \qquad (2.27)$$

where $\Delta\vartheta$ is the maximum allowable temperature rise according to Table 2.14. Polyester–imide and polyamide–imide coat can provide operating temperature of 200^0C.

2.5.2 Commonly Used Insulating Materials

Magnet wire insulation summary is given in Table 2.15. Insulation specifications according to MWS Wire Industries, Westlake Village, CA are listed in Table 2.16.

Heat-sealable Kapton® polyimide films are used as primary insulation on magnet wire at temperatures 220∘C to 240°C [154]. These films are coated with or laminated to Teflon® fluorinated ethylene propylene (FEP) fluoropolymer, which acts as a high-temperature adhesive. The film is applied in tape form by helically wrapping it over and heat-sealing it to the conductor and to itself.

The highest operating temperatures (over 600^0C) can be achieved using *nickel clad copper* or *palladium–silver* conductor wires and ceramic insulation [34].

2.5.3 Impregnation

After coils are wound, they must be somehow secured in place so as to avoid conductor movement. Two standard methods are used to secure the conductors of electrical machines in place:

- *Dipping* the whole component into a varnish-like material, and then baking off its solvent
- *Trickle impregnation* method, which uses heat to cure a catalyzed resin that is dripped onto the component

Polyester, epoxy, or silicon resins are most often used as impregnating materials for treatment of stator or rotor windings. Silicon resins of high thermal endurance are able to withstand $\vartheta_{max} > 225°$C.

Recently, a new method of conductor securing that does not require any additional material and uses very low energy input has emerged [132]. The solid conductor wire (usually copper) is coated with a *heat- and/or solvent-activated adhesive*. The adhesive, which is usually a polyvinyl butyral, utilizes a low-temperature thermoplastic resin [132]. This means that the bonded adhesive can come apart after a certain minimum temperature is reached, or it again comes in contact with the solvent. Normally, this temperature is much lower than the thermal rating of the base insulation layer. The adhesive is activated by either passing the wire through a solvent while winding or heating the finished coil as a result of passing electric current through it.

The conductor wire with a heat-activated adhesive overcoat costs more than the same class of nonbondable conductor. However, a less than two second current pulse is required to bond the heat-activated adhesive layer, and bonding machinery costs about half as much as trickle impregnation machinery [132].

Polyester, epoxy, or silicon resins are used most often as impregnating materials for the treatment of stator windings. Silicon resins of high thermal endurance are able to withstand $\vartheta_{max} > 225°C$.

2.6 Principles of Superconductivity

Superconductivity (SC) is a phenomenon occurring in certain materials at low temperatures, characterized by the complete absence of electrical resistance and the damping of the interior magnetic field (Meissner effect). The *critical temperature* for SCs is the temperature at which the electrical resistivity of an SC drops to zero. Some critical temperatures of metals are aluminum (Al) $T_c = 1.2$ K, tin (Sn) $T_c = 3.7$ K, mercury (Hg) $T_c = 4.2$ K, vanadium (V) $T_c = 5.3$ K, lead (Pb) $T_c = 7.2$ K, niobium (Nb) $T_c = 9.2$ K. Compounds can have higher critical temperatures, e.g., $T_c = 92$ K for $YBa_2Cu_3O_7$, and $T_c = 133$ K for $HgBa_2Ca_2Cu_3O_8$. Superconductivity was discovered by the Dutch scientist H. Kamerlingh Onnes in 1911 (Nobel Prize in 1913). Onnes was the first person to liquefy helium (4.2 K) in 1908.

The superconducting state is defined by three factors (Fig. 2.13):

1. Critical temperature T_c;
2. Critical magnetic field H_c;
3. Critical current density J_c.

Maintaining the superconducting state requires that both the magnetic field and the current density, as well as the temperature, remain below the critical values, all of which depend on the material.

The phase diagram $T_c H_c J_c$ shown in Fig. 2.13 demonstrates the relationship between T_c, H_c, and J_c. When considering all three parameters, the plot represents a critical surface. For most practical applications, SCs must be able to carry high currents and withstand high magnetic field without reverting to their normal state.

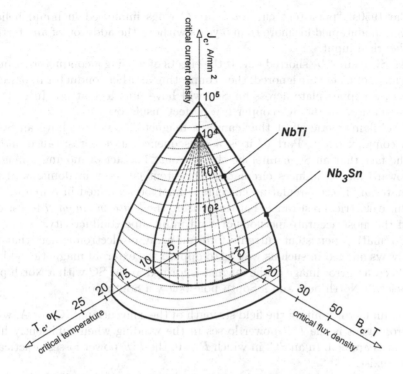

Fig. 2.13. Phase diagram $T_c H_c J_c$.

Meissner effect (sometimes called the Meissner—Ochsenfeld effect) is the expulsion of a magnetic field from an SC.

When a thin layer of insulator is sandwiched between two SCs, until the current becomes critical, electrons pass through the insulator as if it does not exist. This effect is called *Josephson effect*. This phenomenon can be applied to the switching devices that conduct on—off operation at high speed.

In *type I superconductors*, the superconductivity is "quenched" when the material is exposed to a sufficiently high magnetic field. This magnetic field, H_c, is called the *critical field*. In contrast, *type II superconductors* have two critical fields. The first is a low-intensity field H_{c1}, which partially suppresses the superconductivity. The second is a much higher critical field, H_{c2}, which totally quenches[2] the superconductivity. The upper critical field of type II superconductors tends to be two orders of magnitude or more above the critical fields of a type I superconductor.

Some consequences of zero resistance are as follows:

- When a current is induced in a ring-shaped SC, the current will continue to circulate in the ring until an external influence causes it to stop. In

---

[2] Quenching is a resistive heating of an SC and a sudden temperature rise.

the 1950s, "persistent currents" in SC rings immersed in liquid helium were maintained for more than 5 years without the addition of any further electrical input.

- An SC cannot be shorted out. If the effects of moving a conductor through a magnetic field are ignored, then connecting another conductor in parallel, e.g., a copper plate across an SC, will have no effect at all. In fact, by comparison to the SC, copper is a perfect insulator.
- The diamagnetic effect that causes a magnet to levitate above an SC is a complex effect. Part of it is a consequence of zero resistance and of the fact that an SC cannot be shorted out. The act of moving a magnet toward an SC induces circulating persistent currents in domains in the material. These circulating currents cannot be sustained in a material of finite electrical resistance. For this reason, the *levitating magnet* test is one of the most accurate methods of confirming superconductivity.
- Circulating persistent currents form an array of electromagnets that are always aligned in such as way as to oppose the external magnetic field. In effect, a mirror image of the magnet is formed in the SC with a North pole below a North pole and a South pole below a South pole.

The main factor limiting the field strength of the conventional (Cu or Al wire) electromagnet is the I^2R power losses in the winding when sufficiently high current is applied. In an SC, in which $R \approx 0$, the I^2R power losses practically do not exist.

Lattice (from the mathematics point of view) is a partially ordered set in which every pair of elements has a unique *supremum* (least upper bound of elements, called their *join*) and an *infimum* (greatest lower bound, called their *meet*). Lattice is an infinite array of points in space, in which each point has surroundings identical to all others. *Crystal structure* is the periodic arrangement of atoms in the crystal.

The only way to describe SCs is to use quantum mechanics. The model used is the *BSC theory* (named after Bardeen, Cooper, and Schrieffer) was first suggested in 1957 (Nobel Prize in 1973) [20]. It states that

- lattice vibrations play an important role in superconductivity;
- electron—phonon interactions are responsible.

Photons are the quanta of electromagnetic radiation. *Phonons* are the quanta of acoustic radiation. They are emitted and absorbed by the vibrating atoms at the lattice points in the solid. Phonons possess discrete energy $E = h\nu$, where $h = 6.626\ 068\ 96(33) \times 10^{-34}$ Js is *Planck constant*. Phonons propagate through a crystal lattice.

Low temperatures minimize the vibrational energy of individual atoms in the crystal lattice. An electron moving through the material at low temperature encounters less of the impedance due to vibrational distortions of the lattice. The Coulomb attraction between the passing electron and the positive ion distorts the crystal structure. The region of increased positive charge density propagates through the crystal as a quantized sound wave called a phonon. The phonon exchange neutralizes the strong electric repulsion between the two electrons due to Coulomb forces. Because the energy of the paired electrons is lower than that of unpaired electrons, they bind together. This is called *Cooper pairing*. Cooper pairs carry the supercurrent relatively unresisted by the thermal vibration of the lattice. Below T_c, pairing energy is sufficiently strong (Cooper pair is more resistant to vibrations), the electrons retain their paired motion and, upon encountering a lattice atom, do not scatter. Thus, the electric resistivity of the solid is zero. As the temperature rises, the binding energy is reduced and goes to zero when $T = T_c$. Above T_c, a Cooper pair is not bound. An electron alone scatters (collision interactions), which leads to ordinary resistivity. Conventional conduction is resisted by thermal vibration within the lattice.

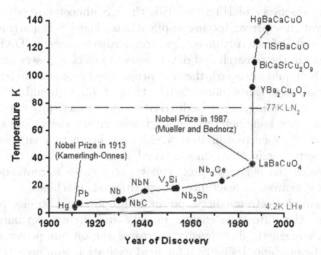

Fig. 2.14. Discovery of materials with successively higher critical temperatures over the last century.

In 1986, J. Georg Bednorz and K. Alex Mueller of IBM Ruschlikon, Switzerland, published results of research [23] showing indications of superconductivity at about 30 K (Nobel Prize in 1987). In 1987 researchers at the University of Alabama at Huntsville (M. K. Wu) and at the University of Houston (C. W. Chu) produced ceramic SCs with a critical temperature ($T_c = 52.5$ K) above the temperature of liquid nitrogen. Discoveries of materials with successively higher critical temperatures over the last century are presented in Fig. ??.

There is no widely accepted temperature that separates *high-temperature superconductors* (HTS) from *low-temperature superconductors* (LTS). Most LTS superconduct at the boiling point of liquid helium (4.2 K = -269^0C at 1 atm). However, all the SCs known before the 1986 discovery of the superconducting oxocuprates would be classified LTS. The barium-lanthanum-cuprate BaLaCuO fabricated by Mueller and Bednorz, with a $T_c = 30$ K $= -243^0$C, is generally considered to be the first HTS material. Any compound that will superconduct at and above this temperature is called HTS. Most modern HTS superconduct at the boiling point of liquid nitrogen (77 K $= -196^0$C at 1 atm).

The most important market for LTS electromagnets are currently *magnetic resonance imaging* (MRI) devices, which enable physicians to obtain detailed images of the interior of the human body without surgery or exposure to ionizing radiation.

All HTS are *cuprates* (copper oxides). Their structure relates to the *perovskite structure* (calcium titanium oxide $CaTiO_3$) with the general formula ABX_3. Perovskite $CaTiO_3$ is a relatively rare mineral occurring in orthorhombic (pseudocubic) crystals[3].

With the discovery of HTS in 1986, the US almost immediately resurrected interest in superconducting applications. The US Department of Energy (DoE) and Defense Advanced Research Projects Agency (DARPA) have taken the lead in the research and development of electric power applications. At 60 to 77 K (liquid nitrogen), thermal properties become more friendly, and cryogenics can be 40 times more efficient than at 4.2 K (liquid helium).

In power engineering, superconductivity can be practically applied to synchronous machines, homopolar machines, transformers, energy storages, transmission cables, fault-current limiters, LSMs, and magnetic levitation vehicles. The use of superconductivity in electrical machines reduces the excitation losses, increases the magnetic flux density, eliminates ferromagnetic cores, and reduces synchronous reactance (in synchronous machines).

The apparent electromagnetic power is proportional to electromagnetic loadings, i.e., the stator line current density and the air gap magnetic flux density. High magnetic flux density increases the output power or reduces the size of the machine. Using SCs for field excitation winding, the magnetic flux density can exceed the saturation magnetic flux of the best ferromagnetic materials. Thus, ferromagnetic cores can be removed.

2.7 Superconducting Wires

The first commercial low LTS wire was developed at *Westinghouse* in 1962. Typical LTS wires are magnesium diboride MgB_2 tapes, NbTi Standard, and Nb_3Sn Standard. LTS wires are still preferred in high field magnets for nuclear magnetic resonance (NMR), *magnetic resonance imaging* (MRI), magnets for accelerators, and fusion magnets. Input cooling power as a function of required temperature is shown in Fig. 2.15.

---

[3] Perovskite (In German "Perovskit") was discovered in the Ural mountains of Russia by G. Rose in 1839 and named for Russian mineralogist, L. A. Perovski (1792—1856).

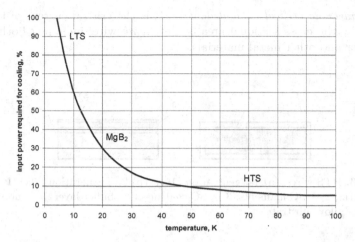

Fig. 2.15. Input cooling power in percent of requirements at 4.2K versus temperature.

Manufacturers of electrical machines need low-cost HTS tapes that can operate at temperatures approaching 77 K for economical generators, motors, and other power devices. The minimum length of a single piece acceptable by the electrical engineering industry is at least 100 m.

2.7.1 Classification of HTS Wires

HTSwires are divided into two categories:

1. *First generation* (1G) superconductors, i.e., *multifilamentary tape con-ductors* BiSrCaCuO (BSCCO) developed up to industrial state. Their properties are reasonable for different use, but prices are still high.
2. *Second generation* (2G) superconductors, i.e., *coated tape conductors*: YBaCuO (YBCO) which offer superior properties.

Fig. 2.16 shows basic 1G and 2G HTS wire tape architecture according to *American Superconductors* [10]. Each type of advanced wire achieves high power density with minimal electrical resistance, but differs in the SC materials manufacturing technology, and, in some instances, its end-use applications.

Parameters characterizing superconductors are

- critical current $I_c \times$ (wire length), Am (200,580 Am in 2008) [196];
- critical current $I_c/$(wire width), A/cm–width (700 A/cm–width in 2009) [10];
- critical current density J_c, A/cm^2;
- engineering critical current density J_e, A/cm^2.

The *engineering critical current density* J_e is the critical current of the wire divided by the cross sectional area of the entire wire, including both superconductor and other metal materials.

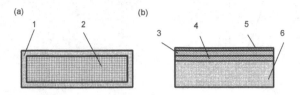

Fig. 2.16. HTS wires: (a) 1st generation (1G); (b) 2nd generation (2G). 1 — silver alloy matrix, 2 — SC filaments, 3 — SC coating, 4 — buffer layer, 5 — noble metal layer, 5 — alloy substrate.

BSCCO 2223 is a commonly used name to represent the HTS material $Bi_{(2-x)}Pb_xSr_2Ca_2Cu_3O_{10}$. This material is used in multifilamentary composite HTS wire and has a typical superconducting transition temperature around 110 K. BSCCO-2223 is proving successful presently but will not meet all industrial requirements in the nearest future. According to *SuperPower* [196], there are clear advantages to switch from 1G to 2G,

- better in-field performance;
- better mechanical properties (higher critical tensile stress, higher bend strain, higher tensile strain);
- better uniformity, consistency, and material homogeneity;
- higher engineering current density;
- lower a.c. losses.

There are key areas where 2G needs to be competitive with 1G in order to be used in the next round of various device prototype projects. Key benchmarks to be addressed are

- long piece lengths;
- critical current over long lengths;
- availability (high throughput, i.e., production volume per year, large deliveries from pilot-scale production);
- comparable cost with 1G.

Commercial quantities of HTS wire based on BSCCO are now available at around five times the price of the equivalent copper conductor. Manufacturers are claiming the potential to reduce the price of YBCO to 50% or even 20% of BSCCO. If the latter occurs, HTS wire will be competitive with copper in many large industrial applications.

Magnesium diboride, MgB_2, is a much cheaper SC than BSCCO and YBCO in terms of dollars per current-carrying capacity × length ($/kA-m).

However, this material must be operated at temperatures below 39K, so the cost of cryogenic equipment is very significant. This can be half the capital cost of the electric machine when cost-effective SC is used. Magnesium diboride, SC MgB_2, might gain niche applications if further developments will be successful.

As of March 2007, the current world record of superconductivity is held by a ceramic SC consisting of thallium, mercury, copper, barium, calcium, strontium, and oxygen with $T_c = 138$ K.

2.7.2 HTS Wires Manufactured by American Superconductors

American Superconductors (AMSC) [10] is the world's leading developer and manufacturer of HTS wires. Two types of HTS wires branded as 344 and 348 have been commercially available since 2005. AMSC HTS wire specifications are given in Table 2.17.

Table 2.17. Specifications of HTS wires manufactured by American Superconductors, Westborough, MA, USA [10]

Specifications	Bismuth based, multifilamentary HTS wire encased in a silver alloy matrix	344 HTS copper stabilized wires, 4.4 mm wide	344 HTS stainless steel stabilized wires, 4.4 mm wide
Grade	BSCC0, 1G	YBCO, 2G	YBCO, 2G
Average thickness, mm	0.21 to 0.23	0.20 ± 0.02	0.15 ± 0.02
Minimum width, mm	3.9		
Maximum width, mm	4.3	4.35 ± 0.05 average	4.33 ± 0.07 average
Minimum double bend diameter at 20°C, mm	100	30	30
Maximum rated tensile stress at 20°C, MPa	65	150	150
Maximum rated wire tension at 20°C, kg	4		
Maximum rated tensile stress at 77K, MPa	65		
Maximum rated tensile strain at 77K, %	0.10	0.3	0.3
Average engineering current density J_e, A/cm^2, at minimum critical current I_c, A	$J_e = 12,700$ $I_c = 115$ $J_e = 13,900$ $I_c = 125$ $J_e = 15,000$ $I_c = 135$ $J_e = 16,100$ $I_c = 145$	$J_e = 8000$ $I_c = 70$	$J_e = 9200$ $I_c = 60$
Continuous piece length, m	up to 800	up to 100	up to 20

Today, AMSC 2G HTS wire manufacturing technology is based on 100-m long, 4-cm wide strips of SC material that are produced in a high-speed, continuous reel–to–reel deposition[4] process, the so-called *rolling assisted biaxially textured substrates* (RABITS). RABITS is a method for creating textured[5] metal substrate for 2G wires by adding a buffer layer between the nickel substrate and YBCO. This is done to prevent the texture of the YBCO from being destroyed during processing under oxidizing atmospheres. This process is similar to the low-cost production of motion picture film in which celluloid strips are coated with a liquid emulsion and subsequently slit and laminated into eight, industry-standard 4.4-mm wide tape-shaped wires (344 SCs). The wires are laminated on both sides with copper or stainless-steel metals to provide strength, durability, and certain electrical characteristics needed in applications. AMSC expects to scale up the 4 cm technology to 1000-m lengths. The company then plans to migrate to 10 cm technology to further reduce manufacturing costs.

Sumitomo Electric Industries, Japan, uses the holmium (Ho) rare element instead of yttrium (Y). According to *Sumitomo*, the HoBCO SC layer allows for higher rate of deposition, high critical current density J_c, and better flexibility of tape than the YBCO SC layer.

2.7.3 HTS Wires Manufactured by SuperPower

SuperPower [196] uses *ion beam assisted deposition* (IBAD), a technique for depositing thin SC films. IBAD combines ion implantation with simultaneous sputtering or another physical vapor deposition (PVD) technique. An ion beam is directed at an angle toward the substrate to grow textured buffer layers. According to *SuperPower*, virtually, any substrate could be used, i.e., high-strength substrates, nonmagnetic substrates, low-cost, off–the–shelf substrates (Inconel, Hastelloy, stainless steel), very thin substrates, resistive substrates (for low a.c. losses), etc. There are no issues with percolation (trickling or filtering through a permeable substance). IBAD can pattern the conductor to very narrow filaments for a low a.c. loss conductor. An important advantage is small grain size in the submicron range.

IBAD MgO—based coated conductor has five thin oxide buffer layers with different functions, such as

1. alumina barrier layer to prevent diffusion of metal element into SC;
2. yttria seed layer to provide good nucleation surface for IBAD MgO;
3. IBAD MgO template layer to introduce biaxial texture;
4. homoepitaxial MgO buffer layer to improve biaxial texture;

---

[4] Deposition (in chemistry) is the settling of particles (atoms or molecules) or sediment from a solution, suspension mixture, or vapor onto a preexisting surface.

[5] Texture in materials science is the distribution of crystallographic orientations of a sample. Biaxially textured means textured along two axes.

5. SrTiO$_3$ (STO) cap layer to provide lattice match between MgO and YBCO and good chemical compatibility.

Tables 2.18 and 2.19 show the cost and selected specifications of HTS 2G tapes manufactured by SuperPower [196].

Table 2.18. Cost per meter and parameters of SCS4050 4 mm HTS tape [196].

SCS4050	\$/m	I_c at 4 mm width, 77 K, self field, A	\$/kA-m	Width with copper stabilizer, mm
2006	100	80	1250	4
2007	65	100	650	4
2010	40	100	400	4

Table 2.19. Cost per meter and parameters of SF12050 12 mm HTS tape [196].

SF12050	\$/m	I_c at 12 mm width, 77 K, self field, A	\$/kA-m	Width without copper stabilizer, mm
2006	150	240	625	12
2007	90	300	400	12

2.8 Laminated Stacks

Most LSMs use laminated armature stacks with rectangular semiopen or open slots. In low-speed industrial applications, the frequency of the armature current is well below the power frequency 50 or 60 Hz so that, from the electromagnetic point of view, laminations can be thicker than 0.5 or 0.6 mm (typical thickness for 50 or 60 Hz, respectively). The laminations are cut to dimensions using stamping presses in mass production or laser cutting machines when making prototypes of LSMs. If the stack is thicker than 50 mm, it is recommended that laminations be grouped into 20 to 40-mm thick packets separated by 4 to 8-mm wide longitudinal cooling ducts. Each of the two external laminations should be thicker than internal laminations to prevent the expansion (swelling) of the stack at toothed edges. The slot pitch of a flat armature core is

$$t_1 = b_{11} + c_t = 2p\tau/s_1 \tag{2.28}$$

where b_{11} is the width of the rectangular slot, c_t is the width of tooth, $2p$ is the number of poles, τ is the pole pitch, and s_1 is the number of slots totally filled with conductors. The shapes of armature slots of flat LSMs are shown in Fig. 2.17.

The laminations are kept together with the aid of seam welds (Fig. 2.18), spot welds, or using bolts and bars pressing the stack (Fig. 2.19).

Stamping and stacking can be simplified by using rectangular laminations, as in Fig. 2.20 [227]. When making a prototype, the laminated stack of a polyphase LSM can also be assembled of E-shaped laminations, the same as those used in manufacturing small single-phase transformers.

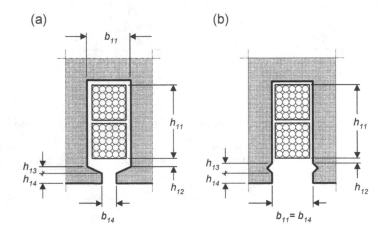

Fig. 2.17. Armature slots of flat LSMs: (a) semiopen, (b) open.

For slotless windings, armature stacks are simply made of rectangular strips of electrotechnical steel.

LSMs for heavy-duty applications are sometimes furnished with finned heat exchangers or water-cooled cold plates that are attached to the yoke of the armature stack.

Armature stacks of tubular motors can be assembled in the following three ways by using

1. star ray arranged long flat slotted stacks of the same dimensions that embrace the excitation system;
2. modules of ring laminations with inner different diameters for teeth and for slots;
3. sintered powders.

The last method seems to be the best from the point of view of performance of the magnetic circuit .

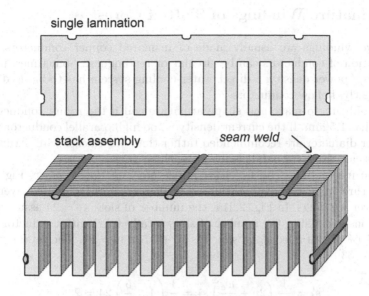

Fig. 2.18. Laminated stack assembled with the aid of seam welds.

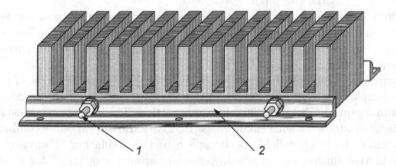

Fig. 2.19. Laminated stack assembled by using: 1 — bolts, and 2 — pressing angle bars.

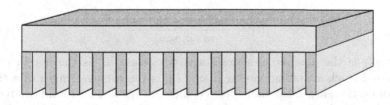

Fig. 2.20. Slotted armature stack of an a.c. linear motor assembled of rectangular laminations.

2.9 Armature Windings of Slotted Cores

Armature windings are usually made of insulated copper conductors. The cross section of conductors can be circular or rectangular. Sometimes, to obtain a high power density, a direct water-cooling system has to be used, and consequently, hollow conductors.

It is difficult to make and shape armature coil if the round conductor is thicker that 1.5 mm. If the current density is too high, parallel conductor wires of smaller diameter are recommended rather than one thicker wire. Armature windings can also have parallel current paths.

Armature windings can be either *single layer* or *double layer*. Fig. 2.21 shows a three-phase, four-pole winding configured both as double-layer and single-layer windings. In Fig. 2.21a, the number of slots $s_1 = 24$ assumed for calculations is equal to the number of slots totally filled with conductors, i.e. 19 plus 50% of half filled slots, which is $19 + 0.5 \times 10 = 24$. The total number of slots is [63]

$$s_1' = \frac{1}{2p}\left(2p + \frac{w_c}{\tau}\right)s_1 = \frac{1}{4}\left(4 + \frac{5}{6}\right)24 = 29$$

where $w_c \leq \tau$ is the coil pitch.

Double-layer armature winding wound with rectangular cross section of conductors of a large-power flat LSM is shown in Fig. 2.22. The coil pitch is equal to three slots, three end slots are half filled, and the end turns are diamond-shaped.

Armature windings of long-core, large-power LSMs can be made of cable. For example, the German *Transrapid* maglev system uses a multistrand, very soft aluminum conductor of 300 mm^2 cross section. The insulation consists of a synthetic elastometer with small dielectric losses. Inner and outer conductive films limit the electric field to the space of the insulation. The cable has a conductive sheath, also consisting of an elastometer mixture, for external protection and electric shielding.

The resistance of a flat armature winding as a function of the winding parameters, dimensions, and electric conductivity is expressed as

$$R_1 = \frac{2(L_i'k_{1R} + l_{1e})N_1}{\sigma s_{w1}a_w a_p} \tag{2.29}$$

where N_1 is the number of series turns per phase, L_i' is the length of the laminated stack including cooling ducts, l_{1e} is the average length of a single end connection, k_{1R} is the skin effect coefficient ($k_{1R} \approx 1$ for round conductors with diameter less than 1 mm and frequency 50 to 60 Hz), σ is the electric conductivity of the conductor at the operating temperature — eqn (2.25), s_{w1} is the cross section of the armature conductor wire, a_w is the number of parallel conductors, and a_p is the number of parallel current paths. For tubular LSMs, $l_{1e} = 0$ and $2L_i'$ in the numerator should be replaced by an

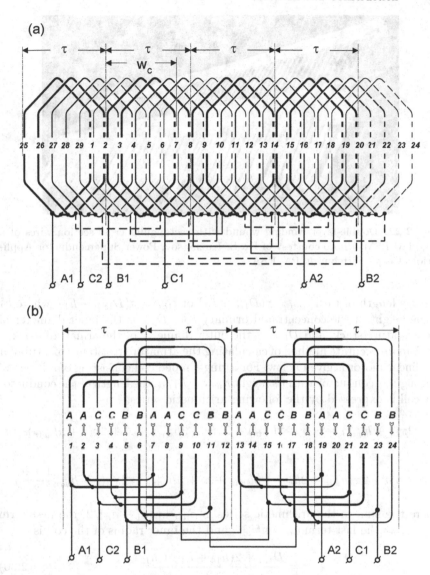

Fig. 2.21. Three-phase, four-pole ($2p = 4$) full pitch windings of an LSM distributed in 24 slots: (a) double-layer winding; (b) single-layer winding.

Fig. 2.22. Double-layer winding wound with rectangular cross section wires of a large PM LSM. Photo courtesy of Dr. S. Kuznetsov, Power Superconductor Applications Corp., Pittsburgh, PA, USA.

average length of turn $l_{1av} = \pi(D_{1in} + h_1)$ or $l_{1av} = \pi(D_{1out} - h_1)$, where h_1 is the height of the concentrated primary coil, D_{1in} is the inner diameter of the armature stack, and D_{1out} is the outer diameter of the armature stack.

A more accurate method of calculating the armature resistance of a tubular a.c. linear motor is given below. For a ring-shaped coil consisting of n layers of rectangular conductor with its height $h_{wir} = h_1/n$, the length of the conductor per coil is expressed by the following arithmetic series:

$$2\pi(r + h_{wir}) + 2\pi(r + 2h_{wir}) + 2\pi(r + 3h_{wir}) + ... + 2\pi(r + nh_{wir})$$

$$= 2\pi r n + 2\pi(h_{wir} + 2h_{wir} + 3h_{wir} + ... + nh_{wir}) = 2\pi n(r + h_{wir}\frac{1+n}{2})$$

where the sum of the arithmetic series is $S_n = n(a_1 + a_n)/2$, the first term $a_1 = h_{wir}$, the last term $a_n = nh_{wir}$, and the inner radius of the coil is

$$r = \frac{D_{1in} + 2(h_{14} + h_{13} + h_{12})}{2} \qquad (2.30)$$

where h_{14} is the slot opening, and h_{13} and h_{12} are dimensions according to Fig. 2.17. For the whole phase winding the length of the conductor per phase is

$$\frac{s_1}{m_1} 2\pi n(r + h_{wir}\frac{1+n}{2})$$

where s_1 is the number of the armature slots, and s_1/m_1 is the number of cylindrical coils per phase in the case of an m_1 phase winding. The resistance of the armature winding per phase is

$$R_1 = \frac{s_1}{m_1}\pi\frac{N_{sl}}{a_w}$$

$$\times \left[D_{1in} + 2(h_{14} + h_{13} + h_{12}) + h_{wir}\left(1 + \frac{N_{sl}}{a_w}\right)\right]\frac{k_{1R}}{\sigma h_{wir}b_{wir}a_w a_p} \quad (2.31)$$

where h_{wir} is the height of the rectangular conductor, b_{wir} is the width of the conductor, σ is the electric conductivity of the conductor, N_{sl} is the number of conductors per slot, a_w is the number of parallel wires, $n = N_{sl}/a_w$ provided that parallel wires are located beside each other, and a_p is the number of parallel current paths. It is recommended that the cross section area of the conductor in the denominator be multiplied by a factor 0.92 to 0.99 to take into account the round corners of rectangular conductors. The bigger the cross section of the conductor, the bigger the factor including round corners. Eqn (2.31) should always be adjusted for the arrangement of conductor wires in slots.

The armature leakage reactance is the sum of the slot leakage reactance X_{1s}, the end connection leakage reactance X_{1e}, and the differential leakage reactance X_{1d} (for higher space harmonics), i.e.,

$$X_1 = X_{1s} + X_{1e} + X_{1d} = 4\pi f\mu_o\frac{L_i N_1^2}{pq_1}(\lambda_{1s}k_{1X} + \frac{l_{1e}}{L_i}\lambda_{1e} + \lambda_{1d} + \lambda_{1t}) \quad (2.32)$$

where N_1 is the number of turns per phase, k_{1X} is the skin-effect coefficient for leakage reactance, p is the number of pole pairs, $q_1 = s_1/(2pm_1)$ is the number of primary slots s_1 per pole per phase, l_{1e} is the length of the primary winding end connection, λ_{1s} is the coefficient of the slot leakage permeance (slot-specific permeance), λ_{1e} is the coefficient of the end turn leakage permeance, and λ_{1d} is the coefficient of the differential leakage. For a tubular LSM, $\lambda_{1e} = 0$, and L_i should be replaced by $L_i = \pi(D_{1in} + h_1)$ or $L_i = \pi(D_{1out} - h_1)$. There is no leakage flux about the end connections as they do not exist in tubular LIMs.

The coefficients of leakage permeances of the slots shown in Fig. 2.17 are:
• for a semiopen slot (Fig. 2.17a)

$$\lambda_{1s} = \frac{h_{11}}{3b_{11}} + \frac{h_{12}}{b_{11}} + \frac{2h_{13}}{b_{11} + b_{14}} + \frac{h_{14}}{b_{14}} \quad (2.33)$$

• for an open slot (Fig. 2.17b)

$$\lambda_{1s} \approx \frac{h_{11}}{3b_{11}} + \frac{h_{12} + h_{13} + h_{14}}{b_{11}} \quad (2.34)$$

The above specific-slot permeances are for single-layer windings. To obtain the specific permeances of slots containing double-layer windings, it is necessary to multiply eqns (2.33) and (2.34) by the factor

$$\frac{3w_c/\tau + 1}{4} \tag{2.35}$$

Such an approach is justified if $2/3 \leq w_c/\tau \leq 1.0$.

The specific permeance of the end turn (overhang) is estimated on the basis of experiments. For double-layer, low-voltage, small- and medium-power motors,

$$\lambda_{1e} \approx 0.34 q_1 (1 - \frac{2}{\pi} \frac{w_c}{l_{1e}}) \tag{2.36}$$

where l_{1e} is the length of a single end turn. Putting $w_c/l_{1e} = 0.64$, eqn (2.36) also gives good results for single-layer windings,

$$\lambda_{1e} \approx 0.2 q_1 \tag{2.37}$$

For double-layer, high-voltage windings,

$$\lambda_{1e} \approx 0.42 q_1 (1 - \frac{2}{\pi} \frac{w_c}{l_{1e}}) k_{w1}^2 \tag{2.38}$$

where the armature winding factor for the fundamental space harmonic $\nu = 1$ is

$$k_{w1} = k_{d1} k_{p1} \tag{2.39}$$

$$k_{d1} = \frac{\sin[\pi/(2m_1)]}{q_1 \sin[\pi/(2m_1 q_1]} \tag{2.40}$$

$$k_{p1} = \sin\left(\frac{\pi}{2} \frac{w_c}{\tau}\right) \tag{2.41}$$

In general,

$$\lambda_{1e} \approx 0.3 q_1 \tag{2.42}$$

for most of the windings.

The specific permeance of the differential leakage flux is

$$\lambda_{1d} = \frac{m_1 q_1 \tau k_{w1}^2}{\pi^2 g k_{C1} k_{sat}} \tau_{d1} \tag{2.43}$$

where the Carter's coefficient including the effect of slotting of the armature stack

$$k_{C1} = \frac{t_1}{t_1 - \gamma_1 g} \tag{2.44}$$

$$\gamma_1 = \frac{4}{\pi} \left[\frac{b_{14}}{2g} \arctan \frac{b_{14}}{2g} - \ln \sqrt{1 + \left(\frac{b_{14}}{2g}\right)^2} \right] \tag{2.45}$$

and the differential leakage factor τ_{d1} is

$$\tau_{d1} = \frac{1}{k_{w1}^2} \sum_{\nu>1} \left(\frac{k_{w1\nu}}{\nu}\right)^2 \tag{2.46}$$

or

$$\tau_{d1} = \frac{\pi^2(10q_1^2+2)}{27} \left[\sin\left(\frac{30^\circ}{q-1}\right)\right]^2 - 1 \tag{2.47}$$

where $k_{w1\nu}$ is the winding factor for $\nu > 1$,

$$k_{w1\nu} = k_{d1\nu}k_{p1\nu} \tag{2.48}$$

$$k_{d1\nu} = \frac{\sin[\nu\pi/(2m_1)]}{q_1 \sin[\nu\pi/(2m_1q_1)]} \tag{2.49}$$

$$k_{p1\nu} = \sin\left(\nu\frac{\pi}{\tau}\frac{w_c}{2}\right) \tag{2.50}$$

Eqns (2.49) and (2.50) express the distribution factor $k_{d1\nu}$ and pitch factor $k_{p1\nu}$ for higher space harmonics $\nu > 1$. The curves of the differential leakage factor τ_{d1} are given in publications dealing with the design of a.c. motors, e.g., [63, 70, 86, 113].

The tooth-top specific permeance

$$\lambda_{1t} \approx \frac{5g_t/b_{14}}{5 + 4g_t/b_{14}} \tag{2.51}$$

should be added to the differential specific permeance λ_{1d} (2.43).

2.10 Slotless Armature Systems

Slotless windings of LSMs for industrial applications can uniformly be distributed on the active surface of the armature core or designed as moving coils without any ferromagnetic stack (air-cored armature). In both cases, the slotless coils can be wound using insulated conductors with round or rectangular cross section (Fig. 2.23) or insulated foil. Fig. 2.24 shows a double-sided PM linear brushless motor with moving inner coil manufactured by Trilogy Systems, Webster, TX, USA [222]. A heavy duty conductor wire with high temperature insulation is used. To improve heat removal, Trilogy Systems forms the winding wire during fabrication into a planar surface at the interface with the aluminum attachment bar (US Patent 4839543, reissue 34674) [P38, P63]. The interface between the winding planar surface and the aluminum flat surface maximizes heat transfer. Once heat is transferred into the attachment bar, it is still important to provide adequate surface area in the

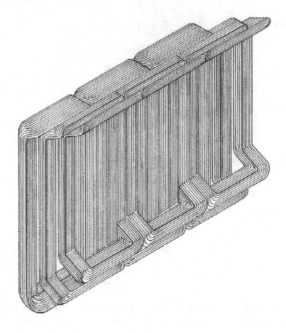

Fig. 2.23. Slotless winding for coreless flat PM LSM according to US Patent 4839543 [P38].

carriage assembly to reject the heat. The use of thermal grease on the coil mounting surface is recommended.

A large-power PM LSM with coreless armature and slotless winding has been proposed for a wheel–on–rail high-speed train [139]. There is a long vertical armature system in the center of the track and train-mounted double-sided PM excitation system.

High-speed maglev trains with SC electromagnets use ground coreless armature windings fixed to or integrated with concrete slabs. A coreless three-phase slotless winding is designed as a double-layer winding [197]. The U-shaped guideway of the Yamanashi Maglev Test Line (Japan) has two three-phase armature windings mounted on two opposite vertical walls of the guide-way (Table 2.20). The on-board excitation system for both the LSM and elec-trodynamic (ELD) levitation is provided by SC electromagnets located on both sides of the vehicle.

Guideways of the Yamanashi Maglev Test Line are classified according to the structure and characteristics of the side wall to which the winding is attached [241]:

(a) Panel type
(b) Side-wall beam type (Fig. 2.25)
(c) Direct attachment type

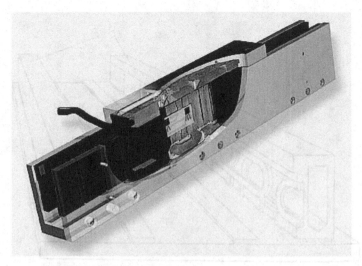

Fig. 2.24. Flat double-sided PM LBM with inner moving coil. Photo courtesy of Trilogy Systems Corporation, Webster, TX, USA.

Table 2.20. Specifications of three-phase armature propulsion winding of SC LSM on Yamanashi Maglev Test Line

Length × width of a coil, m	1.42 × 0.6
Coil pitch, m	0.9
Number of layers	2
Number of turns for the Northern Line	8 (front layer) 10 (back layer)
Number of turns for the Southern Line	7 (front layer) 8 (back layer)
Rated voltage, kV	22 Northern Line 11 Southern Line
Conductor	Al

Panel type windings are produced in the on-site manufacturing yards. Coils are attached to the side of the reinforced concrete panel. The panel is attached to the cast-in-place concrete side walls with the aid of bolts. This type is applied to elevated bridges and high-speed sections.

A *side-wall beam* shown in Fig. 2.25 is a box-type girder. Styrofoam is embedded in the hollow box-type beam to reduce the mass and facilitate the construction. Ground coils are attached to the beam in the manufacturing yard. The beam is then transported to the site and erected on the preproduced support. This type is also applied to elevated bridges and high-speed sections.

In the *direct attachment* type guideway, the coils of the winding are attached direct to the side wall of a cast-in-place reinforced concrete structure.

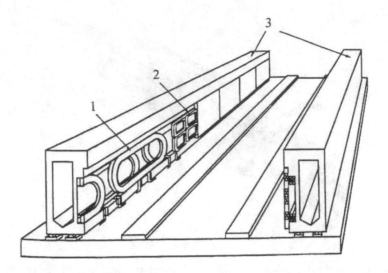

Fig. 2.25. Side-wall beam-type guideway of the Yamansahi Maglev Test Line: 1 — armature (propulsion) winding, 2 — levitation and guidance coil, 3 — twin beams. Courtesy of Central Japan Railway Company and Railway Technical Research Institute, Tokyo, Japan.

This type, unlike the panel or side-wall beam type, cannot respond to structural displacements [241]. The direct attachment guidaway is applied mainly to tunnel sections situated on solid ground.

The resistance of the slotless armature winding can be calculated using eqn (2.29). It is recommended that the finite element method (FEM) be used for calculating the inductance.

2.11 Electromagnetic Excitation Systems

Electromagnetic excitation systems, i.e., salient ferromagnetic poles with d.c. winding, are used in large-power LSMs. For example, the German *Transrapid* system has on board mounted electromagnets (Fig. 2.26), which are used both for propulsion (excitation system of the LSM) and electromagnetic (EML) levitation (attraction forces). Electromagnet modules are approximately 3-m long [234]. There are 12 poles per one module. To deliver electric power to the vehicle, linear generator windings are integrated with each excitation pole. Electromagnet modules are fixed to the levitation frame with the aid of maintenance-free joints. The necessary freedom of movement is obtained by the use of vibration-damping elastic bearings.

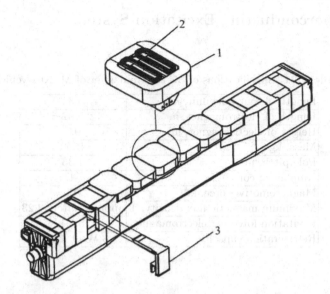

Fig. 2.26. Electromagnetic excitation system with salient poles of *Transrapid* maglev vehicle: 1 — d.c. excitation winding, 2 — linear generator winding, 3 — air gap sensor. Courtesy of Thyssen Transrapid System, München, Germany.

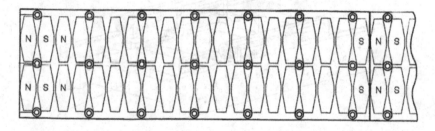

Fig. 2.27. Hexagonal assembly of PMs of LBM. Courtesy of Anorad Corporation, Hauppauge, NY, USA.

2.12 Permanent Magnet Excitation Systems

To minimize the thrust ripple in LSMs or LBMs with slotted armature stack, PMs need to be skewed. The *skew* is approximately equal to one tooth pitch of the armature — eqn (2.28). Instead of skewed assembly (Fig. 1.9) of rectangular PMs, Anorad Corporation proposes to use hexagonal PMs, the symmetry axis of which is perpendicular to the direction of motion (Fig. 2.27) [12].

There are practically no power losses in PM excitation systems (except higher harmonic losses), which do not require any forced cooling or heat exchangers.

2.13 Superconducting Excitation Systems

Table 2.21. Specifications of SC electromagnet of MLX01 vehicle

Length of four pole unit, m	5.32
Length of electromagnet, m	1.07
Height of electromagnet, m	0.5
Mass, kg	1500
Pole pitch, m	1.35
Number of coils	4
Magnetomotive force, kA	700
Maximum magnetic flux density, T	approximately 4.23
Levitation force per electromagnet	115.5 kN
Refrigeration capacity	8 W at 4.3 K

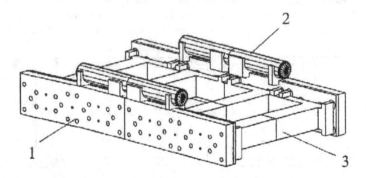

Fig. 2.28. MLX01 SC electromagnet and bogie frame. 1 — SC electromagnet, 2 — tank, 3 — bogie frame. Courtesy of Central Japan Railway Company and Railway Technical Research Institute, Tokyo, Japan.

Superconducting (SC) excitation systems are recommended for high power LSMs which can be used in high speed ELD levitation transport.

In 1972, an experimental SC maglev test vehicle ML100 was built in Japan. Since 1977, when the Miyazaki Maglev Test Center on Kyushu Island opened, maglev vehicles ML and MLU with SC LSMs and electrodynamic suspension systems have been systematically tested. Air-cored armature winding has been installed in the form of a guideway on the ground. In 1990, a new 18.4 km Yamanashi Maglev Test Line (near Mount Fuji) for electrodynamic levitation vehicles with SC LSMs was constructed. In 1993, a test run of MLU002N (Miyazaki) started and since 1995 vehicles MLX01 (Yamanashi Maglev Test Line) have been tested.

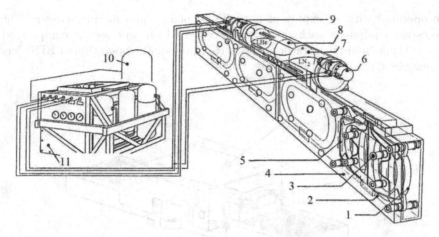

Fig. 2.29. Structure of the MLX01 SC electromagnet and on-board refrigeration system: 1 — SC winding (inner vessel), 2 — radiation shield plate, 3 — support, 4 — outer vessel, 5 — cooling pipe, 6 — 80 K refrigerator, 7 — liquid nitrogen reservoir, 8 — liquid helium reservoir, 9 — 4 K refrigerator, 10 — gaseous helium buffer tank, 11 — compressor unit. Courtesy of Central Japan Railway Company and Railway Technical Research Institute, Tokyo, Japan.

Fig. 2.28 shows the bogie frame and Fig. 2.29 shows the structure of the LTS electromagnet of the MLX01 vehicle manufactured by Toshiba, Hitachi, and Mitsubishi. The bogie frame (Sumitomo Heavy Industries) is laid under the vehicle body. The LTS electromagnet (Table 2.21) is wound with a BSCCO wire and enclosed by and integrated with a stainless inner vessel (Fig. 2.29). The winding is made of *niobium– titanium alloy* wire, which is embedded in a *copper matrix* in order to improve the stability of superconductivity. Permanent flow of current without losses is achieved by keeping the coils within a cryogenic temperature (4.2 K or −269°C) using *liquid helium*. The inner vessel is covered with a radiation shield plate on which a cooling pipe is crawled and liquid nitrogen is circulated inside the pipe to eliminate radiation heat [224]. The shield plate is kept at liquid nitrogen temperature, the boiling point of which is about 77 K.

These components are covered with an outer aluminum vacuum vessel (room temperature) and an insulating material that is packed in the space between the inner vessel and outer vessel [224]. The space is maintained in a high-vacuum range to prolong the life of the insulation. There are four sets of inner vessels (LTS electromagnet) per one outer vessel. The on-board 4 K Gifford–McMahon/Joule–Thomson refrigerator for helium, 80 K Gifford–McMahon refrigerator for nitrogen, liquid helium reservoir, and liquid nitrogen reservoir are incorporated inside the tank on the top of the outer vessel. This refrigerator reliquefies the helium gas evaporated as a result of heat generation inside the LTS electromagnet. For commercial use, electromagnets should be operated with no supply of both liquid helium and liquid nitrogen. The necessary equipment such as the compressor, which supplies the compressed gas to the helium refrigerator and control units for the operation of LTS electromagnets, are located inside the bogie.

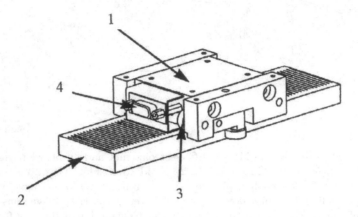

Fig. 2.30. PM HLSM. 1 — forcer, 2 — platen, 3 — mechanical or air bearing, 4 — umbilical cable with power and air hose. Courtesy of Normag Northern Magnetics Inc., Santa Clarita, CA, USA.

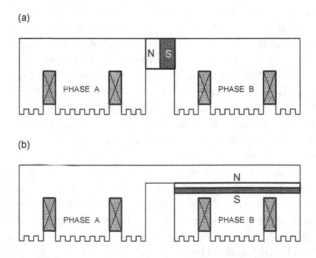

Fig. 2.31. Longitudinal sections of forcers of HLSMs: (a) symmetrical forcer, (b) asymmetrical forcer.

2.14 Hybrid Linear Stepping Motors

The PM HLSM manufactured by Normag Northern Magnetics, Inc., Santa Clarita, CA, USA is shown in Fig. 2.30 [161].

Magnetic circuits of forcers of HLSMs are made of high permeable electrotechnical steels. The thickness of lamination is about 0.2 mm as the input frequency is high. Forcers are designed as symmetrical, i.e., with the PM joining the two stacks (Fig. 2.31a) or two PMs (Fig. 1.19) or asymmetrical, i.e., with PM located in one stack (Fig. 2.31b) designs. The asymmetrical design is easier for assembly.

Examples

Example 2.1

An LSM rated at 60 Hz input frequency is fed from a 50 Hz power supply at the same voltage. Losses in the armature teeth at 60 Hz are $\Delta P_{Fet60} = 420$ W and losses in the armature core (yoke) are $\Delta P_{Fec60} = 310$ W. The hysteresis losses amount to 75% of the total core losses. Find the hysteresis, eddy current and total losses in the armature core at 50 Hz.

Solution

Hysteresis losses in armature teeth at 60 Hz

$$\Delta P_{ht60} = 0.75 \Delta P_{Fet60} = 0.75 \times 420 = 315 \text{ W}$$

Hysteresis losses in armature core (yoke) at 60 Hz

$$\Delta P_{hc60} = 0.75 \Delta P_{Fec60} = 0.75 \times 310 = 232.5 \text{ W}$$

Total hysteresis losses at 60 Hz

$$\Delta P_{h60} = \Delta P_{ht60} + \Delta P_{hc60} = 315 + 232.5 = 547.5 \text{ W}$$

The magnetic flux density is proportional to the EMF, which in turn is proportional to the frequency. Therefore, the magnetic flux density at 60 Hz magnetic flux density at 50 Hz ratio is

$$b = \frac{f_{60}}{f_{50}} = \frac{60}{50} = 1.2$$

On the basis of eqn (2.2) hysteresis losses in teeth at 50 Hz are

$$\Delta P_{ht50} = \frac{f_{50}}{f_{60}} b^2 \Delta P_{ht60} = \frac{50}{60} \times 1.2^2 \times 315 = 378 \text{ W}$$

Hysteresis losses in armature core at 50 Hz are

$$\Delta P_{hc50} = \frac{f_{50}}{f_{60}} b^2 \Delta P_{hc60} = \frac{50}{60} \times 1.2^2 \times 232.5 = 279 \text{ W}$$

Total hysteresis losses at 50 Hz

$$\Delta P_{h50} = \Delta P_{ht50} + \Delta P_{hc50} = 378 + 279 = 657 \text{ W}$$

Eddy-current losses in armature teeth at 60 Hz

$$\Delta P_{et60} = \Delta P_{Fet60} - \Delta P_{ht60} = 420 - 315 = 105 \text{ W}$$

Eddy-current losses in armature core (yoke) at 60 Hz

$$\Delta P_{ec60} = \Delta P_{Fec60} - \Delta P_{hc60} = 310 - 232.5 = 77.5 \text{ W}$$

Total eddy-current losses at 60 Hz

$$\Delta P_{e60} = \Delta P_{et60} + \Delta P_{ec60} = 105 + 77.5 = 182.5 \text{ W}$$

On the basis of eqn (2.2), eddy-current losses in armature teeth at 50 Hz are

$$\Delta P_{et50} = \left(\frac{f_{50}}{f_{60}}\right)^2 b^2 \Delta P_{et60} = \left(\frac{50}{60}\right)^2 \times 1.2^2 \times 105 = 105 \text{ W}$$

Eddy-current losses in armature core at 50 Hz are

$$\Delta P_{ec50} = \left(\frac{f_{50}}{f_{60}}\right)^2 b^2 \Delta P_{et60} = \left(\frac{50}{60}\right)^2 \times 1.2^2 \times 77.5 = 77.5 \text{ W}$$

Total eddy-current losses at 50 Hz

$$\Delta P_{e50} = \Delta P_{et50} + \Delta P_{ec50} = 105 + 77.5 = 182.5 \text{ W}$$

If the frequency is reduced from 60 to 50 Hz, hysteresis losses increase while eddy-current losses remain the same.

Total losses in the armature stack at 60 Hz

$$\Delta P_{Fe60} = \Delta P_{h60} + \Delta P_{e60} = 547.5 + 182.5 = 730 \text{ W}$$

Total losses in the armature stack at 50 Hz

$$\Delta P_{Fe50} = \Delta P_{h50} + \Delta P_{e50} = 657 + 182.5 = 839.5 \text{ W}$$

If the frequency is reduced form 60 to 50 Hz, the total losses in the armature stack increase, i.e.,

$$\frac{\Delta P_{Fe50}}{\Delta P_{Fe50}} = \frac{839.5}{730} = 1.15$$

because the magnetic flux density at the same number of turns and voltage increases as the frequency decreases.

Example 2.2

Find the losses in the armature core of a single-sided 6-pole, 18-slot, 100-Hz, 3-phase PM LSM. The dimensions of the armature core are as follows:

- pole pitch $\tau = 48.0$ mm
- coil pitch $w_c = 48.0$ mm
- effective length of armature stack $L_i = 96.0$ mm
- height of primary core (yoke) $h_{1c} = 10.0$ mm
- armature tooth width $c_t = 7.0$ mm
- armature tooth height $h_t = 23.5$ mm
- armature slot width $b_{11} = 9.0$ mm
- armature slot opening $b_{14} = 3.0$ mm

The magnetic flux density in armature teeth is $B_{1t} = 1.678$ T, flux density in the armature core (yoke) $B_{1c} = 1.468$, specific core losses at 50 Hz and 1 T are $\Delta p_{1/50} = 1.07$ W/kg, stacking factor $k_i = 0.96$, and the factors in eqn (2.3) accounting for the increase in losses due to metallurgical and manufacturing processes are $k_{adt} = 1.6$ and $k_{adc} = 3.2$.

Solution

Slot pitch according to eqn (2.28)

$$t_1 = 9.0 + 8.0 = 16 \text{ mm}$$

Number of slots half-filled with armature winding according to eqn (2.9)

$$s_1' = \frac{1}{6}\left(6 + \frac{48.0}{48.0}\right) \times 18 = 21$$

Length of stack with half-filled slots

$$L_{stack} = s_1' t_1 + c_t = 21 \times 0.016 + 0.007 = 0.343/\text{m}$$

Mass of armature teeth

$$m_{1t} = 7800(s_1'+1)c_t h_t L_i k_i = 7800(21+1) \times 0.007 \times 0.0235 \times 0.096 \times 0.96 = 2.60 \text{ kg}$$

Mass of armature core (yoke)

$$m_{1c} = 7800 L_{stack} h_{1c} L_i k_i = 7800 \times 0.343 \times 0.01 \times 0.096 \times 0.96 = 2.47 \text{ kg}$$

Core losses at $f = 100$ Hz calculated according to eqn (2.3)

$$\Delta P_{Fe} = 1.07\left(\frac{100}{50}\right)^{4/3}(1.6 \times 1.678^2 \times 2.60 + 3.2 \times 1.468^2 \times 2.47) = 77.4 \text{ W}$$

Example 2.3

Find the operating point of an NdFeB grade N35H PM with rectangular cross section placed in a mild steel magnetic circuit with air gap. The parameters of the PM are

- remanent magnetic flux desnity at 20° C $B_{r20} = 1.18$ T
- coercive force at 20° C $H_{c20} = 900$ kA/m
- temperature coefficeint for B_r $\alpha_B = -0.11$ %/$^\circ$C
- temperature coefficeint for H_c $\alpha_H = -0.45$ %/^{o}oC

The PM temperature is $\theta = 100$ °C; the dimensions of PM are $w_M = 42$ mm, $l_M = 105$ mm, $h_M = 5.0$ mm; the air gap is $g = 1.6$ mm, and the PM leakage flux coefficient is $\sigma_{lM} = 1.2$.

Solution

Remanence and coercivity at $\theta = 100$ °C according to eqns (2.21) and (2.22)

$$B_r = 1.18 \left[1 + \frac{-0.11}{100}(100 - 20) \right] = 1.076 \text{ T}$$

$$H_c = 900 \left[1 + \frac{-0.45}{100}(100 - 20) \right] = 576 \text{ kA/m}$$

Relative recoil magnetic permeability

$$\mu_{rrec} = \frac{1.076}{0.4\pi \times 10^{-6} \times 576\ 000} = 1.487$$

Equivalent air gap

$$g_{eq} = g + \frac{h_M}{\mu_{rrec}} = 1.6 + \frac{5.0}{1.487} = 4.96 \text{ mm}$$

Permeance of the air gap according to eqn (B.2)

$$G_g = 0.4\pi \times 10^{-6} \frac{0.042 \times 0.105}{0.00496} = 1.117 \times 10^{-7} \text{ H}$$

Permeance for total magnetic flux

$$G_t = \mu_0 \frac{w_M l_M}{g_{eq}} \sigma_{lM} = 0.4\pi \times 10^{-6} \frac{0.042 \times 0.105}{0.00496} \times 1.2 = 1.340 \times 10^{-7} \text{ H}$$

Line representing the total permeance, which intersects the demagnetization curve (Fig. 2.32)

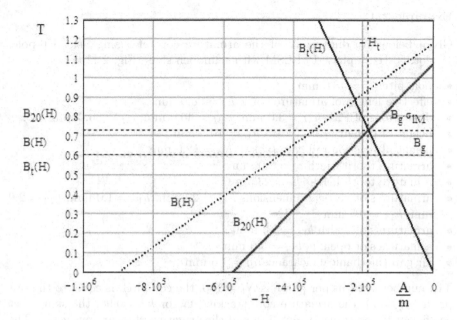

Fig. 2.32. Operating point ($B_g = 0.608$ T, $H_t = 185\ 694.5$ A/m) of PM on demagnetization curve. Numerical example 2.3.

$$B_t(H) = \mu_0 \frac{h_M}{g} H \quad \text{T}$$

Magnetic field intensity corresponding to B_t

$$H_t = \frac{B_r}{\mu_0 h_M / g + B_r / H_c}$$

$$= \frac{1.076}{0.4\pi \times 10^{-6} \times 0.005/0.0016 + 1.076/576\ 000} = 185\ 694.5 \text{ A/m}$$

For $H_t = 185\ 694.5$ A/m, the corresponding magnetic flux density $B_t = 0.729$ T. The air gap magnetic flux density as obtained from the PM diagram (Fig. 2.32)

$$B_g = B_r \left(1 - \frac{H_t}{H_c}\right) \frac{1}{\sigma_{lM}} = 1.076 \left(1 - \frac{185\ 694.5}{576\ 000}\right) \frac{1}{1.2} = 0.608 \text{ T}$$

This point ($B_g = 0.608$ T, $H_t = 185\ 694.5$ A/m) is the intersection point of the demagnetization curve $B(H)$ and permeance line $B_t(H)$ in Fig. 2.32. The air gap magnetic flux

$$\Phi_g = B_g w_M l_M = 0.608 \times 0.042 \times 0.105 = 0.00268 \text{ Wb}$$

Example 2.4

Given below are dimensions of the armature core of a single-sided 6-pole, 18-slot, 50-Hz, 3-phase PM LSM with semiopen slots (Fig. 2.17a:

- pole pitch $\tau = 48.0$ mm
- effective length of armature stack $L_i = 96.0$ mm
- dimensions of PM: $h_M = 3.0$ mm, $w_M = 40.0$ mm, $l_M = 96.0$ mm
- height of primary core (yoke) $h_{1c} = 10.0$ mm
- height of reaction rail core (yoke) $h_{2c} = 12.0$ mm
- armature tooth width $c_t = 8.0$ mm
- armature tooth height $h_t = 23.5$ mm
- armature slot vertical dimensions $h_{11} = 20$ mm, $h_{12} = 1.0$ mm, $h_{13} = 2.0$ mm, $h_{14} = 0.5$ mm
- armature slot width $b_{11} = 8.0$ mm
- armature slot opening $b_{14} = 3.0$ mm
- air gap (mechanical clearance) $g = 1.5$ mm

The number of turns per phase is $N_1 = 180$, the coil pitch is equal to the slot pitch ($w_c = \tau$), the armature core stacking factor $k_i = 0.96$, the skin effect coefficient for reactance $k_{1X} \approx 1.0$, and the armature slots are unskewed. The saturation factor of magnetic circuit is $k_{sat} = 1.153$.

Find the stator winding leakage reactance X_1.

Solution

The number of slots per pole is $Q_1 = 18/6 = 3$, number of slots per pole per phase $q_1 = 18/(3 \times 6) = 1$, winding distribution factor $k_{d1} = \sin[\pi/(2 \times 3)]/(1 \times \sin[\pi/(2 \times 3 \times 1)]) = 1$, winding pitch factor $k_{p1} = \sin[\pi \times 0.048/(2 \times 0.048)] = 1$, winding factor $k_{w1} = 1 \times 1 = 1$, and slot pitch $t_1 = c_t + b_{11} = 8.0 + 8.0 = 16$ mm.

The coefficient of leakage permeance for semi-open slot according to eqn (2.33)

$$\lambda_{1s} = \frac{20.0}{3 \times 8.0} + \frac{1.0}{8.0} + \frac{2 \times 2.0}{8.0 + 3.0} + \frac{0.5}{3.0} = 1.489$$

Estimated length of single-sided end turn

$$l_{1e} = 1.3\tau + 0.004 = 1.3 \times 0.048 + 0.004 = 0.066 \text{ m}$$

The coefficient of leakage permeance for end turns according to eqn (2.36)

$$\lambda_{1e} = 0.34 \times 1.0 \left(1 - \frac{2}{\pi} \frac{0.048}{0.066}\right) = 0.184$$

where $w_c = \tau = 0.048$ m. The armature winding differential leakage factor calculated according to eqn (2.46) for $\nu = 2000$ is $\tau_{d1} = 0.096456$. This coefficient can also be calculated on the basis of simplified formula (2.47), i.e.,

$$\tau_{d1} = \frac{\pi^2(10 \times 1.0^2 + 2)}{27} \left[\sin\left(\frac{30^o}{1.0}\right) \right]^2 - 1 = 0.09662$$

Carter's coefficient according to eqn (2.44)

$$k_{C1} = \frac{0.016}{0.016 - 0.5587 \times 0.016} = 1.0553$$

where $\gamma_1 = 0.5587$ has been obtained from eqn (2.45). The specific permeance for differential leakage flux according to eqn (2.43)

$$\lambda_{1d} = \frac{3 \times 1 \times 0.048 \times 1.0^2}{\pi^2 \times 0.0015 \times 1.0553 \times 1.153} 0.096456 = 0.771$$

The coefficient of tooth-top permeance according to eqn (2.51)

$$\lambda_{1t} = \frac{5 \times 1.5/3.0}{5 + 4 \times 1.5/3.0} = 0.357$$

Leakage reactance of the armature winding according to eqn (2.32) in which $k_{1X} = 1.0$

$$X_1 = 4\pi \times 50 \times 0.4\pi \times 10^{-6} \frac{0.096 \times 180^2}{3 \times 1} \left(1.489 \times 1.0 + \frac{0.066}{0.096} 0.184 + 0.771 + 0.357 \right)$$

$$= 2.246 \ \Omega$$

3

Theory of Linear Synchronous Motors

3.1 Permanent Magnet Synchronous Motors

3.1.1 Magnetic Field of the Armature Winding

The *time-space distribution* of the magnetomotife force (MMF) of a symmetrical polyphase winding with distributed parameters fed with a balanced system of currents can be expressed as

$$\mathcal{F}(x,t) = \frac{N_1\sqrt{2}I_a}{\pi p}\sin(\omega t)\sum_{\nu=1}^{\infty}\frac{1}{\nu}k_{w1\nu}\cos\left(\nu\frac{\pi}{\tau}x\right)$$

$$+\frac{N_1\sqrt{2}I_a}{\pi p}\sin\left(\omega t - \frac{1}{m_1}2\pi\right)\sum_{\nu=1}^{\infty}\frac{1}{\nu}k_{w1\nu}\cos\nu\left(\frac{\pi}{\tau}x - \frac{1}{m_1}2\pi\right)+\dots$$

$$+\frac{N_1\sqrt{2}I_a}{\pi p}\sin\left(\omega t - \frac{m_1-1}{m_1}2\pi\right)\sum_{\nu=1}^{\infty}\frac{1}{\nu}k_{w1\nu}\cos\nu\left(\frac{\pi}{\tau}x - \frac{m_1-1}{m_1}2\pi\right)$$

$$=\frac{1}{2}\sum_{\nu=1}^{\infty}\mathcal{F}_{m\nu}\left\{\sin\left[\left(\omega t - \nu\frac{\pi}{\tau}x\right)+(\nu-1)\frac{2\pi}{m_1}\right]\right.$$

$$\left.+\sin\left[\left(\omega t + \nu\frac{\pi}{\tau}x\right)-(\nu+1)\frac{2\pi}{m_1}\right]\right\} \tag{3.1}$$

where I_a is the armature phase current, m_1 is the number of phases, p is the number of pole pairs, N_1 is the number of series turns per phase, $k_{w1\nu}$ is the winding factor, $\omega = 2\pi f$ is the angular frequency, τ is the pole pitch, and

- for the forward-traveling field

$$\nu = 2m_1 k + 1, \qquad k = 0,1,2,3,4,5,\dots \tag{3.2}$$

- for the backward-traveling field

$$\nu = 2m_1 k - 1, \qquad k = 1, 2, 3, 4, 5 \ldots \tag{3.3}$$

The magnitude of the νth harmonics of the armature MMF is

$$\mathcal{F}_{m\nu} = \frac{2m_1\sqrt{2}}{\pi p} N_1 I_a \frac{1}{\nu} k_{w1\nu} = m_1 [\mathcal{F}_{m\nu}]_{m_1=1} \tag{3.4}$$

where $[\mathcal{F}_{m\nu}]_{m_1=1}$ is the magnitude of the armature MMF of a single-phase winding. The winding factor for the higher-space harmonics $\nu > 1$ is the product of the distribution factor, $k_{d1\nu}$, and the pitch factor, $k_{p1\nu}$, as given by eqns (2.48), (2.49), and (2.50) in Chapter 2. See also eqns (2.39), (2.40), and (2.41) given in Chapter 2, expressing $k_{w1\nu} = k_{w1}$, $k_{d1\nu} = k_{d1}$, and $k_{p1\nu} = k_{p1}$ for $\nu = 1$.

Assuming that $\omega t \mp \nu\pi x/\tau = 0$, the *linear synchronous speed* of the νth harmonic wave of the MMF is

$$v_{s\nu} = \mp 2f\tau \frac{1}{\nu} \tag{3.5}$$

For a three-phase winding, the time space distribution of the MMF is

$$\mathcal{F}(x,t) = \frac{1}{2} \sum_{\nu=1}^{\infty} \mathcal{F}_{m\nu} \left\{ \sin \left[\left(\omega t - \nu\frac{\pi}{\tau}x \right) + (\nu - 1)\frac{2\pi}{3} \right] \right.$$

$$\left. + \sin \left[\left(\omega t + \nu\frac{\pi}{\tau}x \right) - (\nu + 1)\frac{2\pi}{3} \right] \right\} \tag{3.6}$$

For a three-phase winding and the fundamental harmonic $\nu = 1$,

$$\mathcal{F}(x,t) = \frac{1}{2} F_m \sin \left(\omega t - \frac{\pi}{\tau}x \right) \tag{3.7}$$

$$F_m = \frac{2m_1\sqrt{2}}{\pi p} N_1 I_a k_{w1} \approx 0.9 \frac{m_1 N_1 k_{w1}}{p} I_a \tag{3.8}$$

the winding factor $k_{w1} = k_{d1} k_{p1}$ is given by eqns (2.39), (2.40), and (2.41), and $v_s = v_{s\nu=1}$ is according to eqn (1.1).

The peak value of the armature line current density or *specific electric loading* is defined as the number of conductors (one turn consists of two conductors) in all phases $2m_1 N_1$ times the peak armature current $\sqrt{2}I_a$ divided by the armature stack length $2p\tau$, i.e.,

$$A_m = \frac{m_1\sqrt{2}N_1 I_a}{p\tau} \tag{3.9}$$

3.1.2 Form Factors and Reaction Factors

The *form factor of the excitation field* is defined as the ratio of the *amplitude of the first harmonic-to-maximum value of the air gap magnetic flux density*, i.e.,

$$k_f = \frac{B_{mg1}}{B_{mg}} = \frac{4}{\pi} \sin \frac{\alpha_i \pi}{2} \tag{3.10}$$

where α_i is the *pole-shoe b_p to pole pitch τ ratio*, i.e.,

$$\alpha_i = \frac{b_p}{\tau} \tag{3.11}$$

The *form factors of the armature reaction* are defined as the ratios of the *first harmonic amplitudes to maximum values of normal components of armature reaction magnetic flux densities* in the d-axis and q-axis, respectively, i.e.,

$$k_{fd} = \frac{B_{ad1}}{B_{ad}} \qquad k_{fq} = \frac{B_{aq1}}{B_{aq}} \tag{3.12}$$

The *direct* or *d*-axis is the center axis of the magnetic pole, while the *quadrature* or *q*-axis is the axis perpendicular (90° electrical) to the *d*-axis. The peak values of the first harmonics B_{ad1} and B_{aq1} of the armature magnetic flux density can be calculated as coefficients of Fourier series for $\nu = 1$, i.e.,

$$B_{ad1} = \frac{4}{\pi} \int_0^{0.5\pi} B(x) \cos x \, dx \tag{3.13}$$

$$B_{aq1} = \frac{4}{\pi} \int_0^{0.5\pi} B(x) \sin x \, dx \tag{3.14}$$

For a salient-pole motor with electromagnetic excitation and the air gap $g \approx 0$ (fringing effects neglected), the d- and q-axis form factors of the armature reaction are

$$k_{fd} = \frac{\alpha_i \pi + \sin \alpha_i \pi}{\pi} \qquad k_{fq} = \frac{\alpha_i \pi - \sin \alpha_i \pi}{\pi} \tag{3.15}$$

For PM excitation systems, the form factors of the armature reaction are [70]

- for surface PMs

$$k_{fd} = k_{fq} = 1 \tag{3.16}$$

- for buried magnets

$$k_{fd} = \frac{4}{\pi} \alpha_i^2 \frac{1}{\alpha_i^2 - 1} \cos \frac{\pi}{2\alpha_i} \qquad k_{fq} = \frac{1}{\pi}(\alpha_i \pi - \sin \alpha_i \pi) \tag{3.17}$$

Formulae for k_{fd} and k_{fq} for inset type PMs and surface PMs with mild steel pole shoes are given in [66].

The *reaction factors* in the d- and q-axis are defined as

$$k_{ad} = \frac{k_{fd}}{k_f} \qquad\qquad k_{aq} = \frac{k_{fd}}{k_f} \qquad\qquad (3.18)$$

The form factors k_f, k_{fd}, and k_{fq} of the excitation field, and armature reaction and reaction factors k_{ad} and k_{aq} for salient-pole synchronous machines according to eqns (3.10), (3.15), and (3.18), are given in Table 3.1.

Table 3.1. Factors k_f, k_{fd}, k_{fq}, k_{ad}, and k_{aq} for salient-pole synchronous machines according to eqns (3.10), (3.15), and (3.18)

Factor	$\alpha_i = b_p/\tau$						
	0.4	0.5	0.6	$2/\pi$	0.7	0.8	1.0
k_f	0.748	0.900	1.030	1.071	1.134	1.211	1.273
k_{fd}	0.703	0.818	0.913	0.943	0.958	0.987	1.00
k_{fq}	0.097	0.182	0.287	0.391	0.442	0.613	1.00
k_{ad}	0.939	0.909	0.886	0.880	0.845	0.815	0.785
k_{aq}	0.129	0.202	0.279	0.365	0.389	0.505	0.785

Assuming $g = 0$, the *equivalent d-axis field MMF* (which produces the same fundamental wave flux as the armature-reaction MMF) is

$$\mathcal{F}_{ad} = k_{ad}F_{ad} = \frac{m_1\sqrt{2}}{\pi}\frac{N_1 k_{w1}}{p}k_{ad}I_a \sin\Psi \qquad (3.19)$$

where I_a is the armature current, and Ψ is the angle between the phasor of the armature current $\mathbf{I}_a$ and the q-axis, i.e., $I_{ad} = I_a \sin\Psi$. Similarly, the *equivalent q-axis field MMF* is

$$\mathcal{F}_{aq} = k_{aq}F_{aq} = \frac{m_1\sqrt{2}}{\pi}\frac{N_1 k_{w1}}{p}k_{aq}I_a \cos\Psi \qquad (3.20)$$

where $I_{aq} = I_a \cos\Psi$. In the theory of synchronous machines with electromagnetic excitation, the MMFs $\mathcal{F}_{excd}$ and $\mathcal{F}_{excq}$ are defined as the *armature MMFs referred to the field excitation winding*.

3.1.3 Synchronous Reactance

For a salient-pole synchronous machine, the d-axis and q-axis *synchronous reactances* are

$$X_{sd} = X_1 + X_{ad} \qquad\qquad X_{sq} = X_1 + X_{aq} \qquad (3.21)$$

where $X_1 = 2\pi f L_1$ is the armature *leakage reactance* according to eqn (2.32), X_{ad} is the d-axis *armature reaction reactance*, also called d-axis *mutual reactance*; and X_{aq} is the q-axis armature reaction reactance, also called q-axis mutual reactance. The reactance X_{ad} is sensitive to the saturation of the magnetic circuit, while the influence of the magnetic saturation on the reactance X_{aq} depends on the field excitation system design. In salient-pole synchronous machines with electromagnetic excitation, X_{aq} is practically independent of the magnetic saturation. Usually, $X_{sd} > X_{sq}$ except for some PM synchronous machines.

The d-axis armature reaction reactance

$$X_{ad} = k_{fd}X_a = 4m_1\mu_o f \frac{(N_1 k_{w1})^2}{\pi p} \frac{\tau L_i}{g'} k_{fd} \qquad (3.22)$$

where μ_o is the magnetic permeability of free space, L_i is the effective length of the stator core, $g' \approx k_C k_{sat}g + h_M/\mu_{rrec}$ is the equivalent air gap in the d-axis, k_C is the Carter's coefficient for the air gap according to eqn (2.44), $k_{sat} > 1$ is the saturation factor of the magnetic circuit, and

$$X_a = 4m_1\mu_o f \frac{(N_1 k_{w1})^2}{\pi p} \frac{\tau L_i}{g'} \qquad (3.23)$$

is the armature reaction reactance of a non-salient-pole (surface configuration of PMs) synchronous machine. Similarly, for the q-axis,

$$X_{aq} = k_{fq}X_a = 4m_1\mu_o f \frac{(N_1 k_{w1})^2}{\pi p} \frac{\tau L_i}{k_C k_{satq}g_q} k_{fq} \qquad (3.24)$$

where g_q is the air gap in the q-axis. For salient-pole excitation systems, the saturation factor $k_{satq} \approx 1$ since the q-axis armature reaction fluxes, closing through the large air spaces between the poles have insignificant effect on the magnetic saturation.

The leakage reactance X_1 consists of the slot, end-connection differential and tooth-top leakage reactances — see eqn (2.32). Only the slot and differential leakage reactances depend on the magnetic saturation due to leakage fields.

3.1.4 Voltage Induced

The no-load *rms* voltage induced (EMF) in one phase of the armature winding by the d.c. or PM excitation flux Φ_f is

$$E_f = \pi\sqrt{2}f N_1 k_{w1}\Phi_f \qquad (3.25)$$

where N_1 is the number of the armature turns per phase, k_{w1} is the armature winding coefficient, $f = v_s/(2\tau)$, and the fundamental harmonic Φ_{f1} of the excitation magnetic flux density Φ_f without armature reaction is

$$\Phi_{f1} = L_i \int_0^\tau B_{mg1} \sin\left(\frac{\pi}{\tau}x\right) dx = \frac{2}{\pi}\tau L_i B_{mg1} \tag{3.26}$$

Similarly, the voltage E_{ad} induced by the d-axis armature reaction flux Φ_{ad} and the voltage E_{aq} induced by the q-axis flux Φ_{aq} are, respectively,

$$E_{ad} = \pi\sqrt{2}fN_1k_{w1}\Phi_{ad} \tag{3.27}$$

$$E_{aq} = \pi\sqrt{2}fN_1k_{w1}\Phi_{aq} \tag{3.28}$$

The EMFs E_f, E_{ad}, E_{aq}, and magnetic fluxes Φ_f, Φ_{ad}, and Φ_{aq} are used in the construction of phasor diagrams and equivalent circuits. The EMF E_i per phase with the armature reaction taken into account is

$$E_i = \pi\sqrt{2}fN_1k_{w1}\Phi_g \tag{3.29}$$

where Φ_g is the air gap magnetic flux under load (excitation flux Φ_f reduced by the armature reaction flux). Including armature leakage flux Φ_{lg},

$$E_i = \pi\sqrt{2}fN_1k_{w1}\Phi \tag{3.30}$$

where $\Phi = \Phi_g - \Phi_{lg}$. At no-load (very small armature current), $\Phi_g \approx \Phi_f$. Including the saturation of the magnetic circuit,

$$E_i = 4\sigma_f fN_1k_{w1}\Phi \tag{3.31}$$

The form factor σ_f of EMFs depends on the magnetic saturation of armature teeth, i.e., the sum of the air gap magnetic voltage drop (MVD) and the teeth MVD divided by the air gap MVD.

3.1.5 Electromagnetic Power and Thrust

The following set of equations stems from the phasor diagram of a salient-pole synchronous motor shown in Fig. 3.1:

$$V_1 \sin\delta = -I_{ad}R_1 + I_{aq}X_{sq}$$

$$V_1 \cos\delta = I_{aq}R_1 + I_{ad}X_{sd} + E_f \tag{3.32}$$

in which δ is the load angle between the terminal phase voltage $\mathbf{V}_1$ and $\mathbf{E}_f$ (q axis). The currents

$$I_{ad} = \frac{V_1(X_{sq}\cos\delta - R_1\sin\delta) - E_fX_{sq}}{X_{sd}X_{sq} + R_1^2} \tag{3.33}$$

$$I_{aq} = \frac{V_1(R_1\cos\delta + X_{sd}\sin\delta) - E_fR_1}{X_{sd}X_{sq} + R_1^2} \tag{3.34}$$

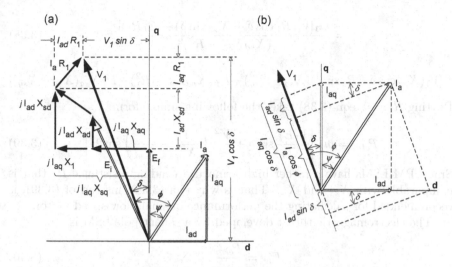

Fig. 3.1. Phasor diagram of an underexcited salient pole synchronous motor: (a) full phasor diagram; (b) auxiliary diagram for calculation of input power.

are obtained by solving the set of eqns (3.32). The *rms* armature current as a function of V_1, E_f, X_{sd}, X_{sq}, δ, and R_1 is

$$I_a = \sqrt{I_{ad}^2 + I_{aq}^2} = \frac{V_1}{X_{sd}X_{sq} + R_1^2} \qquad (3.35)$$

$$\times \sqrt{[(X_{sq}\cos\delta - R_1\sin\delta) - \frac{E_f}{V_1}X_{sq}]^2 + [(R_1\cos\delta + X_{sd}\sin\delta) \; \frac{E_f}{V_1}R_1]^2}$$

The phasor diagram (Fig. 3.1) can also be used to find the input power [70], i.e.,

$$P_{in} = m_1 V_1 I_a \cos\phi = m_1 V_1 (I_{aq}\cos\delta - I_{ad}\sin\delta) \qquad (3.36)$$

Putting eqns (3.32) into eqn (3.36),

$$P_{in} = m_1[I_{aq}E_f + I_{ad}I_{aq}X_{sd} + I_{aq}^2 R_1 - I_{ad}I_{aq}X_{sq} + I_{ad}^2 R_1]$$

$$= m_1[I_{aq}E_f + R_1 I_a^2 + I_{ad}I_{aq}(X_{sd} - X_{sq})] \qquad (3.37)$$

Because the armature core loss has been neglected, the electromagnetic power is the motor input power minus the armature winding loss $\Delta P_{1w} = m_1 I_a^2 R_1 = m_1(I_{ad}^2 + I_{aq}^2)R_1$. Thus,

$$P_{elm} = P_{in} - \Delta P_{1w} = m_1[I_{aq}E_f + I_{ad}I_{aq}(X_{sd} - X_{sq})]$$

$$= \frac{m_1[V_1(R_1 \cos \delta + X_{sd} \sin \delta) - E_f R_1)]}{(X_{sd}X_{sq} + R_1^2)^2} \qquad (3.38)$$

$$\times [V_1(X_{sq} \cos \delta - R_1 \sin \delta)(X_{sd} - X_{sq}) + E_f(X_{sd}X_{sq} + R_1^2) - E_f X_{sq}(X_{sd} - X_{sq})]$$

Putting $R_1 = 0$, eqn (3.38) takes the following simple form,

$$P_{elm} = m_1 \left[\frac{V_1 E_f}{X_{sd}} \sin \delta + \frac{V_1^2}{2} \left(\frac{1}{X_{sq}} - \frac{1}{X_{sd}} \right) \sin 2\delta \right] \qquad (3.39)$$

Small PM LSMs have a rather high armature winding resistance R_1 that is comparable with X_{sd} and X_{sq}. That is why eqn (3.38), instead of (3.39), is recommended for calculating the performance of small, low-speed motors.

The electromagnetic thrust developed by a salient-pole LSM is

$$F_{dx} = \frac{P_{elm}}{v_s} \qquad N \qquad (3.40)$$

Neglecting the armature winding resistance ($R_1 = 0$),

$$F_{dx} = \frac{m_1}{v_s} \left[\frac{V_1 E_f}{X_{sd}} \sin \delta + \frac{V_1^2}{2} \left(\frac{1}{X_{sq}} - \frac{1}{X_{sd}} \right) \sin 2\delta \right] \qquad (3.41)$$

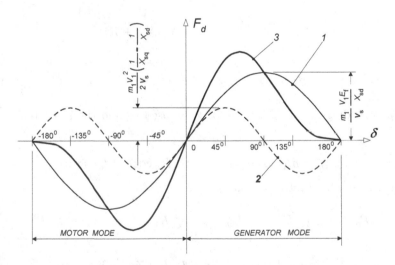

Fig. 3.2. Thrust-angle characteristics of a salient-pole synchronous machine with $X_{sd} > X_{sq}$. 1 — synchronous thrust F_{dsyn} developed by the machine, 2 — reluctance thrust F_{drel}, 3 — resultant thrust F_d.

In a salient-pole synchronous motor, the electromagnetic thrust has two components (Fig. 3.2)

$$F_{dx} = F_{dxsyn} + F_{dxrel} \tag{3.42}$$

where the first term,

$$F_{dxsyn} = \frac{m_1}{v_s} \frac{V_1 E_f}{X_{sd}} \sin \delta \tag{3.43}$$

is a function of both the input voltage V_1 and the excitation EMF E_f. The second term,

$$F_{dxrel} = \frac{m_1 V_1^2}{2v_s} \left(\frac{1}{X_{sq}} - \frac{1}{X_{sd}} \right) \sin 2\delta \tag{3.44}$$

depends only on the voltage V_1 and also exists in an unexcited machine ($E_f = 0$) provided that $X_{sd} \neq X_{sq}$. The thrust F_{dxsyn} is called the *synchronous thrust*, and the thrust F_{dxrel} is called the *reluctance thrust*. The proportion between X_{sd} and X_{sq} strongly affects the shape of curves 2 and 3 in Fig. 3.2. For surface configurations of PMs, $X_{sd} \approx X_{sq}$ (if the magnetic saturation is neglected) and

$$F_{dx} \approx F_{dxsyn} = \frac{m_1}{v_s} \frac{V_1 E_f}{X_{sd}} \sin \delta \tag{3.45}$$

3.1.6 Minimization of *d*-axis Armature Current

For zero *d*-axis armature current, all current $I_a = I_{aq}$ is torque producing. The *d*-axis armature current is zero ($I_{ad} = 0$) if the numerator of eqn (3.33) is zero, i.e.,

$$V_1 X_{sq} \cos \delta - V_1 R_1 \sin \delta - E_f X_{sq} = 0 \tag{3.46}$$

or

$$(-V_1 R_1 \sin \delta)^2 = (E_f X_{sq} - V_1 X_{sq} \cos \delta)^2$$

After putting $\sin^2 \delta = 1 - \cos^2 \delta$, the following 2nd order linear equation is obtained,

$$[(V_1 X_{sq})^2 + (V_1 R_1)^2] \cos^2 \delta - 2V_1 E_f X_{sq}^2 \cos \delta + (E_f X_{sq})^2 - (V_1 R_1)^2 = 0 \tag{3.47}$$

or

$$ax^2 + bx + c = 0 \tag{3.48}$$

where $x = \cos \delta$, $a = (V_1 X_{sq})^2 + (V_1 R_1)^2$, $b = -2V_1 E_f X_{sq}^2$, and $c = (E_f X_{sq})^2 - (V_1 R_1)^2$. Eqn (3.47) or eqn (3.48) has two roots:

$$\delta_1 = \arccos(x_1) = \arccos\left(\frac{-b - \sqrt{\Delta}}{2a}\right) \qquad (3.49)$$

$$\delta_2 = \arccos(x_2) = \arccos\left(\frac{-b + \sqrt{\Delta}}{2a}\right) \qquad (3.50)$$

where the discriminant of quadratic equation $\Delta = b^2 - 4ac$. To obtain $I_{ad} = 0$, at least one root must be a real number. Sometimes both two roots are complex numbers. In this case, for motoring mode, most often the EMF E_f, is greater than the terminal voltage V_1. It means that the number of turns per phase must be reduced.

3.1.7 Thrust Ripple

The *thrust ripple* can be expressed as the *rms* thrust ripple $\sqrt{\sum F_{dx\nu}^2}$ weighted to the mean value of the thrust F_{dx} developed by the LSM, i.e.,

$$f_r = \frac{1}{F_{dx}}\sqrt{\sum_\nu F_{dx\nu}^2} \qquad (3.51)$$

The thrust ripple of an LSM consists of three components: (1) *detent thrust* (cogging thrust), i.e., interaction between the excitation flux and variable permeance of the armature core due to slot openings, (2) distortion of sinusoidal or trapezoidal distribution of the magnetic flux density in the air gap and, (3) phase current commutation and current ripple.

Case Study 3.1. Single-Sided PM LSM

A flat, short-armature, single-sided, three-phase LSM has a long reaction rail with surface configuration of PMs. High-energy sintered NdFeB PMs with the remanent magnetic flux density $B_r = 1.1$ T and coercive force $H_c = 800$ kA/m have been used. The armature magnetic circuit has been made of cold-rolled steel laminations Dk-66 (Table 2.3) The following design data are available: armature phase windings are Y-connected, number of pole pairs $p = 4$, pole pitch $\tau = 56$ mm, air gap in the d-axis (mechanical clearance) $g = 2.5$ mm, air gap in the q-axis $g_q = 6.5$ mm, effective width of the armature core $L_i = 84$ mm, width of the core (back iron) of the reaction rail $w = 84$ mm, height of the armature core (yoke) $h_{1c} = 20$ mm, length of the overhang (one side) $l_e = 90$ mm, number of armature turns per phase $N_1 = 560$, number of parallel wires $a_w = 2$, number of armature slots $s_1 = 24$, width of the armature tooth $c_t = 8.4$ mm, height of the core (yoke) of the reaction rail $h_{2c} = 12$ mm, dimensions of the armature open rectangular slot: $h_{11} = 26.0$ mm, $h_{12} = 2.0$ mm, $h_{13} = 3.0$ mm, $h_{14} = 1.0$ mm, $b_{14} = 10.3$ mm (see Fig. 2.17b), stacking coefficient of the armature core $k_i = 0.96$, conductivity of armature wire $\sigma_{20} = 57 \times 10^6$ S/m at 20^0C, diameter of armature wire $d_{wir} = 1.02$ mm, temperature of armature winding 75^0C, height of the PM $h_M = 4.0$ mm, width of the PM $w_M = 42.0$ mm, length of the PM (in the direction of armature conductors) $l_M = 84$ mm, width of the pole shoe $b_p = w_M = 42.0$ mm, and friction coefficient for rollers at constant speed $\mu_v = 0.01$.

The coil pitch w_c of the armature winding is equal to the pole pitch τ (full pitch winding).

The LSM is fed from a VVVF inverter, and the ratio of the line voltage to input frequency $V_{1L}/f = 10$. The PM LSM has been designed for continuous duty cycle to operate with the load angle δ corresponding to the maximum efficiency. The current density in the armature winding normally does not exceed $j_a = 3.0$ A/mm^2 (natural air cooling).

Calculate the steady-state performance characteristics for $f = 20, 15, 10$ and 5 Hz.

Solution

For $V_{1L}/f = 10$, the input line voltages are 200 V for 20 Hz, 150 V for 15 Hz, 100 V for 10 Hz, and 50 V for 5 Hz. Steady-state characteristics have been calculated using analytical equations given in Chapters 1,2, and 3. The volume of the PM material per $2p\tau$ is $p \times 2h_M \times w_M \times l_M = 4 \times 2 \times 4 \times 42 \times 84 = 112,896$ mm^3.

Given below are parameters independent of the frequency, voltage, and load angle δ:

- number of slots per pole per phase $q_1 = 1$
- winding factor $k_{w1} = 1.0$
- pole pitch measured in slots $= 3$
- pole shoe to pole pitch ratio $\alpha_i = 0.75$
- coil pitch measured in slots $= 3$
- coil pitch measured in millimeters $w_c = 56$ mm
- armature slot pitch $t_1 = 18.7$ mm
- width of the armature slot $b_{11} = b_{12} = b_{14} = 10.3$ mm
- number of conductors in each slot $N_{sl} = 280$
- conductors cross section area to slot area (slot fill factor) $k_{fill} = 0.2252$
- Carter's coefficient $k_C = 1.2$
- form factor of the excitation field $k_f = 1.176$
- form factor of the d-axis armature reaction $k_{fd} = 1.0$
- form factor of the q-axis armature reaction $k_{fd} = 1.0$
- reaction factors $k_{ad} = k_{aq} = 0.85$
- coefficient of leakage flux $\sigma_l = 1.156$
- permeance of the air gap $G_g = 0.1370 \times 10^{-5}$ H
- permeance of the PM $G_M = 0.1213 \times 10^{-5}$ H
- permeance for leakage fluxes $G_{lM} = 0.2144 \times 10^{-6}$ H
- magnetic flux corresponding to the remanent magnetic flux density $\Phi_r = 0.3881 \times 10^{-2}$ Wb

- relative recoil magnetic permeability $\mu_{rrec} = 1.094$
- PM edge line current density $J_M = 800,000.00$ A/m
- mass of the armature core $m_{1c} = 5.56$ kg
- mass of the armature teeth $m_{1t} = 4.92$ kg
- mass of the armature conductors $m_{Cu} = 15.55$ kg
- friction force $F_r = 1.542$ N

Then, resistances and reactances independent of magnetic saturation have been calculated,

- armature winding resistance $R_1 = 2.5643$ Ω at 75°C
- armature winding leakage reactance $X_1 = 4.159$ Ω at $f = 20$ Hz
- armature winding leakage reactance $X_1 = 1.0397$ Ω at $f = 5$ Hz
- armature reaction reactance $X_{ad} = X_{aq} = 4.5293$ Ω at $f = 20$ Hz
- armature reaction reactance $X_{ad} = X_{aq} = 1.1323$ Ω at $f = 5$ Hz
- specific slot leakage permeance $\lambda_{1s} = 1.3918$
- specific leakage permeance of end connections $\lambda_{1e} = 0.2192$
- specific tooth-top leakage permeance $\lambda_{1t} = 0.1786$
- specific differential leakage permeance $\lambda_{1d} = 0.21$
- coefficient of differential leakage $\tau_{d1} = 0.0965$

Note that for calculating X_{ad} of an LSM with surface PMs, the nonferromagnetic air gap is the gap between the ferromagnetic cores of the armature and reaction rail. The relative magnetic permeability of NdFeB PMs is very close to unity.

Steady-state performance characteristics have been calculated as functions of the load angle δ. The load angle of synchronous motors can be compared to the slip of induction motors, which is also a measure of how much the motor is loaded. Magnetic saturation due to main flux and leakage fluxes has been included. An LSM should operate with maximum efficiency. Maximum efficiency usually corresponds to the d-axis current $I_{ad} \approx 0$. Table 3.2 shows fundamental steady-state performance characteristics for $f = 20, 15, 10,$ and 5 Hz. In practice, the maximum efficiency corresponds to small values (close to zero) of the angle Ψ, i.e., the angle between the phasor of the armature current $\mathbf{I}_a$ and the q-axis, which means that the d-axis armature current I_{ad} for maximum efficiency is very small, or $I_{ad} = 0$ (Section 3.1.6 and Chapter 6).

Table 3.3 contains calculation results for maximum efficiency and two extreme frequencies $f = 20$ Hz and $f = 5$ Hz, including magnetic saturation. The magnetic flux density in armature teeth for maximum efficiency is well below the saturation magnetic flux density (over 2.1 T for Dk-69 laminations). This value is close to the saturation value for higher load angles $\delta \geq 60^0$. Also, the armature current I_a and, consequently, line current density A_m and current density j_a increase with the load angle δ.

Table 3.2. Steady-state performance characteristics of a flat three-phase, four-pole LSM with surface PMs and $\tau = 56$ mm

δ deg	Ψ deg	P_{out} W	F_x N	F_z N	I_a A	η —	$\cos\phi$ —
$f = 20$ Hz, $V_{1L-L} = 200$ V, $v_s = 2.24$ m/s							
−20.0	58.69	−670.0	−300.4	992	5.91	0.5987	0.1960
−10.0	80.18	−178.0	−85.2	1004	5.01	0.0312	0.0032
1.0	70.51	308.7	145.3	1108	4.27	0.6582	0.3172
10.0	40.75	739.5	342.1	1273	4.10	0.8223	0.6326
12.0	33.77	839.1	387.7	1319	4.16	0.8351	0.6976
15.0	23.53	991.1	457.2	1393	4.32	0.8463	0.7823
20.0	7.20	1253.0	577.2	1544	4.78	**0.8515**	0.8894
22.4	0.11	1383.0	636.4	1628	5.09	0.8493	0.9238
30.0	−18.73	1803.0	829.1	1931	6.41	0.8287	0.9807
40.0	−36.89	2374.0	1091.0	2461	8.79	0.7805	0.9985
60.0	−72.75	3562.0	1644.0	5511	20.23	0.5212	0.9754
$f = 15$ Hz, $V_{1L-L} = 150$ V, $v_s = 1.68$ m/s							
−20.0	64.25	−406.1	−243.8	977	5.78	0.3703	0.1002
−10.0	84.85	−63.5	−44.1	1024	4.79	0.6035	0.0845
1.0	65.31	276.4	172.3	1160	3.94	0.6714	0.4018
10.0	33.47	576.8	355.3	1350	3.75	0.8146	0.7258
15.0	15.03	749.1	460.4	1487	4.00	**0.8332**	0.8658
19.4	0.21	903.4	554.6	1628	4.44	0.8320	0.9420
20.0	−1.64	924.6	567.6	1646	4.51	0.8309	0.9491
30.0	−27.01	1278.0	783.6	2055	6.24	0.7898	0.9986
40.0	−44.18	1616.0	990.7	2595	8.66	0.7203	0.9973
00.0	−75.94	1911.0	1179.0	5190	18.64	0.4104	0.9615
$f = 10$ Hz, $V_{1L-L} = 100$ V, $v_s = 1.12$ m/s							
−20.0	72.91	−167.7	153.2	990	5.30	0.2779	0.0507
−10.0	87.43	15.5	17.8	1085	4.23	0.0974	0.2176
1.0	55.89	217.4	202.2	1265	3.30	0.6973	0.5462
10.0	20.14	380.5	351.4	1479	3.15	0.8068	0.8648
12.0	11.97	416.2	384.1	1535	3.24	**0.8118**	0.9138
15.0	0.36	469.0	432.5	1625	3.46	0.8108	0.9642
20.0	−16.30	554.5	510.9	1795	4.05	0.7920	0.9979
30.0	−39.74	708.8	652.5	2196	5.81	0.7140	0.9856
40.0	−55.12	822.6	757.8	2709	8.12	0.6060	0.9654
45.0	−61.75	851.9	785.4	3029	9.50	0.5404	0.9576
60.0	−80.56	634.9	592.4	4507	15.50	0.2525	0.9363
$f = 5$ Hz, $V_{1L-L} = 50$ V, $v_s = 0.56$ m/s							
−20	87.20	−8.2	−20.5	1149	3.70	0.0866	0.2957
−10	73.13	45.8	86.8	1278	2.75	0.4260	0.4520
1	35.88	102.2	190.0	1469	1.99	0.7424	0.7999
8.6	0.10	134.3	248.8	1630	2.00	**0.7859**	0.9885
10	−6.18	139.4	258.3	1661	2.06	0.7822	0.9978
15	−25.54	155.4	287.8	1781	2.44	0.7478	0.9831
20	−40.01	167.4	309.8	1911	2.99	0.6891	0.9397
30	−59.02	176.4	326.9	2201	4.36	0.5346	0.8745
40	−71.27	160.6	298.8	2521	5.94	0.3650	0.8547
60	−88.42	21.2	467.9	3254	9.53	0.0292	0.8795

Table 3.3. Calculation results for maximum efficiency with magnetic saturation taken into account

Quantity	$f = 20$ Hz	$f = 5$ Hz
Load angle δ	20°	8.6°
Angle between armature current I_a and q-axis Ψ	7.2°	0.103°
Output power P_{out}, W	1253	134.3
Output power to armature mass, W/kg	48.16	5.16
Input power P_{in}, W	1472	170.9
Electromagnetic thrust F_{dx}, N	578.7	250.4
Thrust F_x, N	577.2	248.8
Normal force F_z, N	1544	1630
Electromagnetic power P_g, W	1296	140.2
Efficiency η	0.8515	0.7859
Power factor $\cos\phi$	0.8894	0.9885
Armature current I_a, A	4.78	2.0
d-axis armature current I_{ad}, A	0.6	0.0
q-axis armature current I_{aq}, A	4.74	2.0
Armature line current density, peak value, A_m, A/m	50,670	21,170
Current density in the armature winding j_a, A/mm^2	2.946	1.231
Air gap magnetic flux density, maximum value B_{mg}, T	0.5244	0.5338
Per phase EMF excited by PMs E_f, V	91.22	23.42
Magnetic flux in the air gap Φ_g, Wb	0.1852×10^{-2}	0.1901×10^{-2}
Armature winding loss ΔP_{1w}, W	175.6	30.65
Armature core loss ΔP_{1Fe}, W	14.45	2.39
Mechanical losses ΔP_m, W	3.45	0.90
Additional losses ΔP_{ad}, W	25.1	2.7
Armature leakage reactance X_1, Ω	4.1268	1.0351
d-axis synchronous reactance X_{sd}, Ω	8.566	2.144
q-axis synchronous reactance X_{sq}, Ω	8.656	2.167
Magnetic flux density in the armature tooth, B_{1t}, T	1.2138	1.2471
Magnetic flux density in the armature core (yoke) B_{1c}, T	0.574	0.5893
Saturation factor of the magnetic circuit k_{sat}	1.0203	1.0208

3.1.8 Magnetic Circuit

The equivalent magnetic circuit shown in Fig. 3.3 has been created on the basis of the following assumptions:

(a) Symmetry axis exists every 180° electrical degrees (one pole pitch).
(b) The magnetic flux density, magnetic field intensity and relative magnetic permeability in every point of each ferromagnetic portion of the magnetic circuit (PMs, cores, teeth) is constant.
(c) The air gap leakage flux is only between the heads of teeth.

(d) The magnetic flux of the armature (primary unit) penetrates only through the teeth and core (yoke).

(e) The equivalent reluctance of teeth per pole pitch is $\Re_t/Q_1$, where $\Re_t$ is the reluctance of a single tooth and Q_1 is the number of teeth (slots) per pole.

Each portion of the magnetic circuit is replaced by equivalent reluctances:

- reluctance of PM

$$\Re_M = \frac{h_M}{\mu_0 \mu_{rrec} w_M L_M} \tag{3.52}$$

- reluctance of air gap

$$\Re_g = \frac{g k_C}{\mu_0 w_M l_M} = \frac{g k_C}{\mu_0 \alpha_i \tau L_M} \tag{3.53}$$

- reluctance of a single tooth

$$\Re_t = \frac{h_t}{\mu_0 \mu_{rt} c_t L_i k_i} \tag{3.54}$$

- reluctance of the armature core (yoke) per pole pitch

$$\Re_{1c} \approx \frac{\tau + h_{1c}}{\mu_0 \mu_{r1c} h_{1c} L_i k_i} \tag{3.55}$$

- reluctance of a the reaction rail core (yoke) per pole pitch

$$\Re_{2c} \approx \frac{\tau + h_{2c}}{\mu_0 \mu_{r2c} h_{2c} L_M} \tag{3.56}$$

- reluctance for the PM leakage flux

$$\Re_{lM} = \frac{1}{G_{lM}} \tag{3.57}$$

in which

$$G_{lM} \approx 2\mu_0(0.52 l_M + 0.26 w_M + 0.308 h_M) \quad \text{if} \quad h_M \leq x_M \tag{3.58}$$

$$G_{lM} \approx 2\mu_0 \left(\frac{h_M l_M}{x_M} + 0.26 w_M + 0.308 h_M \right) \quad \text{if} \quad h_M > x_M \tag{3.59}$$

- reluctance for the air gap leakage flux

$$\Re_{lg} \approx \frac{1}{\mu_0} \frac{5 + 4 g k_C/b_{14}}{5 g k_C/b_{14}} \frac{1}{L_i} \tag{3.60}$$

In the foregoing equations (3.52) to (3.60), μ_{rrec} is the relative recoil magnetic permeability of the PM, μ_{rt} is the relative magnetic permeability of the armature tooth, μ_{r1c} is the relative magnetic permeability of the armature core (yoke), μ_{r2c} is the relative magnetic permeability of the reaction rail (core), h_M is the height of the PM per pole, w_M is the width of the PM, and l_M is the length of the PM in the direction perpendicular to the plane of the traveling field, h_{1c} is the height of the armature core (yoke), h_{2c} is the height of the reaction rail core, b_{14} is the armature slot opening, k_C is Carter's coefficient, τ is the pole pitch, L_i is the effective length of the armature stack (in the direction perpendicular to laminations), and x_M is the distance between adjacent PMs. The magnetic flux Φ_f is excited by the MMF $\mathcal{F}_M = H_c h_M$ of PMs, the armature reaction MMF $\mathcal{F}_{ad}$ is given by eqn (3.19), Φ_f is the PM excitation flux (flux at no load), Φ_g is the air gap magnetic flux, and Φ is the magnetic flux linked with the primary winding (air gap flux Φ_g reduced by the air gap leakage flux Φ_{lg}, if included).

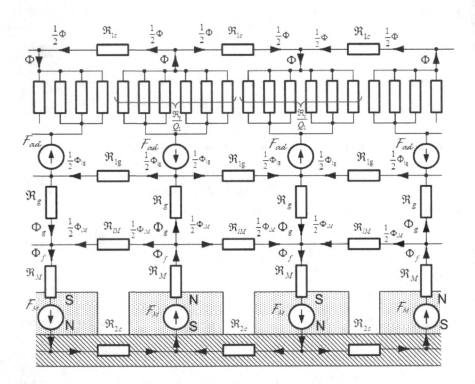

Fig. 3.3. Magnetic circuit of an LSM with surface PMs. Symbols are described in the text.

Reluctances for leakage fluxes $\mathfrak{R}_{lM}$, according to eqns (3.57), (3.58), and (3.59), have been calculated by dividing the magnetic field into simple solids.

The reluctance (3.58) is a parallel connection of the reluctances of two one-quarters of a cylinder (B.16), two half-cylinders (B.15), and four one-quarters of a sphere (B.21). The reluctance (3.59) is a parallel connection of the reluctances of two prisms (B.13), two half-cylinders (B.15), and four one-quarters of a sphere (B.21).

The following Kirchhoff's equations can be written for the magnetic circuit presented in Fig. 3.3:

$$2\Phi \frac{R_t}{Q_1} + \frac{1}{2}\Phi\Re_{1c} - \frac{1}{2}\Phi_{lg}\Re_{lg} = -2\mathcal{F}_{ad} \tag{3.61}$$

$$\Phi_g = \Phi + \Phi_{lg} \tag{3.62}$$

$$\Phi_f = \Phi_g + \Phi_M \tag{3.63}$$

$$2\Phi_f\Re_M + \frac{1}{2}\Phi_{lM}\Re_{lM} + \frac{1}{2}\Phi_f\Re_{2c} = 2\mathcal{F}_M \tag{3.64}$$

$$2(\mathcal{F}_M - \mathcal{F}_{ad}) = 2\Phi_f\Re_M + 2\Phi_g\Re_g + 2\Phi\frac{R_t}{Q_1} + \frac{1}{2}\Phi\Re_{1c} + \frac{1}{2}\Phi_f\Re_{2c} \tag{3.65}$$

The solution to these equations gives magnetic fluxes in the following form,

- magnetic flux excited by PMs

$$\Phi_f = \frac{2(\mathcal{F}_M - \mathcal{F}_{ad} - 4\mathcal{F}_{ad}\Re_g/\Re_{lg})A + 2(\mathcal{F}_M + \mathcal{F}_{ad}\Re_{lM}/\Re_{lg})(2/\Re_{lM})B}{AC + BD} \tag{3.66}$$

- magnetic flux linked with the armature winding

$$\Phi = \frac{2(\mathcal{F}_M - \mathcal{F}_{ad} - 4\mathcal{F}_{ad}\Re_g/\Re_{lg})D - 2(\mathcal{F}_M + \mathcal{F}_{ad}\Re_{lM}/\Re_{lg})(2/\Re_{lM})C}{AC + BD} \tag{3.67}$$

- air gap magnetic flux

$$\Phi_g = \Phi A + \frac{4\mathcal{F}_{ad}}{\Re_{lg}} \tag{3.68}$$

The leakage fluxes result from eqns (3.62) and (3.63), i.e.,

$$\Phi_{lg} = \Phi_g - \Phi \tag{3.69}$$

$$\Phi_{lM} = \Phi_f - \Phi_g \tag{3.70}$$

In the above eqns (3.66) to (3.68),

$$A = 1 + \frac{1}{\Re_{lg}} \left(2\frac{\Re_t}{Q_1} + \frac{1}{2}\Re_{1c} \right) \tag{3.71}$$

$$B = 2A\Re_g + 2\frac{\Re_t}{Q_1} + \frac{1}{2}\Re_{1c} \tag{3.72}$$

$$C = 2\Re_{PM} + \frac{1}{2}\Re_{2c} \tag{3.73}$$

$$D = 1 + 4\frac{\Re_M}{\Re_{lM}} + \frac{\Re_{2c}}{\Re_{lM}} \tag{3.74}$$

The coefficient of PMs leakage flux

$$\sigma_{lM} = 1 + \frac{\Phi_{lM}}{\Phi_f} \tag{3.75}$$

The coefficient of total leakage flux

$$\sigma_l \approx 1 + \frac{\Phi_{lM} + \Phi_{lg}}{\Phi_f} \tag{3.76}$$

If $\Re_{lg} \to \infty$ and $\Re_{lM} \to \infty$, then $A \to 1$. It means that leakage fluxes $\Phi_{lg} \to 0$ and $\Phi_{lM} \to 0$. Thus, the magnetic flux

$$\Phi_f = \Phi_g = \Phi = \frac{2(\mathcal{F}_M - \mathcal{F}_{ad})}{2\left(\Re_M + \Re_g + \frac{\Re_t}{Q_1}\right) + 0.5(\Re_{1c} + \Re_{2c})} \tag{3.77}$$

3.1.9 Direct Calculation of Thrust

The thrust of a PM LSM can be calculated directly on the basis of the electromagnetic field distribution [152]. Fig. 3.4 shows a single-sided PM LSM in the Cartesian coordinate system. The problem can be simplified to a two-dimensional (2D) field distribution where currents (y direction) in the armature winding are perpendicular to the laminations, and magnetic flux density has only two components, i.e., tangential component B_x and normal component B_z.

The 2D electromagnetic field distribution will be found on the basis of the following assumptions:

(a) The armature core is an isotropic and slotless cube with its magnetic permeability tending to infinity and electric conductivity tending to zero.
(b) The armature winding is represented by an infinitely thin current sheet distributed uniformly at the active surface of the armature core.
(c) The armature currents flow only in the direction perpendicular to the xz plane, i.e., in the y direction.
(d) The width of the PM is $b_p < \tau$.

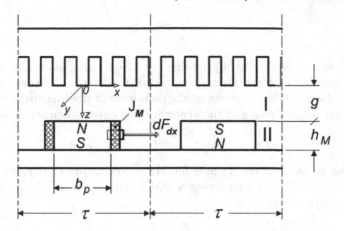

Fig. 3.4. Model of a single-sided LSM with surface configuration of PMs for 2D electromagnetic field analysis.

(e) Isotropic PMs are magnetized in the normal direction (z coordinate) and have zero electric conductivity.

(f) Each PM is represented by an equivalent coil embracing the PM and carrying a fictitious surface current that produces an equivalent magnetic flux.

(g) The magnetic permeability of the space between PMs is equal to that of PMs.

(h) The core of the reaction rail is an isotropic cube with its magnetic permeability tending to infinity and electric conductivity tending to zero.

Simplifications given by assumptions (a) and (b) can be corrected by replacing the air gap g between the armature core and PMs by an equivalent air gap $g' = k_C g$, where k_C is Carter's coefficient. The time-space distribution of the armature line current density for the fundamental space harmonic can be obtained by taking the first derivative of the primary MMF distribution with respect to the x coordinate. According to eqns (3.7) and (3.8),

$$a(x,t) = \frac{d\mathcal{F}(x,t)}{dx} = -\frac{m_1\sqrt{2}}{p\tau}N_1 I_a k_{w1}\cos(\omega t - \beta x) \tag{3.78}$$

or

$$a(x,t) = -Re[A_m e^{j\omega t - \beta x}] = -Re[A_m e^{j\omega t} e^{-j\beta x}] \tag{3.79}$$

where the peak value of the line current density

$$A_m = \frac{m_1\sqrt{2}N_1 k_{w1} I_a}{p\tau} \tag{3.80}$$

and the constant

$$\beta = \frac{\pi}{\tau} \qquad (3.81)$$

The line current density (3.80) obtained as $d\mathcal{F}(x,t)/dx$ has in numerator the effective number of turns $N_1 k_{w1}$, instead of the number of turns N_1 — see eqn (3.9). Eqn 3.9) is according to the definition of the line current density.

The space distribution of the armature line current density is simply

$$a(x) = -Re[A_m e^{-j\beta x}] = -A_m \cos \beta x \qquad (3.82)$$

According to assumption (f) and for 2D problem, the equivalent edge line current density [152] representing a PM is the same as H_c for the linear demagnetization curve, i.e.,

$$J_M = \frac{B_r}{\mu_0 \mu_{rrec}} \qquad \text{A/m} \qquad (3.83)$$

where B_r is the remanent magnetic flux density, and μ_{rrec} is the relative recoil magnetic permeability of a PM (Chapter 2 and Appendix A). If, say, $B_r = 1.1$ T and $\mu_{rrec} = 1.05$, the equivalent edge line current density $J_M \approx 0.834 \times 10^6$ A/m.

The PM magnetic flux density equal to the remanent flux density B_r is constant over the whole width $b_p = \alpha_i \tau$ of the PM and takes its sign according to the polarity of PMs. Such a periodical function $b(x) = \sum_{\nu=1}^{\infty} b_\nu \sin(\nu \pi x / \tau)$ can be resolved into Fourier series the coefficient b_ν of which, for the fundamental harmonic

$$\frac{2}{\tau} \int_{0.5(\tau - b_p)}^{0.5(\tau + b_p)} B_r \sin(\beta x) dx = B_r \frac{4}{\pi} \sin \frac{\alpha_i \pi}{2} = B_r k_f \qquad (3.84)$$

is equal to the product of B_r times the form factor of the excitation field according to eqn (3.10). Thus, the equivalent current density distribution representing the PM for $\nu = 1$ is

$$j_M(x) = k_f \frac{B_r}{\mu_0 \mu_r} \sin \beta x = k_f J_M \sin(\beta x) \qquad (3.85)$$

Assumption (g) can be partially justified due to the fact that the relative magnetic permeability of the NdFeB PM is usually 1.0 to 1.1, i.e., close to the relative magnetic permeability of free space.

The 2D electromagnetic field distribution excited by the armature winding in both regions I and II (Fig. 3.4) can be described by Laplace's equation

$$\frac{\partial^2 \mathbf{A}}{\partial x^2} + \frac{\partial^2 \mathbf{A}}{\partial z^2} = 0 \qquad (3.86)$$

where $\mathbf{A}$ is the magnetic vector potential defined as $\mathbf{B} = curl\mathbf{A}$ ($\mathbf{B} = \nabla \mathbf{A}$).

Using the method of separation of variables, the solution to eqn (3.86) for the fundamental space harmonic can have the following form:

$$\mathbf{A}_y(x, z, t) = e^{j(\omega t - \beta x)}(Ae^{-\beta z} + Be^{\beta z}) \tag{3.87}$$

According to eqn (3.82), the space distribution of the primary line current density is according to cosinusoidal law. Hence, the line current density given by eqn (3.87) can be expressed as a real number, i.e.,

$$A_y(x, z) = \cos(\beta x)(Ae^{-\beta z} + Be^{\beta z}) \tag{3.88}$$

According to assumption (c), the magnetic vector potential can only have one component A_y in the y direction so that, in the Cartesian coordinate system,

$$\mathbf{B}_x = -\frac{\partial \mathbf{A}_y}{\partial z} \qquad \mathbf{B}_z = \frac{\partial \mathbf{A}_y}{\partial x} \tag{3.89}$$

The components of the magnetic flux density in region I and region II are

$$B_x(x, z) = -\frac{\partial A_y(x, z)}{\partial z} = \beta \cos(\beta x)(Ae^{-\beta z} - Be^{\beta z}) \tag{3.90}$$

$$B_z(x, z) = \frac{\partial A_y(x, z)}{\partial x} = -\beta \sin(\beta x)(Ae^{-\beta z} + Be^{\beta z}) \tag{3.91}$$

On the basis of assumption (a), the magnetic permeability of the primary stack tends to infinity. Thus, at $z = 0$,

$$\frac{B_{xI}(x, z = 0)}{\mu_0} = a(x)$$

or, on the basis of eqn (3.82),

$$\beta(A_I - B_I) = -\mu_0 A_m \tag{3.92}$$

where the amplitude of the line current density A_m is according to eqn (3.80).

At $z = g$,

$$\frac{B_{xI}(x, z = g)}{\mu_0} = \frac{B_{xII}(x, z = g)}{\mu_0 \mu_{rrec}} \qquad \text{and} \qquad B_{zI}(x, z = g) = B_{zII}(x, z = g)$$

or

$$\mu_{rrec}(A_I e^{-\beta g} - B_I e^{\beta g}) = A_{II} e^{-\beta g} - B_{II} e^{\beta g} \tag{3.93}$$

$$A_I e^{-\beta g} + B_I e^{\beta g} = A_{II} e^{-\beta g} + B_{II} e^{\beta g} \tag{3.94}$$

At $z = g + h_M$,

$$B_x(x, z = g + h_M) = 0$$

or

$$A_{II}e^{-\beta(g+h_M)} - B_{II}e^{\beta(g+h_M)} = 0 \qquad (3.95)$$

The foregoing boundary conditions allow for finding all constants A_I, A_{II}, B_I, and B_{II} (four equations). For example, the constants A_{II} and B_{II} are

$$A_{II} = C'e^{\beta(g+h_M)} \qquad B_{II} = C'e^{-\beta(g+h_M)}$$

where

$$C' = \frac{1}{\mu_{rrel}\cosh(\beta h_M)\sinh(\beta g) + \sinh(\beta h_M)\cosh(\beta g)}) \frac{\mu_0\mu_{rrel}}{2\beta} A_m$$

Putting the constants A_{II} and B_{II} into eqns (3.91), the space distribution of the first harmonic of the normal component of the armature magnetic flux density is

$$B_{zII}(x,z) = -2\beta C' \sin(\beta x)\cosh[\beta(g+h_M-z)] \qquad (3.96)$$

The thrust for the fundamental harmonic can be found on the basis of the Lorentz equation. The force increment acting on an edge with the coordinate $x = b_p/2$ and line current density J_M is $dF_{dx} = B_{zII}(x = 0.5b_p, z)J_M L_i dz$. For $2 \times 2p$ edges and neglecting the "−" sign

$$F_{dx} = 8pL_i J_M \beta C' \int_g^{g+h_M} \sin\alpha_i\frac{\pi}{2}\cosh\left[\beta(g+h_M-z)\right]dz \qquad (3.97)$$

$$\int_q^{g+h_M} \cosh\left[\beta(h_M+g-z)\right]dz = \frac{1}{\beta}\sinh(\beta h_M) = \frac{\tau}{\pi}\sinh(\beta h_M)$$

Finally,

$$F_{dx} = \frac{4}{\pi}p\tau L_i B_r A_m \sin\left(\frac{\alpha_i\pi}{2}\right)\frac{\tanh(\beta h_M)}{\mu_{rrel}\sinh(\beta g) + \tanh(\beta h_M)\cosh(\beta g)}) \qquad (3.98)$$

The above eqn (3.98) has been derived and verified by H. Mosebach [152] and then developed further for armature line current waveforms other than sinusoidal [153].

3.2 Motors with Superconducting Excitation Coils

The model of a *coreless LSM with SC electromagnets* is shown in Fig. 3.5 [14, 62, 124, 190]. The following assumptions have been made to find the 2D distribution of magnetic and electric field components:

(a) The armature winding is represented by an infinitely long (x direction) and infinitely thin (z direction) current sheet that is distributed uniformly at $z = g$ (xy surface).

(b) The distribution of electromagnetic field in the x direction is periodical with the period equal to 2τ.

(c) The armature currents flow only in the direction perpendicular to the xz plane, i.e., in the y direction.

(d) The air-cored excitation winding is represented by an infinitely thin current sheet distributed uniformly at the $z = 0$ of the xy surface.

(e) The electromagnetic field does not change in the y direction.

(f) End effects due to the finite length of windings (x direction) are neglected.

(g) Only the fundamental space harmonic $\nu = 1$ of the field distribution in the x direction is taken into account.

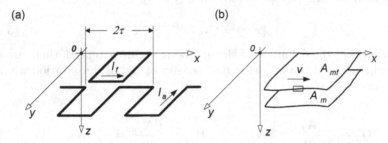

Fig. 3.5. Model of an air-cored LSM with SC excitation system for the electromagnetic field analysis: (a) winding layout, (b) armature and field excitation current sheets.

The armature winding can be represented by the following space-time distribution of the line current density (current sheet) expressed as a complex number

$$\mathbf{a}(x, t) = A_m e^{j(\omega t - \beta x)} \tag{3.99}$$

and the field excitation winding can be described by the following space-time distribution of the complex line current density

$$\mathbf{a}_f(x, t) = A_{mf} e^{j(\omega t - \beta x - \epsilon)} \tag{3.100}$$

where the peak values of line current densities are

- for the armature winding — see eqn (3.80)

$$A_m = \frac{m_1 \sqrt{2} N_1 k_{w1} I_a}{p\tau} = \frac{2m_1 \sqrt{2} N_{1p} k_{w1} I_a}{\tau} \tag{3.101}$$

- for the field excitation winding (d.c. current excitation)

$$A_{mf} = \frac{2N_f k_{wf} I_f}{p\tau} = \frac{4N_{fp} k_{wf} I_f}{\tau} \tag{3.102}$$

The number of series armature turns per phase is $N_1 = 2pN_{1p}$ where N_{1p} is the number of armature series turns per phase per pole, and the number of field series turns $N_f = 2pN_{fp}$ where N_{fp} is the number of field turns per pole.

The so-called *force angle* $\epsilon = 90^0 \mp \Psi$ is the angle between phasors of the excitation flux $\mathbf{\Phi}_f$ in the d-axis and the armature current $\mathbf{I}_a$.

The 2D distribution of the magnetic vector potential of the field excitation winding is described by the Laplace's equation

$$\frac{\partial^2 \mathbf{A}_{fy}}{\partial x^2} + \frac{\partial^2 \mathbf{A}_{fy}}{\partial z^2} = 0 \tag{3.103}$$

The general solution to eqn (3.103) for $0 \leq z \leq g$ is

$$\mathbf{A}_{fy}(x, z) = C_f e^{j(\omega t - \beta x - \epsilon)} e^{-\beta z} \tag{3.104}$$

On the basis of the definition of the magnetic vector potential, there are only two components of the magnetic flux density of the field excitation winding, i.e.,

$$\mathbf{B}_{fx} = -\frac{\partial \mathbf{A}_{fy}}{\partial z} = \beta \mathbf{A}_{fy} \qquad \mathbf{B}_{fz} = \frac{\partial \mathbf{A}_{fy}}{\partial x} = -j\beta \mathbf{A}_{fy} \tag{3.105}$$

According to Ampère's circuital law applied to the field excitation current sheet $(z = 0)$,

$$2 \int \mathbf{H}_{fx}(x, z = 0) dx = \int \mathbf{a}(x, t) dx$$

and using the first eqn (3.105),

$$\mathbf{H}_{fx}(x, z = 0) = \frac{\mathbf{B}_{fx}(x, z = 0)}{\mu_0} = -\frac{1}{\mu_0} \frac{\partial \mathbf{A}_{fy}(x, z = 0)}{\partial z} = \frac{\beta}{\mu_0} \mathbf{A}_{fy}(x, z = 0)$$

The constant C_f in eqn (3.104) is

$$C_f = \frac{\mu_0 A_{mf}}{2\beta}$$

and

$$\mathbf{A}_{fy}(x, z) = \frac{\mu_0}{2\beta} A_{mf} e^{j(\omega t - \beta x - \epsilon)} e^{-\beta z} \tag{3.106}$$

The components $\mathbf{B}_{fx}$ and $\mathbf{B}_{fz}$ can be found on the basis of eqns (3.105) and (3.106).

The magnetic vector potential of the armature winding has a similar form as eqn (3.106), i.e.,

$$\mathbf{A}_y(x, z) = \frac{\mu_0}{2\beta} \mathbf{A}_m e^{j(\omega t - \beta x)} e^{-\beta(g - z)} \qquad (3.107)$$

The forces in the x and z direction per unit area can be found on the basis of the Lorentz equation, i.e.,

- tangential force

$$f_{dx} = \frac{1}{2} Re[\mathbf{a}(x, t)\mathbf{B}_{fz}^*] = -\frac{1}{4}\mu_0 A_m A_{mf} e^{-\beta g} \sin \epsilon \qquad \text{N/m}^2 \qquad (3.108)$$

- normal force

$$f_{dz} = -\frac{1}{2} Re[\mathbf{a}(x, t)\mathbf{B}_{fx}^*] = -\frac{1}{4}\mu_0 A_m A_{mf} e^{-\beta g} \cos \epsilon \qquad \text{N/m}^2 \qquad (3.109)$$

Multiplying by the area $2p\tau$ of the SC electromagnet,

- the electromagnetic thrust

$$F_{dx} = -F_{max} \sin \epsilon \qquad (3.110)$$

- the normal repulsive force

$$F_{dz} = -F_{max} \cos \epsilon \qquad (3.111)$$

where the peak force

$$F_{max} = 4\mu_0 m_1 p \sqrt{2} N_{1p} k_{w1} N_{fp} k_{wf} \frac{L_i}{\tau} I_a I_f e^{-\beta g} \qquad (3.112)$$

Case Study 3.2. Single-Sided Air-Cored LSM

Given are the following design data of a single-sided air-cored LSM with SC excitation winding (Figs 1.17 and 3.5): $m_1 = 3$, $p = 2$, $N_{fp}k_{wf}I_f = 700 \times 10^3$ A, $k_{w1} = 1.0$, $I_a = 1000$ A, $g = 0.1$ m, $\tau = 1.35$ m, and $L_i = 1.07$ m. Find:

(a) The maximum force as a function of the number of armature turns $2 \le N_{1p} \le 20$
(b) The electromagnetic thrust F_{dx} and normal force F_{dz} as functions of the force angle ϵ.

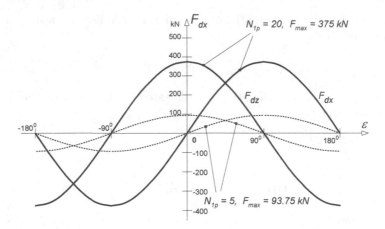

Fig. 3.6. Electromagnetic thrust F_{dx} and normal force F_{dz} as functions of the force angle ϵ for typical parameters of an air-cored LSM.

Solution

The maximum force for $N_{1p} = 2$ is

$$F_{max} = 4 \times 0.4\pi \times 10^{-6} \times 3 \times 2 \times \sqrt{2} \times 2 \times 1.0 \times 700 \times 10^{3} \times 1000.0$$

$$\times \frac{1.07}{1.35} e^{-0.1\pi/1.35} = 37,501.4 \text{ N} \approx 37.5 \text{ kN}$$

For $N_{1p} = 5$, $F_{max} = 93.75$ kN; for $N_{1p} = 10$, $F_{max} = 187.5$ kN; for $N_{1p} = 15$, $F_{max} = 281.25$ kN; and for $N_{1p} = 20$, $F_{max} = 375.0$ kN. The "−" sign has been neglected.

The forces $F_{dx} = F_{max} \sin \epsilon$ and $F_{dz} = F_{max} \cos \epsilon$ as functions of ϵ are plotted in Fig. 3.6.

3.3 Double-Sided LSM with Inner Moving Coil

The analysis of electromagnetic field in a double-sided PM LSM (Fig. 2.24) will be performed on the basis of the following assumptions:

(a) All regions are extended to infinity in the $\pm x$ direction.
(b) Magnetic permeability of ferromagnetic core (return path for magnetic flux) tends to infinity.
(c) Isotropic PMs are magnetized in the normal direction (z coordinate) and have zero electric conductivity.
(d) PMs are represented by an equivalent line current density varying periodically with the period of $2\tau/\nu$, where $\nu = 1, 3, 5, \ldots$.

(e) The model is linear.
(f) Armature reaction is negligible.

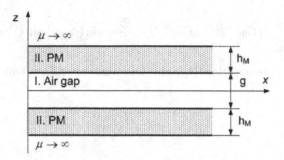

Fig. 3.7. A layer model of a double-sided PM LSM with inner moving coil ($I_a = 0$).

A layer model is shown in Fig. 3.7. The armature current is assumed to be zero ($I_a = 0$), i.e., there is no armature reaction. Thus, the electromagnetic field in the air is described by Laplace's equation, and in PMs by Poisson equation [77, 228], i.e.,

- for $-0.5g \leq z \leq 0.5g$

$$\frac{\partial^2 \mathbf{A}_I(x, z)}{\partial x^2} + \frac{\partial^2 \mathbf{A}_I(x, z)}{\partial z^2} = 0 \qquad (3.113)$$

- for $-0.5g + h_m \leq z \leq 0.5g + h_M$

$$\frac{\partial^2 \mathbf{A}_{II}(x, z)}{\partial x^2} + \frac{\partial^2 \mathbf{A}_{II}(x, z)}{\partial z^2} = -\mu_{rec}\mathbf{j}_M(x) \qquad (3.114)$$

where $\mu_{rec} = \mu_0\mu_{rrec}$ is the recoil permeability of PM; $\mathbf{A}_I$ and $\mathbf{A}_{II}$ are magnetic vector potentials in the air and PMs, respectively; and $j_M(x)$ is the PM equivalent line current density, which can be written in the following scalar form:

$$j_M(x) = \sum_{\nu=1,3,...}^{\infty} J_M \frac{4}{\nu\pi} \sin\left(\nu\alpha_i\frac{\pi}{2}\right) \sin(\beta_\nu x) \qquad (3.115)$$

where $\nu = 1, 3, 5, \ldots$ are higher space harmonics, τ is pole pitch, J_M is given by eqn (3.83), and

$$\beta_\nu = \nu\beta = \nu\frac{\pi}{\tau} \qquad (3.116)$$

See also eqns (3.81) and (3.85) in which $\nu = 1$. Using the method of variable separation, general solutions to eqns (3.113) and (3.114) are

$$A_I = \sum_{\nu=1,3,...}^{\infty} \left(C_1 e^{-\beta_\nu z} + C_2 e^{\beta_\nu z}\right) \sin\left(\beta_\nu x\right) \tag{3.117}$$

$$A_{II} = \sum_{\nu=1,3,...}^{\infty} \left[C_3 e^{-\beta_\nu z} + C_4 e^{\beta_\nu z} + \frac{4B_r \tau}{\nu^2 \pi^2} \sin\left(\beta_\nu x\right)\right] \sin\left(\beta_\nu x\right) \tag{3.118}$$

On the basis of assumption (b), the following boundary condition can be written,

$$z = 0 \qquad H_I = 0$$

$$z = \pm 0.5g \qquad H_{xI} = H_{xII} \qquad \text{and} \qquad B_{zI} = B_{zII} \tag{3.119}$$

$$z = \pm(0.5g + h_M) \qquad H_{xII} = 0$$

From the above boundary conditions (3.119), constants C_1, C_2, C_3, and C_4 can be found, i.e.,

$$C_1 = C_2 e^{-\beta_\nu g} \tag{3.120}$$

$$C_2 = \frac{4B_r \tau}{\nu^2 \pi^2} \sin\left(\beta_\nu x\right)$$

$$\times \frac{1}{(e^{-\beta_\nu g} + 1) + \frac{1}{\mu_0(e^{2\beta_\nu h_M} - 1)}\left[\mu_{rec}(-e^{-\beta_\nu g} + 1)\right](e^{2\beta_\nu h_M} - 1)} \tag{3.121}$$

$$C_3 = C_4 e^{2\beta_\nu h_M} \tag{3.122}$$

$$C_4 = \mu_{rec}(-e^{-\beta_\nu g} + 1)\frac{1}{e^{2\beta_\nu h_M} - 1}C_2 \tag{3.123}$$

The magnetic flux density components can be found from the definition of the magnetic vector potential $\mathbf{B} = \nabla \times \mathbf{A}$. For example, the normal component of the magnetic flux density in the middle of the air gap is

$$B_{zI}(x) = -\frac{\partial A}{\partial x}$$

$$= -\sum_{\nu=1,3,...}^{\infty} \beta_\nu \left(C_1 e^{0.5\beta_\nu g} + C_2 e^{-0.5\beta_\nu g}\right)\cos(\beta_\nu x) \tag{3.124}$$

Eqn (3.124) does not include the armature field ($I_a = 0$), but only the PM excitation field.

3.4 Variable Reluctance Motors

The variable reluctance LSM, called simply the *linear reluctance motor* (LRM), does not have any excitation system, so the EMF $E_f = 0$. The thrust is expressed by eqn (3.44) and is proportional to the input voltage squared V_1^2, the difference $X_{sd} - X_{sq}$ between the d- and q-axis synchronous reactances and $\sin(2\delta)$ where δ is the load angle (between the terminal voltage V_1 and the q-axis). Including the stator winding resistance R_1, the thrust is

$$F_{dxrel} = \frac{P_{elm}}{v_s} = \frac{m_1 V_1^2}{2v_s} \frac{X_{sd} - X_{sq}}{(X_{sd}X_{sq} + R_1^2)^2}[(X_{sd}X_{sq} - R_1^2)\sin 2\delta$$

$$+R_1(X_{sd} + X_{sq})\cos 2\delta - R_1(X_{sd} - X_{sq})] \tag{3.125}$$

where P_{elm} is according to eqn (3.38) for $E_f = 0$.

For the same load, the input current of a reluctance LSM is higher than that of a PM LSM since the EMF induced in the armature winding by the field excitation system is zero — eqns (3.33) and (3.34). Correspondingly, it affects efficiency because of higher power loss dissipated in the armature winding. The thrust can be increased by magnifying the ratio X_{sd}/X_{sq}. However, this in turn involves a heavier magnetizing current, resulting in further increase in the input current due to a high reluctance of the magnetic circuit in the q-axis.

3.5 Switched Reluctance Motors

The inductance of a *linear switched reluctance motor* (Fig. 1.21) can be approximated by the following function (see also eqn 5.9):

$$L(x) = \frac{1}{2}(L_{max} + L_{min}) - \frac{1}{2}(L_{max} - L_{min})\cos\left(\frac{\pi}{\tau}x\right)$$

$$= L_0 - \frac{1}{2}(k_L - 1)L_{min}\cos\frac{\pi}{\tau}x \tag{3.126}$$

where

$$L_0 = \frac{1}{2}(L_{max} + L_{min}) \tag{3.127}$$

$$k_L = \frac{L_{max}}{L_{min}} \tag{3.128}$$

L_{max} is the maximum inductance (armature and platen poles are aligned), and L_{min} is the minimum inductance (complete misalignment of armature and platen poles).

If the magnetic saturation is negligible (magnetization characteristic is linear), the electromagnetic thrust (in the x-direction) according to eqns (1.11) and (1.13) is

$$F_{dx} = \frac{dW}{dx} = \frac{1}{2}I_a^2\frac{dL(x)}{dx} = F_{max}I_a^2\sin\left(\frac{\pi}{\tau}x\right) \qquad (3.129)$$

where the maximum force at $I_a = 1$ A

$$F_{max} = \frac{1}{4}\frac{\pi}{\tau}(k_L - 1)L_{min} \qquad (3.130)$$

The electromagnetic thrust of a linear switched reluctance motor is directly proportional to the armature current squared I_a^2 and the ratio k_L of maximum to minimum inductance, and it is inversely proportional to the pole pitch τ. Eqns (3.129) and (3.130) do not take into account the current waveform shape and current turn-on and turn-off instants.

The thrust changes direction at the aligned position of the armature and the reaction rail poles. If the reaction rail continues past the aligned position, the attractive force between the armature and reaction rail poles produces retarding (braking) force. An accurate current turn-off instant provides elimination of braking force.

Sensitivity analysis of the effect of geometrical parameters on the performance of a linear switched reluctance motor, especially on the thrust profile, is given, e.g., in [9].

Examples

Example 3.1

A single-sided 6-pole, Y-connected PM LSM with surface configuration of PMs is fed with 230 V line-to-line voltage at 50 Hz. The number of turns per phase is $N_1 = 180$, winding factor $k_{w1} = 1$, pole pitch $\tau = 48$ mm, armature winding resistance per phase $R_1 = 1.614$ Ω, armature winding leakage reactance per phase $X_1 = 2.246$ Ω, d-axis armature reaction reactance $X_{ad} = 2.710$ Ω, q-axis armature reaction reactance $X_{aq} = 2.515$ Ω, magnetic flux density in the armature teeth is the same as that in core (yoke), i.e., $B_{1t} = B_{1c} = 1.468$ T, mass of armature teeth $m_{1t} = 2.973$ kg, mass of armature core (yoke) $m_{1c} = 2.473$ kg, the magnetic flux of PMs linked with the armature winding under load is $\Phi = 2.124\times10^{-3}$ Wb, specific armature core losses $\delta p_{1/50} = 1.07$ W/kg, windage losses $\Delta P_{wind} = 10$ W, mechanical losses in linear bearings $\Delta P_m = 80$ W, and losses in PMs $\Delta P_{PM} = 24.6$ W. The armature coils are wound with two ($a_w = 2$) parallel copper conductors AWG20 (0.812 mm diameter). Find the parameters at load angle $\delta = 12^o$ and steady-state performance characteristics.

Solution

The phase voltage is $V_1 = 230/\sqrt{3} = 132.8$ V. The EMF per phase induced in the armature winding

$$E_f = \pi\sqrt{2} \times 50 \times 180 \times 1.0 \times 2.124 \times 10^{-3} = 84.95 \text{ V}$$

Synchronous reactances

$$X_{sd} = 2.246 + 2.710 = 4.956 \ \Omega$$

$$X_{sq} = 2.246 + 2.515 = 4.761 \ \Omega$$

For $\delta = 12°$, the d- and q-axis armature currents according to eqns (3.33) and (3.34) are

$$I_{ad} = \frac{132.8[4.761 \times \cos(12°) - 1.614 \times \sin(12°)] - 84.95 \times 4.761}{4.956 \times 4.761 + 1.614^2} = 6.466 \text{ A}$$

$$I_{aq} = \frac{132.8[1.614 \times \cos(12°) + 4.956 \times \sin(12°)] - 84.95 \times 1.614}{4.956 \times 4.761 + 1.614^2} = 7.99 \text{ A}$$

The total armature current

$$I_a = \sqrt{6.466^2 + 7.99^2} = 10.8 \text{ A}$$

The angle between the armature current and the q-axis

$$\psi = \arcsin\left(\frac{6.466}{10.8}\right) = 38.98° = 0.68 \text{ rad}$$

Armature winding losses

$$\Delta P_{1w} = 3 \times 10.8^2 \times 1.614 = 511.6 \text{ W}$$

Losses in armature teeth and core (yoke) according to eqn (2.3),

$$\Delta P_{Fe} = 1.07 \times \left(\frac{50}{50}\right)^{4/3} (1.8 \times 1.468^2 \times 2.973 + 3.0 \times 1.468^2 \times 2.473) = 29.4 \text{ W}$$

where $k_{adt} = 1.8$ and $k_{adc} = 3.0$. Total losses

$$\Delta P = 511.6 + 29.4 + 80 + 10 + 24.6 = 655.6 \text{ W}$$

Electromagnetic power calculated on the basis of eqn (3.38)

$$P_{elm} = 3[10.8\cos(12°) \times 84.95 + 6.466 \times 7.99(4.956 - 4.761)] = 2066.4 \text{ W}$$

Input power

$$P_{in} = P_{elm} + \Delta P_{1w} + \Delta P_{Fe} = 2066.4 + 511.6 + 29.4 = 2607.5 \text{ W}$$

Output power

$$P_{out} = P_{in} - \Delta P = 2607.5 - 655.6 = 1951.8 \text{ W}$$

Input apparent power

$$S_{in} = 3 \times 132.8 \times 10.8 = 4095.0 \text{ VA}$$

Power factor $\cos\phi$

$$\cos\phi = \frac{P_{in}}{S_{in}} = \frac{2607.5}{4095.0} = 0.637; \qquad \phi = \text{arccos}\,\phi = 50.45°$$

Phase voltage across input terminals of the armature winding calculated according to eqn (3.32),

$$V_1 = \frac{1}{\cos\phi}(E_f + I_{ad}X_{sd} + I_{aq}R_1)$$

$$= \frac{1}{0.637}(84.95 + 6.466 \times 4.956 + 7.99 \times 1.614) = 132.8 \text{ V}$$

Linear synchronous speed

$$v_s = 2 \times 50 \times 0.048 = 4.8 \text{ m/s}$$

Electromagnetic thrust (force in the x-direction)

$$F_{dx} = \frac{2066.4}{4.8} = 430.5 \text{ N}$$

Output power calculated on the basis of electromagnetic power

$$P_{out} = P_{elm} - \Delta P_m - \Delta P_{wind} - \Delta P_{PM} 2066.4 - 80.0 - 10.0 - 24.6 = 1951.8 \text{ W}$$

Useful thrust

$$F_x = \frac{1951.8}{4.8} = 406.6 \text{ N}$$

Efficiency

$$\eta = \frac{1951.8}{2607.5} = 0.749$$

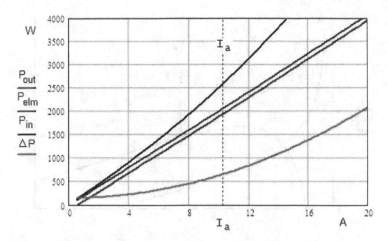

Fig. 3.8. Output power P_{out}, electromagnetic power P_{elm}, input power P_{in}, and total losses ΔP plotted against armature current I_a.

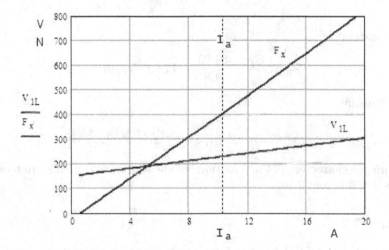

Fig. 3.9. Line voltage V_{1L} and useful thrust F_x plotted against armature current I_a.

Current density in the armature winding

$$j_a = \frac{10.8}{2 \times (\pi \times 0.000812^2/4)} = 9.93 \times 10^6 \text{ A/m}^2$$

Line current density (peak value)

$$A_m = \frac{3\sqrt{2}180 \times 10.8}{3 \times 0.048} = 54\ 514 \text{ A/m}$$

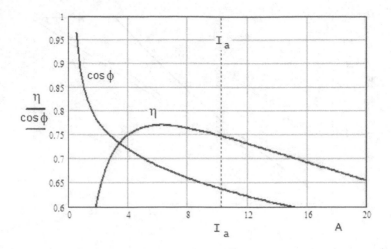

Fig. 3.10. Efficiency η and power factor $\cos\phi$ plotted against armature current I_a.

EMF constant

$$k_E = \frac{E_f}{v_s} = \frac{84.95}{4.8} = 17.697 \text{ Vs/m}$$

Force constant

$$k_F = \frac{F_{dx}}{I_a} = \frac{430.5}{10.8} = 41.881 \text{ N/A}$$

Performance characteristics as functions of the armature currents are plotted in Figs 3.8, 3.9, and 3.10.

Example 3.2

A single-sided PM LSM has the following resistances and reactances: $R_1 = 2.8$ Ω, $X_1 = 3.8$ Ω, $X_{ad} = 4.76$ Ω, $X_{aq} = 4.57$ Ω. At frequency $f = 50$ Hz and phase voltage $V_1 = 220$ V, the EMF per phase is $E_f = 200$ V. The armature windings are Y-connected. The pole pitch is $\tau = 40$ mm. Find the armature current I_a, electromagnetic power P_{elm}, electromagnetic thrust F_{dx}, and primary winding losses at zero d-axis armature current.

Solution

Synchronous reactances in the d- and q-axis

$$X_{sd} = 3.8 + 4.76 = 8.56 \text{ }\Omega; \qquad X_{sq} = 3.8 + 4.57 = 8.37 \text{ }\Omega$$

The discriminant of the quadratic equation (3.48) is

$$\Delta = b^2 - 4ac = (-6.165 \times 10^6)^2 - 4 \times 3.77 \times 10^6 \times 2.423 \times 10^6 = 1.212 \times 10^6 \text{ V}^6/\text{A}^2$$

where

$$a = (V_1 X_{sq})^2 + (V_1 R_1)^2 = (220 \times 8.37)^2 + (220 \times 2.8)^2 = 3.77 \times 10^6 \text{ V}^3/\text{A}$$

$$b = -2V_1 E_f X_{sq}^2 = -2 \times 220 \times 200 \times 8.37^2 = -6.165 \times 10^6 \text{ V}^3/\text{A}$$

$$c = (E_f X_{sq})^2 - (V_1 R_1)^2 = (200 \times 8.37)^2 - (220 \times 2.8)^2 = 2.423 \times 10^6 \text{ V}^3/\text{A}$$

Roots of the second-order equation (3.48)

$$x_1 = \frac{6.165 \times 10^6 - \sqrt{1.212 \times 10^6}}{2 \times 3.77 \times 10^6} = 0.657$$

$$x_2 = \frac{6.165 \times 10^6 + \sqrt{1.212 \times 10^6}}{2 \times 3.77 \times 10^6} = 0.978$$

$$\delta_1 = \arccos(x_1) = 48.94° \qquad \delta_2 = \arccos(x_2) = 11.95°$$

For $\delta_2 = 11.95°$, the armature currents according to eqns (3.33) and (3.34) are

$$I_{ad} = \frac{220[8.37 \times \cos(11.95°) - 2.8 \times \sin(11.95°)] - 200 \times 8.37}{8.56 \times 8.37 + 2.8^2} = 0 \text{ A}$$

$$I_{aq} = \frac{220[2.8 \times \cos(11.95°) + 8.56 \times \sin(11.95°)] - 200 \times 2.8}{8.56 \times 8.37 + 2.8^2} = 5.44 \text{ A}$$

The total armature current is torque producing, i.e.,

$$I_a = \sqrt{0^2 + 5.44^2} = 5.44 \text{ A}$$

The angle between the primary current and q-axis must be zero, i.e.,

$$\psi = \arcsin\left(\frac{I_{ad}}{I_a}\right) = 0°$$

Electromagnetic power at $I_{ad} = 0$ is calculated on the basis of eqn (3.38):

$$P_{elm} = 3[5.44 \times 200 + 0 \times 5.44(8.56 - 8.37)] = 3265 \text{ W}$$

Linear synchronous speed

$$v_s = 2 \times 50 \times 0.04 = 4.0 \text{ m/s}$$

Electromagnetic force at $I_{ad} = 0$

$$F_{dx} = \frac{3265}{4.0} = 816.3 \text{ N}$$

Primary winding losses at $I_{ad} = 0$

$$\Delta P_{1w} = 3 \times 5.44 \times 2.8 = 248.8 \text{ W}$$

Example 3.3

Given below are dimensions of a 400-N, 6-pole, 18-slot, 50-Hz, 3-phase PM LSM:

- pole pitch $\tau = 48.0$ mm
- effective length of armature stack $L_i = 96.0$ mm
- dimensions of PMs $h_M = 3.0$ mm, $w_M = 40.0$ mm, $l_M = 96.0$ mm
- height of armature core $h_{1c} = 10.0$ mm
- height of reaction rail core $h_{2c} = 12.0$ mm
- armature tooth width $c_t = 8.0$ mm
- armature tooth height $h_t = 23.5$ mm
- armature slot opening $b_{14} = 3.0$ mm
- air gap (mechanical clearance) $g = 1.5$ mm

Materials:

- Vacodym 510 NdFeB PMs with $B_{r20} = 1.35$ T, $H_{c20} = 1015$ kA/m at 20° C, $\alpha_B = -0.115$ %/°oC, $\alpha_H = -0.4$ %/°oC
- armature core stacked of M19 silicon steel, thickness 0.35 mm (Fig. 2.2)
- solid core of reaction rail made of steel 4340 (Fig. 2.3)

The number of turns per phase is $N_1 = 180$, the coil pitch is equal to the slot pitch ($w_c = \tau$), the armature core stacking factor $k_i = 0.96$, the estimated coefficient of PM leakage flux $\sigma_{lM} \approx 1.3$, and the exptected temperature of PMs is $\vartheta_{PM} = 50°C$. The armature slots are unskewed. The rated armature current is $I_a = 11.65$ A at $\psi = 40.5° = 0.707$ rad.

Find the magnetic fluxes Φ_f, Φ_g, and Φ according to eqns (3.66), (3.68) and (3.67), including armature reaction.

Solution

The number of slots per pole is $Q_1 = 18/6 = 3$, number of slots per pole per phase $q_1 = 18/(3 \times 6) = 1$, winding distribution factor $k_{d1} = \sin(\pi/2 \times 3)/(1 \times \sin(\pi/2 \times 3 \times 1) = 1$, winding pitch factor $k_{p1} = \sin(\pi \times 0.048/(2 \times 0.048) = 1$, winding factor $k_{w1} = 1 \times 1 = 1$, slot pitch $t_1 = 6 \times 0.048/18 = 0.016$ m, pole shoe width to pole pitch ratio $\alpha_i = 40/48 = 0.833$, form factor of the excitation field $k_f = (4/\pi) \sin(0.833 \times \pi/2) = 1.23$, reaction factor in the d-axis $k_{ad} = 1/1.23 = 0.813$, distance between neighboring surface PMs $x_M = 48.0 - 40.0 = 8.0$ mm.

Remanence of the PM at $\vartheta_{PM} = 50°C$ according to eqn (2.21)

$$B_r = 1.35 \left[1 + \frac{-0.115}{100}(50 - 20) \right] = 1.303 \text{ T}$$

Cooercivity of the PM at $\vartheta_{PM} = 50°C$ according to eqn (2.22)

$$H_c = 1015 \left[1 + \frac{-0.4}{100}(50 - 20) \right] = 893 \text{ kA/m}$$

Relative recoil magnetic permeability

$$\mu_{rrec} = \frac{1.303}{0.4\pi \times 10^{-6} \times 893\ 000} = 1.161$$

Air gap magnetic flux density according to eqn (2.20)

$$B_g = \frac{1.303}{1 + \mu_{rrec} \times 1.5/3.0} = 0.8245 \text{ T}$$

Air gap magnetic flux density including PM leakage flux

$$B_g = 0.8245\frac{1}{1.3} = 0.634 \text{ T}$$

It has been estimated that the PM leakage flux coefficient is $\sigma_{lM} \approx 1.3$. Similar value must obtained after calculating the magnetic fluxes. The magnetic flux in the air gap

$$\Phi_g = \alpha_i B_g \tau L_i = 0.833 \times 0.634 \times 0.048 \times 0.096 = 0.00243 \text{ Wb}$$

Magnetic flux density in primary teeth

$$B_{1t} = \frac{B_g t_1}{c_t k_i} = \frac{0.634 \times 0.016}{0.008 \times 0.96} = 1.322 \text{ T}$$

The corresponding magnetic field intensity read on the magnetization curve B-H of M19 silicon steel is $H_{1t} = 477.7$ A/m. The relative magnetic permeability of teeth

$$\mu_{rt} = \frac{1.322}{0.4\pi \times 10^{-6} \times 477.7} - 2202$$

Cross section of a tooth including stacking coefficient k_i

$$S_t = 0.008 \times 0.096 \times 0.96 = 7.373 \times 10^{-4} \text{ m}^2$$

Reluctance of a tooth according to eqn (3.54)

$$\Re_t = \frac{0.0235}{0.4\pi \times 10^{-6} \times 2202 \times 7.373 \times 10^{-4}} = 1.152 \times 10^4 \text{ 1/H}$$

Magnetic voltage drop along the armature tooth

$$V_{\mu 1t} = \Phi_g \frac{\Re_t}{Q_1} = 0.00271\frac{1.152 \times 10^4}{3} = 9.35 \text{ A}$$

Magnetic flux density in the armature core (yoke)

$$B_{1c} \approx \frac{\Phi_g}{2}\frac{1}{h_{1c}L_i k_i} = \frac{0.00243}{2}\frac{1}{0.01 \times 0.096 \times 0.96} = 1.321 \text{ T}$$

The corresponding magnetic field intensity read on the B-H curve of M19 silicon steel is $H_{1c} = 475.15$ A/m. The relative magnetic permeability of the armature core (yoke)

$$\mu_{r1c} = \frac{1.321}{0.4\pi \times 10^{-6} \times 475.15} = 2212$$

Cross section of the armature core including the stacking coefficient k_i

$$S_{1c} = 0.01 \times 0.096 \times 0.96 = 9.216 \times 10^{-4} \text{ m}^2$$

Reluctance of the armature core according to eqn (3.55)

$$\mathfrak{R}_{1c} = \frac{0.048 + 0.5 \times 0.01}{0.4\pi \times 10^{-6} \times 2212 \times 9.216 \times 10^{-4}} = 2.069 \times 10^4 \text{ 1/H}$$

Magnetic voltage drop along the armature core (yoke)

$$V_{\mu 1c} = \frac{1}{2}\Phi_g \mathfrak{R}_{1c} = \frac{1}{2}0.00243 \times 2.069 \times 10^4 = 25.2 \text{ A}$$

Magnetic flux density in the reaction rail core

$$B_{2c} \approx \frac{\Phi_g}{2}\sigma_{lM}\frac{1}{h_{2c}L_i} = \frac{0.00243}{2}1.3\frac{1}{0.012 \times 0.096} = 1.374 \text{ T}$$

The magnetic field intensity corresponding to $B_{2c} = 1.374$ T is $H_{2c} = 726.1$ A/m The magnetization curve of solid steel 4340 is plotted in Fig. 2.3.
Relative magnetic permeability of the reaction rail core

$$\mu_{r2c} = \frac{1.374}{0.4\pi \times 10^{-6} \times 726.1} = 514.7$$

Cross section of the reaction rail

$$S_{2c} = h_{2c}l_M = 0.012 \times 0.096 = 1.152 \times 10^{-3} \text{ m}^2$$

Reluctance of the secondary core according to eqn (3.56)

$$\mathfrak{R}_{2c} = \frac{0.048 + 0.5 \times 0.012}{0.4\pi \times 10^{-6} \times 514.8 \times 1.152 \times 10^{-3}} = 7.247 \times 10^4 \text{ 1/H}$$

Magnetic voltage drop along the reaction rail core

$$V_{\mu 2c} = \frac{1}{2}\Phi_g \sigma_{lM} \mathfrak{R}_{2c} = \frac{1}{2}0.00243 \times 1.3 \times 7.247 \times 10^4 = 114.7 \text{ A}$$

Reluctance of PM according to eqn (3.52)

$$\mathfrak{R}_M = \frac{0.003}{0.4\pi \times 10^{-6} \times 1.161 \times 0.04 \times 0.096} = 5.356 \times 10^5 \text{ 1/H}$$

Magnetic voltage drop across the PM

$$V_{\mu PM} = \Phi_g \sigma_{lM} \Re_M = 0.00243 \times 1.3 \times 5.356 \times 10^5 = 1695.8 \text{ A}$$

Carter's coefficient according to eqns (2.44) and (2.45)

$$\gamma = \frac{4}{\pi} \left[\frac{0.003}{0.0015} \arctan \left(0.5 \frac{0.003}{0.0015} \right) - \ln \sqrt{1 + \left(\frac{0.003}{0.0015} \right)^2} \right] = 0.5587$$

$$k_C = \frac{0.016}{0.016 - 0.5587 \times 0.0015} = 1.0553$$

Air gap cross section

$$S_g = \alpha_i \tau L_i = 0.833 \times 0.048 \times 0.096 = 3.838 \times 10^{-3} \text{ m}^2$$

Reluctance of the air gap according to eqn (3.53)

$$\Re_g = \frac{0.0015 \times 1.0553}{0.4\pi \times 10^{-6} \times 3.838 \times 10^{-3}} = 3.282 \times 10^5 \text{ 1/H}$$

Magnetic voltage drop across the air gap

$$V_{\mu g} = \Phi_g \Re_g = 0.00243 \times 3.282 \times 10^5 = 799.0 \text{ 1/H}$$

Total MMF per pole pair

$$\mathcal{F} = 2(V_{\mu g} + V_{\mu PM} + V_{\mu t}) + V_{\mu 1c} + V_{\mu 2c}$$

$$= 2(799.0 + 1695.8 + 9.35) + 25.2 + 114.7 = 5147.2 \text{ A}$$

Saturation factor of magnetic circuit according to eqn (B.26)

$$k_{sat} = 1 + \frac{2 \times 9.35 + 25.2 + 114.7}{2 \times 799.0} = 1.099$$

Reluctance for PM leakage flux ($h_M < x_M$) according to eqns (3.57) and (3.59)

$$G_{lM} \approx 2 \times 0.4\pi \times 10^{-6}(0.52 \times 0.096 + 0.26 \times 0.04 + 0.308 \times 0.003) = 1.539 \times 10^{-7} \text{ H}$$

$$\Re_{lM} = \frac{1}{1.539 \times 10^{-7}} = 6.5 \times 10^6 \text{ 1/H}$$

Reluctance for the air gap leakage flux according to eqn (3.60)

$$\Re_{lg} = \frac{1}{0.4\pi \times 10^{-6}} \frac{5 + 4 \times 0.0015 \times 1.0553/0.003}{5 \times 0.0015 \times 1.0553/0.003} \frac{1}{0.096} = 2.23 \times 10^7 \ 1/\text{H}$$

Constants expressed by eqns (3.71) to 3.74)

$$A = 1 + \frac{2}{2.23 \times 10^7} \left(2\frac{1.152 \times 10^4}{3} + \frac{1}{2}2.069 \times 10^4 \right) = 1.002$$

$$B = 2 \times 1.002 \times 3.282 \times 10^5 + 2\frac{1.152 \times 10^4}{3} + \frac{1}{2}2.069 \times 10^4 = 6.754 \times 10^5 \ \Omega$$

$$C = 2 \times 5.356 \times 10^5 + \frac{1}{2}7.247 \times 10^4 = 1.107 \times 10^6 \ \Omega$$

$$D = 1 + 4\frac{5.356 \times 10^5}{6.5 \times 10^6} + \frac{7.247 \times 10^4}{6.5 \times 10^6} = 1.341$$

Equivalent MMF excited by one pole of PM

$$\mathcal{F}_M = H_c h_M = 893\,000 \times 0.003 = 2679 \ \text{A}$$

The d-axis armature reaction MMF according to eqn (3.19)

$$\mathcal{F}_{ad} = 0.813\frac{3\sqrt{2}}{\pi \times 3}180 \times 1.0 \cdot 11.65 \sin(40.5^\circ) = 491.1 \ \text{A}$$

Denominator in eqns (3.66), (3.67), and (3.68)

$$AC + BD = 1.002 \times 1.107 \times 10^6 + 6.754 \times 10^5 \times 1.341 = 2.015 \times 10^6 \ \Omega$$

Magnetic flux of PMs (per pole pair) according to eqn (3.66)

$$\Phi_f = \frac{2}{2.015 \times 10^6} \left(2679 - 491.1 - 4 \times 491.1\frac{3.282 \times 10^5}{2.23 \times 10^7} \right) \times 1.002$$

$$+\frac{2}{2.015 \times 10^6} \left(2679 + 491.1\frac{6.5 \times 10^6}{2.23 \times 10^7} \right) \frac{1}{6.5 \times 10^6} \times 6.754 \times 10^5 = 2.73 \times 10^{-3}\text{Wb}$$

Magnetic flux linked with the armature winding according to eqn (3.67)

$$\Phi = \frac{2}{2.015 \times 10^6} \left(2679 - 491.1 - 4 \times 491.1\frac{3.282 \times 10^5}{2.23 \times 10^7} \right) \times 1.341$$

$$-\frac{2}{2.015 \times 10^6} \left(2679 + 491.1\frac{6.5 \times 10^6}{2.23 \times 10^7} \right) \frac{1}{6.5 \times 10^6} \times 1.107 \times 10^6 = 1.92 \times 10^{-3} \ \text{Wb}$$

Air gap magnetic flux according to eqn (3.68)

$$\Phi_g = 1.92 \times 10^{-3} \times 1.002 + \frac{4 \times 491.1}{2.23 \times 10^7} = 2.011 \times 10^{-3} \text{ Wb}$$

Air gap leakage flux according to eqn (3.69)

$$\Phi_{lg} = 2.011 \times 10^{-3} - 1.92 \times 10^{-3} = 9.102 \times 10^{-5} \text{ Wb}$$

PM leakage flux according to eqn (3.70)

$$\Phi_{lM} = 2.73 \times 10^{-3} - 2.011 \times 10^{-3} = 7.19 \times 10^{-4} \text{ Wb}$$

Coefficient of leakage flux according to eqn (3.76)

$$\sigma_l \approx 1 + \frac{\Phi_{lM} + \Phi_{lg}}{\Phi_f} = 1 + \frac{7.19 \times 10^{-4} + 9.102 \times 10^{-5}}{2.73 \times 10^{-3}} \approx 1.3$$

For $\mathcal{F}_{ad} = 0$ (no armature reaction), magnetic fluxes are $\Phi_f = 3.217 \times 10^{-3}$ Wb, $\Phi_g = 2.664 \times 10^{-3}$ Wb, $\Phi = 2.660 \times 10^{-3}$ Wb, $\Phi_{lg} = 4.292 \times 10^{-6}$ Wb, $\Phi_{lM} = 5.53 \times 10^{-4}$ Wb. The coefficient of leakage flux decreases, i.e., $\sigma_l = 1.173$.

If $\Re_{lg} \to \infty$ and $\Re_{lPM} \to \infty$ (no leakage fluxes) and $\mathcal{F}_{ad} = 0$ (no armature reaction), then, according to eqn (3.77)

$$\Phi_f = \Phi_g = \Phi$$

$$= \frac{2 \times 2679}{2\left(1.152 \times 10^4/3 + 5.356 \times 10^5 + 3.282 \times 10^5\right) + 0.5(2.069 + 7.247) \times 10^4}$$

$$= 3.008 \times 10^{-3} \text{ Wb}$$

Example 3.4

Specifications of a 4-pole, single-sided, PM LSM are given in Tables 3.2 and 3.3, i.e., $\delta = 8.6^0$, $\tau = 56$ mm, $b_p = 42$ mm, $L_i = 84$ mm, $B_r = 1.1$ T, $H_c = 800\,000$ A/m, $B_{mg} = 0.5338$ T, $N_1 = 560$, $k_{w1} = 1.0$, $g = 2.5$ mm, $k_C = 1.2$, $h_M = 4$ mm, $I_a = 2.0$, $X_1 = 1.0397$ Ω, $X_{ad} = X_{aq} = 1.1323$ Ω. The input frequency is $f = 5$ Hz, and input line-to-line voltage is $V_{1L} = 50$ V.

Find the electromagnetic thrust F_{dx} using eqns (3.98) and (3.40).

Solution

The relative recoil magnetic permeability is $\mu_{rrec} = 0.5338/(0.4\pi \times 10^{-6} \times 800\,000) = 1.094$, synchronous speed $v_s = 2 \times 5.0 \times 0.056 = 0.56$ m/s, phase voltage $V_1 = 50/\sqrt{3} = 28.87$, pole shoe width to pole pitch ratio $\alpha_i = 42/56 = 0.75$ and the constant according to eqn (3.81) is $\beta = \pi/0.056 = 56.1 \ 1/m$.

The equivalent edge line current density according to eqn (3.83)

$$J_M = \frac{1.1}{0.4\pi \times 10^{-6} \times 1.094} = 800\ 000 \ \text{A/m}$$

The line current density according to eqn (3.80)

$$A_m = \frac{3\sqrt{2} \times 560 \times 1.0 \times 2.0}{4 \times 0.056} = 21\ 213 \ \text{A/m}$$

The electromagnetic thrust F_{dx} according to eqn (3.98)

$$F_{dx} = \frac{4}{\pi} 4 \times 0.084 \times 1.1 \times 21\ 213 \sin\left(\frac{\pi 0.75}{2}\right)$$

$$\times \frac{\tanh(56.1 \times 0.040)}{1.094 \sinh(56.1 \times 0.0025 \times 1.2) + \tanh(56.1 \times 0.040)\sinh(56.1 \times 0.0025 \times 1.2)}$$

$$= 278.0 \ \text{N}$$

The form factor of the excitation field as given by eqn (3.10)

$$k_f = \frac{4}{\pi} \sin\left(\frac{0.75\pi}{2}\right) = 1.176$$

The amplitude of the first harmonic of the magnetic flux density in the air gap

$$B_{mg1} = 1.176 \times 0.5338 = 0.628 \ \text{T}$$

The magnetic flux

$$\Phi_f = \frac{2}{\pi} 0.056 \times 0.084 \times 0.628 = 1.88 \times 10^{-3} \ \text{Wb}$$

The EMF

$$E_f = \pi\sqrt{2} \times 5.0 \times 560 \times 1.0 \times 1.88 \times 10^{-3} = 23.4 \ \text{V}$$

Synchronous reactances

$$X_{sd} = 1.0397 + 1.1323 = 2.172 \ \Omega \qquad X_{sq} = 1.0397 + 1.1323 = 2.172 \ \Omega$$

The electromagnetic thrust F_{dx} according to eqn (3.41)

$$F_{dx} = \frac{3}{0.56}\left[\frac{28.87 \times 23.4}{2.172}\sin(8.6^0) + \frac{28.87^2}{2}\left(\frac{1}{2.172} - \frac{1}{2.172}\right)\sin(2 \times 8.6^0)\right]$$

$$= 249 \ \text{N}$$

The thrust $F_{dx} = 278$ N given by eqn (3.98) is greater than the thrust $F_{dx} = 249$ N given by eqn (3.41).

Example 3.5

A linear switched reluctance motor has the following parameters: number of armature turns per phase $N = 180$, air gap between the armature core and reaction rail at aligned position $g = 0.8$ mm, width of armature pole equal to the reaction rail pole $b_p = 12$ mm, pole pitch $\tau = 32$ mm, width of armature stack $L_i = 40$, mm and height of the reaction rail pole $h_{rlp} = 8$ mm.

Find approximate distribution of the inductance and electromagnetic thrust F_{dx} along the pole pitch for armature currents $I_a = 5$, 10 and 15 A.

Assumption: Magnetic saturation, leakage and fringing fluxes have been neglected.

Solution

Air gap reluctance in aligned position

$$\Re_{ga} = \frac{2g}{\mu_0 b_p L_i} = \frac{2 \times 0.0008}{0.4\pi \times 10^{-6} \times 0.012 \times 0.04} = 2.653 \times 10^6 \ 1/\text{H}$$

Air gap reluctance in unaligned position

$$\Re_{gu} = \frac{2(g + h_{rlp})}{\mu_0 b_p L_i} = \frac{2(0.0008 + 0.008)}{0.4\pi \times 10^{-6} \times 0.012 \times 0.04} = 29.18 \times 10^6 \ 1/\text{H}$$

Maximum inductance (aligned position)

$$L_{max} = \frac{1}{\Re_{ga}} N^2 = \frac{1}{2.653 \times 10^6} 180^2 = 0.0122 \ \text{H}$$

Minimum inductance (unaligned position)

$$L_{min} = \frac{1}{\Re_{gu}} N^2 = \frac{1}{29.18 \times 10^6} 180^2 = 0.0011 \ \text{H}$$

Maximum-to-minimum inductance ratio according to eqn (3.128)

$$k_L = \frac{0.0122}{0.0011} = 11$$

Constant value of inductance according to eqn (3.127)

$$L_0 = \frac{1}{2}(0.0122 + 0.0011) = 0.0067 \ \text{H}$$

Variation of inductance with pole pitch according to eqn (3.126) is shown in Fig. 3.11. Maximum electromagnetic thrust at $I_a = 1$ A according to eqn (3.130)

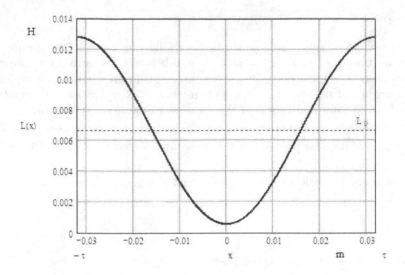

Fig. 3.11. Variation of inductance with armature position.

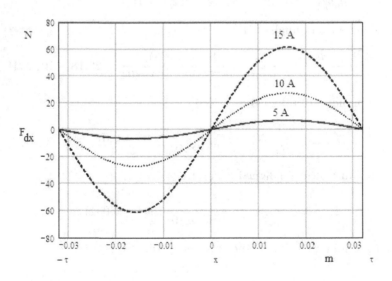

Fig. 3.12. Variation of electromagnetic thrust with armature position.

$$F_{max} = \frac{1}{4}\frac{\pi}{0.032}0.0011 = 0.273 \text{ N}$$

Variation of electromagnetic thrust with current and pole pitch according to eqn (3.126) is shown in Fig. 3.12. For $I_a = 5$ A, the thrust at $0.5\tau = 16$ mm is $F_{dx} = 6.813$ N; for $I_a = 10$ A, the thrust at $0.5\tau = 16$ mm is $F_{dx} = 27.25$ N; and for $I_a = 15$ A, the thrust at $0.5\tau = 16$ mm is $F_{dx} = 61.32$ N.

The presented method of approximate calculation of inductance and electromagnetic thrust does not include magnetic saturation, fringing and leakage flux, and can be used only for very rough estimation of the performance of an SR linear motor. Fringing and leakage fluxes can be included by using field plotting or dividing the magnetic field into simple solids as shown in Appendix B.

4

FEM Analysis

Linear-motion electromagnetic systems have been widely studied using electric circuit approach, e.g., [24, 63, 71, 142, 160]. Their operation can be simulated more accurately on the basis of electromagnetic field analysis [48, 101, 173]. Accurate estimation of electromagnetic parameters is very important. The inductance of the stator windings and the propulsion force (thrust) can be calculated from the field differential parameters such as magnetic potential and magnetic flux density distributions.

The finite element method (FEM) analysis allows for calculating the magnetic field distribution in all regions, i.e., air gap, reaction rail, armature core and armature winding. To effectively use the FEM for electromagnetic field analysis, an adequate discretization of the analyzed regions should be done. The most important is the selection of an appropriate mesh in each region using the so-called *mesh generators*. A certain number of mesh generation attempts leads to the required refinement of each region. After calculating the integral parameters of the magnetic field, the dynamic characteristics can be simulated (Appendix D) using a set of electrical and mechanical balance equations, which are solved simultaneously.

This chapter deals with the mathematical modeling of the electromagnetic field in PM LSMs using the 2D and 3D FEM. The 2D computer program has been based on the magnetic vector potential. For the 3D field simulation, the FEM computations have been done using total and reduced magnetic scalar potentials.

4.1 Fundamental Equations of Electromagnetic Field

4.1.1 Magnetic Field Vector Potential

Neglecting the displacement currents, which are of no significance at power frequencies (50 or 60 Hz), convection currents, and motion of polarized dielectric, Maxwell equations can be written in the following simplified form:

$$\nabla \times \mathbf{H} = \mathbf{J}; \qquad \nabla \times \mathbf{E} = -\nabla \times (\mathbf{B} \times \mathbf{v}) \qquad (4.1)$$

where $\mathbf{B}$ is the vector of magnetic flux density, $\mathbf{H}$ is the vector of magnetic field intensity, $\mathbf{E}$ is the vector of electric field intensity, and $\mathbf{v}$ is the vector of linear velocity. The electric current density

$$\mathbf{J} = \sigma \mathbf{E} \qquad (4.2)$$

where σ is the electrical conductivity, $-\partial \mathbf{B}/\partial t = 0$ and $-\nabla \times (\mathbf{B} \times \mathbf{v})$ represent the motion of PMs with respect to the armature system.

The magnetic vector potential $\mathbf{A}$ satisfies the equations

$$\nabla \cdot \mathbf{A} = 0 \quad \text{and} \quad \mathbf{B} = \nabla \times \mathbf{A} \qquad (4.3)$$

Assuming nonlinear magnetic permeability $\mu(B)$ of the ferromagnetic material, the magnetic field intensity $\mathbf{H} = \mathbf{B}/\mu(B)$ can be calculated on the basis of the flux density ($\mathbf{B}$) distribution [26]. However, to obtain the magnetic flux density ($\mathbf{B}$) values, the nonlinear Poisson's differential equation must be solved, i.e.,

$$\nabla \times \left(\frac{1}{\mu(B)} \nabla \times \mathbf{A} \right) = \mathbf{J} \qquad (4.4)$$

which governs the magnetic field in an LSM.

Excluding the PM domains and strongly anisotropic domain, ferromagnetic materials can be assumed isotropic. If they are modeled as magnetically linear materials, the above Poisson's equation (4.4) governs the whole analyzed domain. For rare earth PMs, their properties can be modeled with magnetization vector $\mathbf{M}$, e.g., see [70].

In the 3D Cartesian coordinate system, the *curl* $= \nabla$ of the magnetic vector potential $\mathbf{A}$ is expressed by the formula

$$\mathbf{B} = \nabla \times \mathbf{A}$$

$$= \left(\frac{\partial A_z}{\partial y} - \frac{\partial A_y}{\partial z} \right) \mathbf{1}_x + \left(\frac{\partial A_x}{\partial z} - \frac{\partial A_z}{\partial x} \right) \mathbf{1}_y + \left(\frac{\partial A_y}{\partial x} - \frac{\partial A_x}{\partial y} \right) \mathbf{1}_z \qquad (4.5)$$

The magnetic flux density distribution can also be calculated after solving the partial differential equation (PDE) with the vector potential in 3D.

In the case of a planar symmetry x, z, i.e., in the 2D system as shown, e.g., in Fig. 3.4, only the A_y component exists. In this case, the simplified differential equation describing the magnetic field is

$$\frac{\partial}{\partial x} \left[\frac{1}{\mu(B)} \frac{\partial A_y}{\partial x} \right] + \frac{\partial}{\partial z} \left[\frac{1}{\mu(B)} \frac{\partial A_y}{\partial z} \right] = J_y \qquad (4.6)$$

The magnetic flux density is calculated as a *curl* of the vector $\mathbf{A}$ ($\nabla \times \mathbf{A}$), i.e.,

$$\mathbf{B} = \nabla \times \mathbf{A} = -\frac{\partial A_y}{\partial z}\mathbf{1}_x + \frac{\partial A_y}{\partial x}\mathbf{1}_z \tag{4.7}$$

In the cylindrical coordinate system r, φ, z, the *curl* of the magnetic vector potential $\mathbf{A}$ can be written in the following generalized form:

$$\mathbf{B} = \nabla \times \mathbf{A}$$

$$= \left(\frac{1}{r}\frac{\partial A_z}{\partial \varphi} - \frac{\partial A_\varphi}{\partial z}\right)\mathbf{1}_r + \left(\frac{\partial A_r}{\partial z} - \frac{\partial A_z}{\partial r}\right)\mathbf{1}_\varphi + \frac{1}{r}\left(\frac{\partial (rA_\varphi)}{\partial r} - \frac{\partial A_r}{\partial \varphi}\right)\mathbf{1}_z \tag{4.8}$$

For cylindrical symmetry, the components A_r and A_z of the magnetic vector potential will vanish. Owing to the only J_φ component of the excitation current density, the A_φ component governs the field in the considered domain, and the elliptic PDE can be expressed as

$$\frac{\partial}{\partial r}\left[\frac{1}{\mu(B)}\left(\frac{\partial A_\varphi}{\partial r} + \frac{A_\varphi}{r}\right)\right] + \frac{\partial}{\partial z}\left[\frac{1}{\mu(B)}\frac{\partial A_\varphi}{\partial z}\right] = -J_\varphi \tag{4.9}$$

Thus, the magnetic flux density can be calculated from the magnetic vector potential, i.e.,

$$\mathbf{B} = \nabla \times \mathbf{A} = -\frac{\partial A_\varphi}{\partial z}\mathbf{1}_r + \left(\frac{\partial A_\varphi}{\partial r} + \frac{A_\varphi}{r}\right)\mathbf{1}_z \tag{4.10}$$

The nonlinear system of difference equations, obtained after discretization of the analyzed domain, can be solved with a variety of iterative conjugate gradient solvers (CGS). According to the third author [211], it is convenient to use the preconditioned conjugate gradient (PCG) code method.

4.1.2 Electromagnetic Forces

Simulation methods offer two techniques for the electromagnetic force calculation:

- Maxwell stress tensor approach
- The virtual work method

For the 2D and 3D problems, the Maxwell stress tensor can be written as [25, 41, 226]

$$\mathbf{F}_e = \int_\Omega \mathbf{f}d\Omega = \oint_\Gamma [T] \cdot d\mathbf{\Gamma} \tag{4.11}$$

The Greek letter $\mathbf{\Gamma}$ denotes the normal vector to the closed surface embracing the 3D region. For the 2D region, it becomes a closed contour.

In the 3D Cartesian coordinate system, the stress tensor can be expressed as

$$[T] = \begin{bmatrix} \mu(B)\left(H_x^2 - \frac{1}{2}H^2\right) & \mu(B)H_xH_y & \mu(B)H_xH_z \\ \mu(B)H_yH_x & \mu(B)\left(H_y^2 - \frac{1}{2}H^2\right) & \mu(B)H_yH_z \\ \mu(B)H_zH_x & \mu(B)H_zH_y & \mu(B)\left(H_z^2 - \frac{1}{2}H^2\right) \end{bmatrix} \quad (4.12)$$

where the magnetic permeability $\mu(B)$ is a nonlinear function of B. For the 2D x-z system with planar symmetry, the stress tensor can be written as

$$[T] = \begin{bmatrix} \mu(B)\left(H_x^2 - \frac{1}{2}H^2\right) & \mu(B)H_xH_z \\ \mu(B)H_zH_x & \mu(B)\left(H_z^2 - \frac{1}{2}H^2\right) \end{bmatrix} \quad (4.13)$$

In the 3D cylindrical coordinate system r, φ, z, Maxwell stress tensor can be expressed in the following matrix form:

$$[T] = \begin{bmatrix} \mu(B)\left(H_r^2 - \frac{1}{2}H^2\right) & \mu(B)rH_rH_\varphi & \mu(B)H_rH_z \\ \mu(B)rH_rH_\varphi & \mu(B)r^2\left(H_\varphi^2 - \frac{1}{2}H^2\right) & \mu(B)rH_\varphi H_z \\ \mu(B)H_rH_z & \mu(B)rH_\varphi H_z & \mu(B)\left(H_z^2 - \frac{1}{2}H^2\right) \end{bmatrix} \quad (4.14)$$

In the case of a cylindrical symmetry, the Maxwell stress tensor is [216]

$$[T] = \begin{bmatrix} \mu(B)\left(H_r^2 - \frac{1}{2}H^2\right) & \mu(B)H_rH_z \\ \mu(B)H_rH_z & \mu(B)\left(H_z^2 - \frac{1}{2}H^2\right) \end{bmatrix} \quad (4.15)$$

4.1.3 Inductances

The magnetic flux that links the N-turn armature coil can be calculated as the sum of the fluxes inside each turn (wire loop) [213]. This is done by integration of the magnetic flux density components bounded by the turn. Using Stokes theorem for the 2D region, the surface integral is replaced with the line integral, i.e.,

$$\Psi = \sum_{k=1}^{N} \int_\Gamma B_n d\Gamma = \sum_{k=1}^{N} \int_l A_t dl \quad (4.16)$$

where A_t is the component of the magnetic vector potential tangential to the line l. The static inductance of the N-turn coil is defined as the flux linkage Ψ divided by the current I in the coil (1.10), i.e.,

$$L_s = \frac{\Psi}{I} \quad (4.17)$$

where Ψ is given by eqn (4.16). The dynamic inductance

$$L_d = \frac{\partial \Psi}{\partial i} \quad (4.18)$$

is more important in the modeling and simulation of transient operation of an LSM.

In most cases, the magnetic flux is calculated by integration over the surface penetrated by the flux. However, when the magnetic potential vector is used [144], the magnetic flux linked with the armature winding can be calculated with the aid of the formula

$$\Psi = \frac{\int \mathbf{A} \cdot \mathbf{J} dV}{I} \tag{4.19}$$

4.1.4 Magnetic Scalar Potential

In regions without electric current, the total magnetic scalar potential ψ is expresses as

$$\nabla \cdot \mu \nabla \psi = 0 \tag{4.20}$$

The scalar magnetic potential has been used, e.g., in the Opera 3D FEM package [162].

If the scalar potential values in the nodes of the FEM mesh are known, the magnetic field intensity vectors can be calculated as

$$\mathbf{H} = -\nabla \psi \tag{4.21}$$

The reduced scalar potential φ is used in regions with currents. In such regions, the resultant magnetic field intensity is a sum of two components:

$$\mathbf{H} = \mathbf{H}_m + \mathbf{H}_S \tag{4.22}$$

where

$$\mathbf{H}_m = -\nabla \varphi \tag{4.23}$$

and

$$\mathbf{H}_S = \int\limits_{\Omega_J} \frac{\mathbf{J} \times \mathbf{R}}{|R|^3} d\Omega_J \tag{4.24}$$

The second component H_S in eqn (4.22) is calculated using Biot—Savart law. Thus, the PDE for the current-carrying regions can be written as

$$\nabla \cdot \mu \nabla \varphi - \nabla \cdot \mu \left(\int\limits_{\Omega_J} \frac{\mathbf{J} \times \mathbf{R}}{|R|^3} d\Omega_J \right) = 0 \tag{4.25}$$

The electromagnetic force developed by an electromagnetic device is calculated by integration of the normal (to the surface S) component of the magnetic flux density over each side of the reaction rail, e.g. ([70], p.104). In the

3D Cartesian coordinate system, the components of the electromagnetic force are

$$F_{dx} = \int_S \left[\frac{1}{\mu} B_x (\mathbf{B} \cdot \mathbf{n}) - \frac{1}{2\mu} |B|^2 n_x \right] dS \tag{4.26}$$

$$F_{dy} = \int_S \left[\frac{1}{\mu} B_y (\mathbf{B} \cdot \mathbf{n}) - \frac{1}{2\mu} |B|^2 n_y \right] dS \tag{4.27}$$

$$F_{dz} = \int_S \left[\frac{1}{\mu} B_z (\mathbf{B} \cdot \mathbf{n}) - \frac{1}{2\mu} |B|^2 n_z \right] dS \tag{4.28}$$

The other integral parameter of the magnetic field — the static inductance of the winding — can be calculated either with the aid of the flux linkage (see eqn (4.17))

$$L_s = \frac{N\Phi}{I} = \frac{N \int_S \mathbf{B} \cdot d\mathbf{s}}{I} \tag{4.29}$$

or the total energy W expressed by eqn (1.12). The magnetic energy can be obtained by performing the integration over the whole analyzed volume V, so that the static inductance is

$$L_s = \frac{2W}{I^2} = \frac{\frac{1}{2} \int_V \mathbf{B} \cdot \mathbf{H} dV}{I} \tag{4.30}$$

4.1.5 Magnetic Energy and Coenergy

When the energy stored in the electric field is negligible as compared with the magnetic field energy, the mechanical force for an arbitrary displacement x of the moving part of an electromechanical device can be calculated on the basis of the magnetic coenergy W' [145], i.e.,

$$F = \frac{\partial W'}{\partial \xi} \tag{4.31}$$

where ξ is a generalized coordinate. See also eqn (1.13). Including nonlinearity, the coenergy [145, 160, 212] is expressed as

$$W' = \int_V \left[\int_0^H B dH \right] dV \tag{4.32}$$

where V is the sectional volume of the calculated region.

The inductance of the stator coil can also be found either on the basis of the energy stored in the coil or from the flux linkage Ψ. The energy stored in the magnetic field is expressed as

$$W = \int\limits_{V} \left[\int\limits_{0}^{B} H \, dB \right] dV \qquad (4.33)$$

Obviously, for a linear magnetic system, the energy must be equal to the coenergy [145]. In this case, when the moving part is partially saturated, the expressions (4.32) and (4.33) have to be used.

When the system is slightly saturated, the average inductance of the stator can be obtained from the magnetic energy W given by eqn (1.11).

4.2 FEM Modeling

In the FEM approach, the considered region Ω is divided into a number of nonoverlapping subdomains $\Omega^{(e)}$ called *elements*, and then an approximation function $u^{(e)}$ over each element is determined [25, 189]. The elements can be of different shape so that different approximation functions can be used. The most popular elements are 3D tetrahedral and 2D triangular elements. Owing to simplifications in the modeling, linear approximation functions are useful.

In the 3D FEM, a cubic approximation is frequently used. After setting up the boundary and interface conditions, the system of linear equations is created [253]. Linear equations can be solved using special numerical methods, which are very effective for coarse and diagonal matrices that are obtained in the FEM algorithm.

4.2.1 3D Modeling in Cartesian Coordinate System

For simulation of the magnetic field distribution in LSMs, the boundary problems for PDE in the 3D or 2D regions should be solved. In a 3D region, the Poisson's PDE for the scalar potential function $\varphi(x, y, z)$ can be written as

$$a_x \frac{\partial^2 \varphi}{\partial x^2} + a_y \frac{\partial^2 \varphi}{\partial y^2} + a_z \frac{\partial^2 \varphi}{\partial z^2} = -g \qquad (4.34)$$

where $g = g(x, y, z)$ is an arbitrary function of variables x, y, z, while a_x, a_y, a_z are constants.

According to the weighted residual method, in order to obtain a variational form of eqn (4.34), it has to be multiplied by the weighted function v, while $a_x = a_y = a_z = 1$. Thus,

$$\int\limits_{\Omega^{(e)}} \left\{ v \left(\frac{\partial^2 \varphi}{\partial x^2} + \frac{\partial^2 \varphi}{\partial y^2} + \frac{\partial^2 \varphi}{\partial z^2} \right) + vg \right\} dV = 0 \qquad (4.35)$$

where $dV = dx\,dy\,dz$ is the volume of an element.

After some mathematical transformations, the variational form of eqn (4.34) for a finite element e can be written as

$$\int_{\Omega^{(e)}} \left[-\left(\frac{\partial v}{\partial x} \frac{\partial \varphi}{\partial x} + \frac{\partial v}{\partial y} \frac{\partial \varphi}{\partial y} + \frac{\partial v}{\partial z} \frac{\partial \varphi}{\partial z} \right) + vg \right] dV + \oint_{\Gamma^{(e)}} v \nabla \varphi \cdot \mathbf{n} ds = 0 \quad (4.36)$$

where Γ is the boundary surface of the element, and $\mathbf{n}$ is the unit vector, normal to the boundary surface Γ.

The surface integral of eqn (4.36) denotes the Neumann boundary condition, which is defined as

$$\nabla \varphi \cdot \mathbf{n} = \left(\frac{\partial \varphi}{\partial \mathbf{n}} \right)_\Gamma = f(\Gamma) \quad \text{for} \quad \Gamma \in \Omega \quad (4.37)$$

The Neumann boundary condition simulates the field lines direction at the boundary points. In addition, the Dirichlet boundary condition

$$\varphi(\Gamma) = f(\Gamma) \quad \text{for} \quad \Gamma \in \Omega \quad (4.38)$$

is assigned by fixing known values of the potential at boundary nodes. When the potential function φ is interpolated with the linear combination of node potentials, i.e., the so-called shape functions N_j, the method of solution becomes simpler. The function φ can be interpolated as

$$\varphi = \sum_{j=1}^{n} \varphi_j N_j \quad (4.39)$$

where n is the total number of nodes in each element.

After combining eqns (4.39) and (4.36) and assuming that the shape functions N_j are equal to the weighted function (Bubnov–Galerkin's method), the following system of equations is obtained for each finite element e:

$$F_i^{(e)} = \sum_{j=1}^{n} \int_{\Omega^e} \left[\left(\frac{\partial N_i}{\partial x} \frac{\partial N_j}{\partial x} + \frac{\partial N_i}{\partial y} \frac{\partial N_j}{\partial y} + \frac{\partial N_i}{\partial z} \frac{\partial N_j}{\partial z} \right) dV \right] \varphi_j$$

$$- \int_{\Omega^{(e)}} g N_i dV - \oint_{\Gamma^{(e)}} N_i q_n ds = 0 \quad (4.40)$$

where $q_n = \nabla \varphi \cdot \mathbf{n}$ is the Neumann boundary condition (4.37), and i, j are numbers of nodes of an element. Thus, for each element e, a system of equations in matrix form can be written, i.e.,

$$[K]^{(e)}[\varphi]^{(e)} = [f^{(e)}] \quad (4.41)$$

where the element $K_{ij}^{(e)}$ of the coefficient matrix $[K]^{(e)}$ is

$$K_{ij}^{(e)} = \int\limits_{\Omega^{(e)}} \left(\frac{\partial N_i}{\partial x}\frac{\partial N_j}{\partial x} + \frac{\partial N_i}{\partial y}\frac{\partial N_j}{\partial y} + \frac{\partial N_i}{\partial z}\frac{\partial N_j}{\partial z} \right) dV \qquad (4.42)$$

and

$$f_i^{(e)} = \int\limits_{\Omega^{(e)}} gN_i dV + \oint\limits_{\Gamma^{(e)}} q_n N_i ds \qquad (4.43)$$

Combining eqns (4.41) for all elements, the global system of equations is obtained:

$$[K]^{n\times n}[\varphi]^{n\times 1} = [f]^{n\times 1} \qquad (4.44)$$

where n is the number of unknown potential values in the nodes of the whole analyzed region.

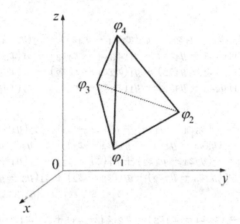

Fig. 4.1. Tetrahedral element in 3D space.

The interpolation functions for a linear tetrahedral element shown in Fig. 4.1 can be written in the following form:

$$\varphi^{(e)}(x, y, z) = \beta_1 + \beta_2 x + \beta_3 y + \beta_4 z \qquad (4.45)$$

where $\beta_1, \beta_2, \beta_3, \beta_4$ are constants of the interpolation function.

Taking into account the values of the potential at the i-th coordinate and β coefficients at the j-th node, the following system of equations is obtained

$$\begin{bmatrix} \varphi_1^{(e)} \\ \varphi_2^{(e)} \\ \varphi_3^{(e)} \\ \varphi_4^{(e)} \end{bmatrix} = \begin{bmatrix} 1 & x_1 & y_1 & z_1 \\ 1 & x_2 & y_2 & z_2 \\ 1 & x_3 & y_3 & z_3 \\ 1 & x_4 & y_4 & z_4 \end{bmatrix} \begin{bmatrix} \beta_1 \\ \beta_2 \\ \beta_3 \\ \beta_4 \end{bmatrix} \implies [\varphi] = [M][\beta] \qquad (4.46)$$

in which

$$[\varphi] = \begin{bmatrix} \varphi_1^{(e)} \\ \varphi_2^{(e)} \\ \varphi_3^{(e)} \\ \varphi_4^{(e)} \end{bmatrix} \qquad [M] = \begin{bmatrix} 1 & x_1 & y_1 & z_1 \\ 1 & x_2 & y_2 & z_2 \\ 1 & x_3 & y_3 & z_3 \\ 1 & x_4 & y_4 & z_4 \end{bmatrix} \qquad [\beta] = \begin{bmatrix} \beta_1 \\ \beta_2 \\ \beta_3 \\ \beta_4 \end{bmatrix} \qquad (4.47)$$

After solving eqn (4.41), the interpolation constants β_1, β_2, β_3, and β_4 in the interpolation function (4.45) are

$$\begin{bmatrix} \beta_1 \\ \beta_2 \\ \beta_3 \\ \beta_4 \end{bmatrix} = \frac{1}{\det[M]} ([M]^D)^T \begin{bmatrix} \varphi_1 \\ \varphi_2 \\ \varphi_3 \\ \varphi_4 \end{bmatrix} = \frac{1}{6V} \begin{bmatrix} m_{11} & m_{21} & m_{31} & m_{41} \\ m_{12} & m_{22} & m_{32} & m_{42} \\ m_{13} & m_{23} & m_{33} & m_{43} \\ m_{14} & m_{24} & x_{34} & x_{44} \end{bmatrix} \begin{bmatrix} \varphi_1 \\ \varphi_2 \\ \varphi_3 \\ \varphi_4 \end{bmatrix} \qquad (4.48)$$

where

$$\begin{cases} m_{11} = x_2(y_3 z_4 - y_4 z_3) - y_2(x_3 z_4 - x_4 z_3) + z_2(x_3 y_4 - x_4 y_3) \\ m_{21} = -x_1(y_3 z_4 - y_4 z_3) + y_1(x_3 z_4 - x_4 z_3) - z_1(x_3 y_4 - x_4 y_3) \\ m_{31} = -x_1(y_2 z_4 - y_4 z_2) + y_1(x_2 z_4 - x_4 z_2) - z_1(x_2 y_4 - x_4 y_2) \\ m_{41} = -x_1(y_2 z_3 - y_3 z_2) + y_1(x_2 z_3 - x_3 z_2) - z_1(x_2 y_3 - x_3 y_2) \end{cases}$$

$$\begin{cases} m_{12} = -(y_3 z_4 - y_4 z_3) + y_2(z_4 - z_3) - z_2(y_4 - y_3) \\ m_{22} = (y_3 z_4 - y_4 z_3) - y_1(z_4 - z_3) + z_1(y_4 - y_3) \\ m_{32} = -(y_2 z_4 - y_4 z_2) + y_1(z_4 - z_2) - z_1(y_4 - y_3) \\ m_{42} = (y_2 z_3 - y_3 z_2) - y_1(z_3 - z_2) + z_1(y_3 - y_2) \end{cases}$$

$$\begin{cases} m_{13} = (x_3 z_4 - x_4 z_3) - x_2(z_4 - z_3) + z_2(x_4 - x_3) \\ m_{23} = -(x_4 z_3 - x_4 z_3) + x_1(z_4 - z_3) - z_1(x_4 - x_3) \\ m_{33} = (x_2 z_4 - x_4 z_2) - x_1(z_4 - z_2) + z_1(x_4 - x_2) \\ m_{43} = -(x_2 z_3 - x_3 z_2) + x_1(z_3 - z_2) - z_1(x_3 - x_2) \end{cases}$$

$$\begin{cases} m_{14} = -(x_3 y_4 - x_4 y_3) + x_2(y_4 - y_3) - y_2(x_4 - x_3) \\ m_{24} = (x_3 y_4 - x_4 y_3) - x_1(y_4 - y_3) + y_1(x_4 - x_3) \\ m_{34} = -(x_2 y_4 - x_4 y_2) + x_1(y_4 - y_2) - y_1(x_4 - x_2) \\ m_{44} = (x_2 y_3 - x_3 y_2) - x_1(y_3 - y_2) + y_1(x_3 - x_2) \end{cases}$$

Thus, the magnetic scalar potential φ of the element e can be expressed as

$$\varphi^{(e)}(x, y, z) = \frac{1}{6V} \left[\sum_{j=1}^{n} \left(\varphi_j^{(e)} (m_{j1} + m_{j2}x + m_{j3}y + m_{j4}z) \right) \right]$$

$$= \sum_{j=1}^{n} \varphi_j^{(e)} N_j^{(e)} \tag{4.49}$$

where

$$N_j = \frac{1}{6V} \left(m_{j1} + m_{j2}x + m_{j3}y + m_{j4}z \right) \tag{4.50}$$

The above shape functions N_j arise in eqns (4.39) and (4.40) to form the final system of eqns (4.41). The coefficients of the matrix $[K]$ and vector $[f]$ can be calculated by using either the numerical or analytical approaches. The analytical approach is more effective than numerical calculations.

4.2.2 2D Modeling of Axisymmetrical Problems

In the case of 2D problems, the differential equations depend on the assumed symmetry. For the planar symmetry, PDEs are similar to those for the 3D field. The only differences are limits of the sums in eqns (4.39) and (4.49). In the 2D cases, the limit number is two, naturally.

In the cylindrical coordinate system r, φ, z, for axisymmetrical problems, the z-axis is the axis of symmetry. Thus, for linear problems, eqn (4.9) can be brought to the form

$$\frac{1}{r} \frac{\partial A_\varphi}{\partial r} + \frac{\partial^2 A_\varphi}{\partial r^2} + \frac{\partial^2 A_\varphi}{\partial z^2} - \frac{A_\varphi}{r^2} = -\mu J_\varphi \tag{4.51}$$

Implementing the same procedure as for the 3D Cartesian coordinate system, the above eqn (4.51) has to be multiplied by a weighted function v, i.e.,

$$\int_{\Omega^{(e)}} \left[v \left(\frac{\partial^2 A_\varphi}{\partial z^2} + \frac{\partial^2 A_\varphi}{\partial r^2} \right) + \frac{v}{r} \frac{\partial A_\varphi}{\partial r} - v \frac{A_\varphi}{r} + v\mu J_\varphi \right] \cdot 2\pi r dr dz = 0 \tag{4.52}$$

Assuming the net values for the function $f(\Gamma)$ in eqn (4.37), the so-called zero Neumann conditions are considered. Thus, eqn (4.52) can be rewritten to obtain

$$\int_{\Omega^{(e)}} \left[-\left(\frac{\partial A_\varphi}{\partial z} \frac{\partial v}{\partial z} + \frac{\partial A_\varphi}{\partial r} \frac{\partial v}{\partial r} \right) \right.$$

$$\left. + \frac{v}{r} \frac{\partial A_\varphi}{\partial r} - v \frac{A_\varphi}{r} + v\mu J_\varphi \right] \cdot 2\pi r dr dz = 0 \tag{4.53}$$

The potential can be interpolated similarly to the linear combination in the 3D region. Using linear functions, the approximation for a triangular element is

$$A^{(e)} = \sum_{j=1}^{3} A_j^{(e)} N_j(r, z) \tag{4.54}$$

Putting eqn (4.54) and $v = \sum_{i=1}^{3} A_i^{(e)} N_i(r, z)$ into eqn (4.53), and after some mathematical transformations of eqn (4.54), the following functional $F_i^{(e)}$ can be obtained, i.e.,

$$F_i^{(e)} = \frac{1}{2} \sum_{j=1}^{3} \left[A_i^{(e)} \int_{\Omega^{(e)}} \left(\frac{\partial N_i}{\partial z} \frac{\partial N_j}{\partial z} + \frac{1}{r^2} \frac{\partial (rN_i)}{\partial r} \frac{\partial (rN_j)}{\partial r} \right) \cdot 2\pi r dr dz \right] A_j^{(e)}$$

$$-\mu \sum_{i=1}^{3} \sum_{j=1}^{3} A_i^{(e)} \left(\int_{\Omega^{(e)}} N_i J^{(e)} 2\pi r dr dz \right) \tag{4.55}$$

where i, j are numbers of local nodes of a given element. The matrix equation for each element e can be expressed as

$$\frac{1}{2} \left[A^{(e)} \right]^T \left[K^{(e)} \right] \left[A^{(e)} \right] = \mu \left[A^{(e)} \right]^T \left[T^{(e)} \right] \tag{4.56}$$

where

$$K_{ij}^{(e)} = \int_{\Omega^{(e)}} \left(\frac{\partial N_i}{\partial z} \frac{\partial N_j}{\partial z} + \frac{1}{r^2} \frac{\partial (rN_i)}{\partial r} \frac{\partial (rN_j)}{\partial r} \right) \cdot 2\pi r dr dz \tag{4.57}$$

$$T_i^{(e)} = \int_{\Omega^{(e)}} N_i J^{(e)} \cdot 2\pi r dr dz \tag{4.58}$$

The interpolation functions for a linear triangular element shown in Fig. 4.2 can be assumed as

$$A^{(e)}(r, z) = \beta_1 + \beta_2 r + \beta_3 z \tag{4.59}$$

where $\beta_1, \beta_2, \beta_3$ are constants of the interpolation function.

Taking into account the values of the potential in nodes of the triangular element, the following matrix equation is obtained:

$$\begin{bmatrix} A_1^{(e)} \\ A_2^{(e)} \\ A_3^{(e)} \end{bmatrix} = \begin{bmatrix} 1 & r_1 & z_1 \\ 1 & r_2 & z_2 \\ 1 & r_3 & z_3 \end{bmatrix} \begin{bmatrix} \beta_1 \\ \beta_2 \\ \beta_3 \end{bmatrix} \implies [A] = [M][\beta] \tag{4.60}$$

in which

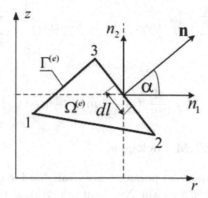

Fig. 4.2. Triangular element in axisymmetric coordinates.

$$[A] = \begin{bmatrix} A_1^{(e)} \\ A_2^{(e)} \\ A_3^{(e)} \end{bmatrix} \qquad [M] = \begin{bmatrix} 1 & r_1 & z_1 \\ 1 & r_2 & z_2 \\ 1 & r_3 & z_3 \end{bmatrix} \qquad [\beta] = \begin{bmatrix} \beta_1 \\ \beta_2 \\ \beta_3 \end{bmatrix} \qquad (4.61)$$

After solving eqn (4.60), the values of the interpolation constants in eqn (4.59) can be determined, i.e.,

$$\begin{bmatrix} \beta_1 \\ \beta_2 \\ \beta_3 \end{bmatrix} = [M]^{-1} \begin{bmatrix} A_1 \\ A_2 \\ A_3 \end{bmatrix} = \frac{1}{2S_c} \begin{bmatrix} m_{11} & m_{21} & m_{31} \\ m_{12} & m_{22} & m_{32} \\ m_{13} & m_{23} & m_{33} \end{bmatrix} \begin{bmatrix} A_1 \\ A_2 \\ A_3 \end{bmatrix} \qquad (4.62)$$

where

$$2S_c = r_1(z_2 - z_3) + r_2(z_3 - z_1) + r_3(z_1 - z_2)$$

$$\begin{cases} m_{11} = r_2 z_3 - z_2 r_3 \\ m_{12} = z_2 - z_3 \\ m_{13} = r_3 - r_2 \end{cases}$$

$$\begin{cases} m_{21} = r_3 z_1 - z_3 r_1 \\ m_{22} = z_3 - z_1 \\ m_{23} = r_1 - r_3 \end{cases}$$

$$\begin{cases} m_{31} = r_1 z_2 - z_1 r_2 \\ m_{32} = z_1 - z_2 \\ m_{33} = r_2 - r_1 \end{cases}$$

Thus, the value of the potential A in each element e can be expressed as

$$A^{(e)}(r, z) = \frac{1}{2S_c} \left[\sum_{j=1}^{3} \left(A_j^{(e)} \left(m_{j1} + m_{j2}r + m_{j3}z \right) \right) \right] = \sum_{j=1}^{3} A_j^{(e)} N_j^{(e)} \quad (4.63)$$

where

$$N_j = \frac{1}{2S_c} \left(m_{j1} + m_{j2}r + m_{j3}z \right) \quad (4.64)$$

4.2.3 Commercial FEM Packages

Nowadays, many commercial packages are available for the FEM simulation of electromagnetic fields, e.g., MagNet from Infolytica Co., Montreal, Canada, Maxwell from Ansoft Co., Pittsburgh, PA, USA; Flux from Magsoft Co., Troy, NY, USA; JMAG from JMAG Group, Tokyo, Japan; and Opera from Vector Fields Ltd., Oxford, UK [13, 33, 193, 229]. All these packages have 2D and 3D solvers for electrostatic, magnetostatic and eddy-current problems. They are mostly based on PDEs. Eqns (4.9) or (4.6) can also be solved with the 2D codes, and this will be shown in Chapter 5 of this book.

Although, each computer package uses its own methodology, fundamental expressions arise from the theory of the electromagnetic field [25, 41, 226]. For example, the electromagnetic force has been obtained from Maxwell's stress tensor, which can be written in the following general form:

$$dF = \left(\frac{1}{2} \left(\mathbf{H} \left(\mathbf{B} \cdot \mathbf{n} \right) + \mathbf{B} \left(\mathbf{H} \cdot \mathbf{n} \right) - \left(\mathbf{H} \cdot \mathbf{B} \right) \mathbf{n} \right) \right) \cdot d\mathbf{\Gamma} \quad (4.65)$$

Each approach has its own merits and, in some cases, one cannot be conveniently substituted by the other.

4.3 Time-Stepping FEM Analysis

Time stepping FEM is a combination of the FEM and state space methods. It is used in the analysis and synthesis of electrical machines to increase the accuracy of simulation of dynamic characteristics.

The 2D problem in PM electrical machines can be described by the following equation [101, 242]:

$$\frac{\partial}{\partial x} \left(\frac{1}{\mu} \frac{\partial A}{\partial x} \right) + \frac{\partial}{\partial z} \left(\frac{1}{\mu} \frac{\partial A}{\partial z} \right) = \sigma \frac{dA}{dt} - J_a - J_M \quad (4.66)$$

where A is the y-component (perpendicular to the plane of laminations) of the magnetic vector potential, μ is the permeability, σ is the electric conductivity, J_a is the current density of the armature (primary) winding, and J_M is the equivalent current density of the PM magnetization vector — see also eqn (3.83).

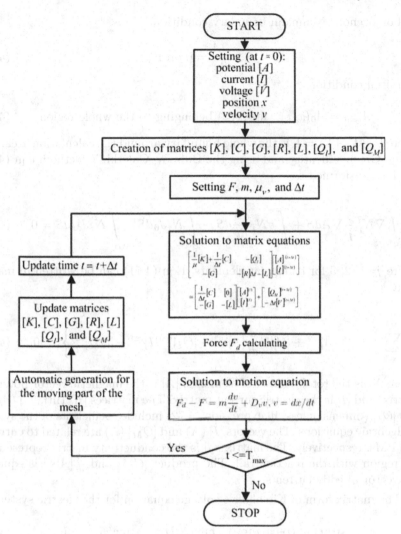

Fig. 4.3. Flowchart of time-stepping FEM analysis.

Eqn (4.66) can be solved using appropriate boundary and initial conditions, i.e.,

- Neumann boundary condition

$$\frac{1}{\mu}\frac{\partial A}{\partial n} = A_1(t) \text{ on } \Gamma_1 \tag{4.67}$$

- homogenous Dirichlet boundary condition

$$A = 0 \text{ on } \Gamma_2 \tag{4.68}$$

- homogenous Neumann boundary condition

$$\frac{\partial A}{\partial n} = 0 \text{ on } \Gamma_3 \qquad (4.69)$$

- initial condition

$$A_{t=0} = A_0(x, z) \quad \text{for } (x, z) \text{ belonging to the whole region} \qquad (4.70)$$

where $\Gamma_1 + \Gamma_2 + \Gamma_3 = \Gamma$ is the whole boundary around the calculation area. Applying Green's identity and using the Bubnov–Galerkin's method, eqn (4.66) can be transformed to

$$\int_S \nabla N_j^T \frac{1}{\mu} \nabla A dS + \int_S \sigma N_j \frac{dA}{dt} dS - \int_S N_j J_0 dS - \int_S N_j J_M dS = 0 \qquad (4.71)$$

where $j = 1, 2, 3$ for triangular elements. Eqn (4.71) can be written in matrix form

$$\sum_{e=1}^{N_e} \left\{ [K]^{(e)} [A]^{(e)} + [C]^{(e)} \frac{d}{dt} [A]^{(e)} - [Q_I]^{(e)} [I]^{(e)} - [Q_M]^{(e)} \right\} = [0] \qquad (4.72)$$

where N_e is the total number of elements, $[A]$ is the magnetic vector potential matrix, and $[I]$ is the electric current matrix. The matrices $[K]$ (m), $[C]$ (S/m), and $[Q_I]$ (dimensionless) in expression (4.72) include coefficients of the system of algebraic equations. The vectors $[I]$ (A) and $[Q_M]$ (A) are related to currents and PMs, respectively. The matrix $[C]$ is the conductivity matrix representing the region with the reaction rail. The product $[C]^{(e)}$ and $\frac{d}{dt}[A]^{(e)}$ is equal to the vector of eddy currents.

The matrix form of Kirchhoff's voltage equation for the electric system is

$$[V]^{(e)} = [R]^{(e)} [I]^{(e)} + [L]^{(e)} \frac{d}{dt} [I]^{(e)} + [G]^{(e)} \frac{d}{dt} [A]^{(e)} \qquad (4.73)$$

where $[R]$, $[L]$, and $[G]$ are resistance, inductance, and conductance matrices, respectively. The conductance matrix $[G]$ relates to the region with the armature winding. The product $[G]^{(e)}$ and $\frac{d}{dt}[A]^{(e)}$ is equal to the voltage (EMF) induced in the armature winding. By applying Euler's backward time difference method and assuming that the derivatives of $[I]$ and $[A]$ are

$$\frac{d}{dt}[I] \approx \frac{[I]^{(t+\Delta t)} - [I]^{(t)}}{\Delta t} \qquad (4.74)$$

$$\frac{d}{dt}[A] \approx \frac{[A]^{(t+\Delta t)} - [A]^{(t)}}{\Delta t} \qquad (4.75)$$

the whole electromechanical system matrix can be expressed as [101]

$$
\begin{bmatrix} [K] + \frac{1}{\Delta t}[C] & -[Q_I] \\ -[G] & -[R]\Delta t - [L] \end{bmatrix} \begin{bmatrix} [A]^{(t+\Delta t)} \\ [I]^{(t+\Delta t)} \end{bmatrix}
$$
$$
= \begin{bmatrix} \frac{1}{\Delta t}[C] & [0] \\ -[G] & -[L] \end{bmatrix} \begin{bmatrix} [A]^{(t)} \\ [I]^{(t)} \end{bmatrix} + \begin{bmatrix} [Q_M]^{(t+\Delta t)} \\ -\Delta t[V]^{(t+\Delta t)} \end{bmatrix} \qquad (4.76)
$$

Compare also [50, 51]. The mechanical balance equation of the system is

$$
F_d - F = m\frac{dv}{dt} + D_v v \qquad (4.77)
$$

where D_v is the mechanical damping constant, $v = dx/dt$ is the linear velocity, F_d is the electromagnetic force, and F is the external force. The electromagnetic force F_d can be calculated using e.g., Maxwell's stress tensor. As the part of the mesh is moving with the displacement of the reaction rail, the moving mesh technique is used to model the movement of the reaction rail [242]. The flowchart of time-stepping FEM analysis is given in Fig. 4.3.

4.4 FEM Analysis of Three-Phase PM LSM

With the growing demand on LSMs, an accurate approach to their design is required. Nowadays, the FEM method is regarded as the most accurate computational tool and necessary, among others, for the calculation of inductances, forces, and operating characteristics. In this section, FEM calculations supported by analytical calculations for PM LSMs are discussed.

4.4.1 Geometry

The presented modular PM LSM can be built both in flat and tubular forms (Figs. 4.4 to 4.7). The flat topology is characterized by the rectangular form of the armature and reaction rail slices. The concentration of magnetic energy is in the air gap between the flat reaction rail (excitation system) with PMs of w_M thickness and the armature of l_M length (Fig. 4.5).

For the flat construction, the dimension l_M perpendicular to the plane of the longitudinal section (Fig. 4.5) is very important because the output power depends on that dimension. The armature and reaction rail are assumed to be of the same width l_M. For comparative analysis of the flat and tubular topologies, their electromagnetic parameters have been calculated.

4.4.2 Specifications of Investigated Prototypes of PM LSMs

Dimensions of investigated prototypes of PM LSMs (Figs. 4.5 and 4.7) are given in Table 4.1. Magnetization curves, i.e., B-H curve and relative magnetic

Fig. 4.4. (see color insert.) Outline of 3-phase flat LSM. 1 — armature coil, 2 — ferromagnetic core, 3 — tooth of the armature segment, 4 — PM, 5 — flat ferromagnetic bar.

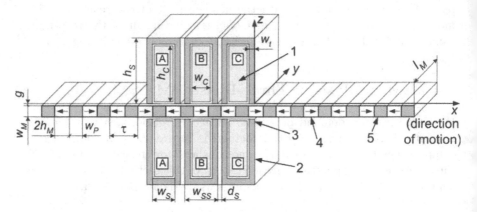

Fig. 4.5. Dimensions of 3-phase flat LSM. 1 — armature coil, 2 — ferromagnetic core, 3 — tooth of the armature segment, 4 — PM, 5 — flat ferromagnetic bar.

permeability curve μ_r verus H of mild steel used for the investigated LSMs are plotted in Fig. 4.8. The electric conductivity of mild steel at 20°C is $\sigma = 4.5 \times 10^6$ S/m.

Owing to the modular construction shown in Figs 4.4 to 4.7, it is possible to design a series of LSMs with different number of phases m_1. The formula for the appropriate distance between segments for a 3-phase motor is [220, 230]

$$\text{If} \quad w_{ss} \leq \frac{4\tau}{3}, \quad \text{then} \quad d_s = k \cdot \frac{\tau}{3} - w_{ss} + \tau, \quad k = 1, 2 \ldots$$

$$\text{else} \quad d_s = \frac{\tau}{3} - w_{ss} + k \cdot \tau, \quad k = 1, 2 \ldots \tag{4.78}$$

where k is the smallest integer number for which $d_s > 0$. For the main dimensions of three-phase LSMs presented in Table 4.1 and magnetization curves

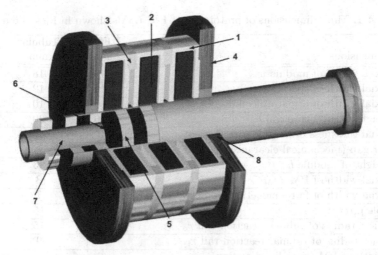

Fig. 4.6. (see color insert.) Cutaway view of 3-phase tubular LSM. 1 armature segment, 2 — armature coil, 3 — nonferromagnetic ring, 4 — armature cover, 5 — PM, 6 — ferromagnetic ring, 7 — nonferromagnetic tube, 8 — linear slide bearing.

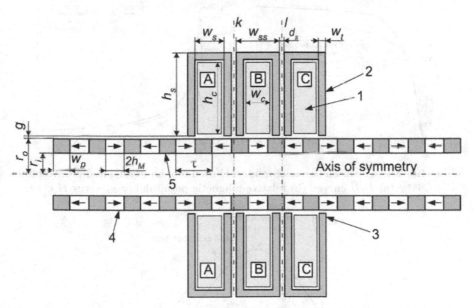

Fig. 4.7. Longitudinal section of 3-phase tubular PM LSM. 1 — armature coil, 2 — ferromagnetic core, 3 — tooth of the armature segment, 4 — PM, 5 — ferromagnetic ring.

Table 4.1. Main dimensions of prototypes of PM LSMs shown in Figs. 4.5 and 4.7

Dimensions	Flat mm	Tubular mm
Axial width of module w_{ss}	18	18
Axial width of space for coil w_s	12	12
Axial width of armature coil w_c	10	10
Height of armature coil h_c	30	30
Axial thickness of module core (leg) w_t	3	3
Air gap (mechanical clearance) g	1	1
Height of module h_s	35	35
Axial width of PM $2h_M$	8	8
Axial width of ferromagnetic core between PMs w_p	7	7
Pole pitch τ	15	15
Outer radius of tubular reaction rail r_o	—	15
Inner radius of tubular reaction rail r_i	—	9
Width of PM (in radial direction)	6	$w_M = r_o - r_i = 6$
Length of PM l_M	48.7	—

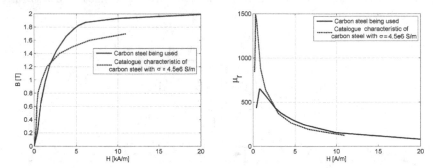

Fig. 4.8. Magnetization curves of mild steel used for the investigated prototypes of PM LSMs: (a) B-H curves; (b) relative magnetic permeability μ_r verus H curves.

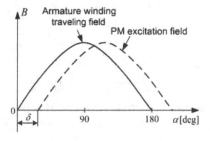

Fig. 4.9. Load angle δ defined as angular displacement between PM magnetic flux density and armature magnetic flux density waveforms (see also Fig. 3.1).

(a)

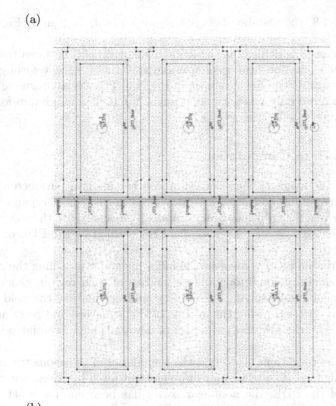

(b)

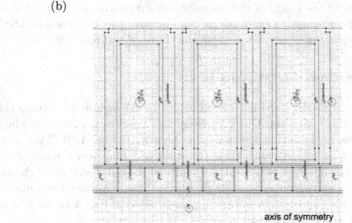

axis of symmetry

Fig. 4.10. Portion of the discretization mesh for 3-phase PM LSMs: (a) flat motor, (b) tubular motor.

shown in Fig. 4.8, the distance between segments is $d_s = 2$ mm (Figs. 4.5 and 4.7).

The *load angle* δ is defined as the angular displacement between the zero crossings of the excitation field waveform and traveling field waveform generated by the armature winding, as shown in Fig. 4.9. The electromagnetic force (thrust) is produced only when the load angle $\delta \neq 0$. The maximum force is for the load angle $\delta = 90°$ (see Fig. 3.2).

4.4.3 Approach to Computation

In calculating the magnetic flux density and other integral parameters, displacement currents and eddy currents have been neglected. The partial saturation of the reaction rail has been included, especially for the maximum allowable current. The electromagnetic field is governed by the PDE of elliptic type (4.51).

The computation process involves the following steps: modeling the geometry, setting boundary conditions and properties of each region, generating the finite element mesh, solving eqn (4.51), and calculating the field integral parameters. After drawing the outline of the armature and reaction rail (Figs. 4.5 and 4.7), the physical properties of materials have been introduced.

The Dirichlet conditions $A_\varphi = 0$ at the boundaries of the geometric model have been predefined. The triangular finite element mesh has been generated. For solution of the PDE, the nonlinear solver has been employed [41, 212]. For each position of the reaction rail, to minimize errors (in differentiation or integration of the magnetic potential), a nonuniform grid has been created, so that its refinement was extremely near the edges of the reaction rail.

4.4.4 Discretization of LSM Area in 2D

The field problems for the LSM have been solved in 2D regions by solving the PDE equation (4.51) with the Dirichlet boundary conditions. An exemplary mesh of finite elements inside the motor is shown in Fig. 4.10. Two nonferromagnetic distance layers with their thickness of 2 mm between armature segments have been inserted (4.6). In the vicinity of the outer surface of the armature core crude nonferromagnetic rings (4-mm thickness) have been placed (Figs. 4.6 and 4.10). The cuts in outer rings are visible in Fig. 4.10, as well as in Figs 4.11 and 4.12.

As previously mentioned, a fine mesh is required in FEM modeling. This mesh should be dense enough to minimize the calculation error. On the other hand, more dense discretization causes longer time of computations (solution to eqn (4.44)). Thus, it is important to optimize the finite element mesh. Modern mesh generators can find a compromise between accuracy and computation time.

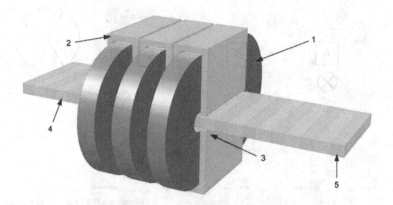

Fig. 4.4 Outline of 3-phase flat LSM. 1 — armature coil, 2 — ferromagnetic core, 3 — tooth of the armature segment, 4 — PM, 5 — flat ferromagnetic bar.

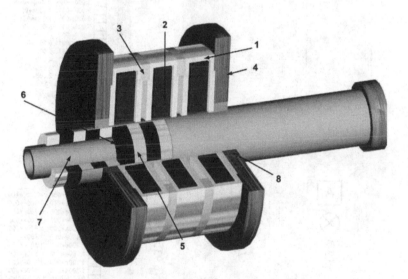

Fig. 4.6 Cutaway view of 3-phase tubular LSM. 1 armature segment, 2—armature coil, 3 — nonferromagnetic ring, 4 — armature cover, 5 — PM, 6 — ferromagnetic ring, 7 — nonferromagnetic tube, 8 — linear slide bearing.

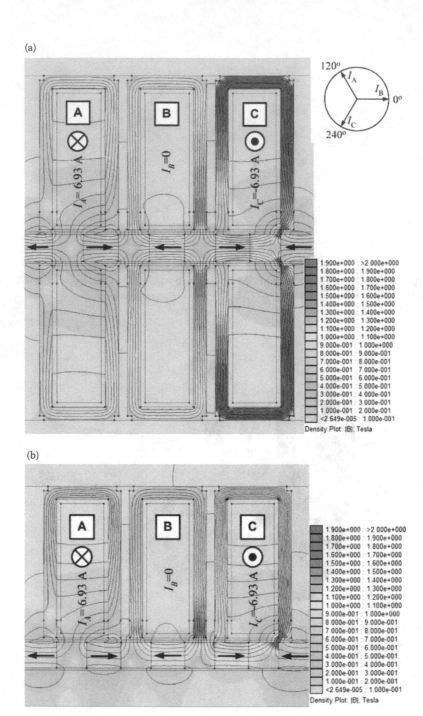

Fig. 4.11 Magnetic field distribution in longitudinal sections of 3-phase PM LSMs at no-load ($\delta = 0$): (a) flat motor, (b) tubular motor.

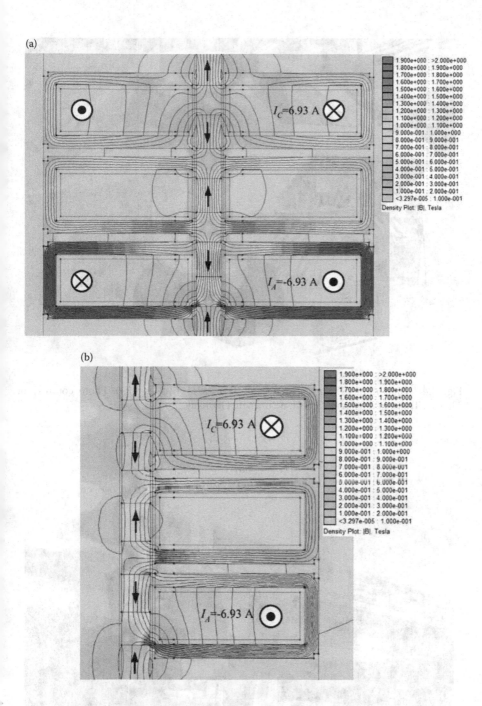

Fig. 4.22 Magnetic flux density maps in the case of excitation by the imaginary components of armature currents: (a) flat motor, (b) tubular motor.

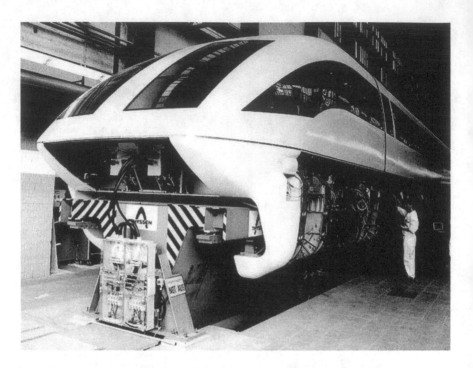

Fig. 8.4 *Transrapid 07 Europa* (Emsland Transrapid Test Facility). Photo courtesy of Thyssen Transrapid System, GmbH, M¨unchen, Germany.

Fig. 8.9 *Transrapid 08* . Photo courtesy of *Thyssen Transrapid System, GmbH,* München, Germany.

Fig. 8.11 *Transrapid* in Shanghai, China.

Fig. 8.13 Yamanashi Maglev Test Line: Ogatayama Bridge over the Chuo Expressway. Courtesy of Central Japan Railway Company and Railway Technical Research Institute, Tokyo, Japan.

Fig. 8.19 Double-cusp-shaped head car (facing Koufu) of the MLX01 Maglev Train at Expo 2005, Aichi Prefecture, Japan. Courtesy of Central Japan Railway Company and Railway Technical Research Institute, Tokyo, Japan.

Fig. 8.25 Prototype of urban maglev vehicle built by General Atomics, San Diego, CA, USA. Photo taken by the first author.

Fig. 8.28 Principle of operation of General Atomics' maglev vehicle. 1 — upper Halbach array levitation magnets, 2 — lower Halbach array levitation magnets, 3 — Litz wire guideway, 4 — LSM armature winding, 5 — propulsion magnets. Photo taken by the first author.

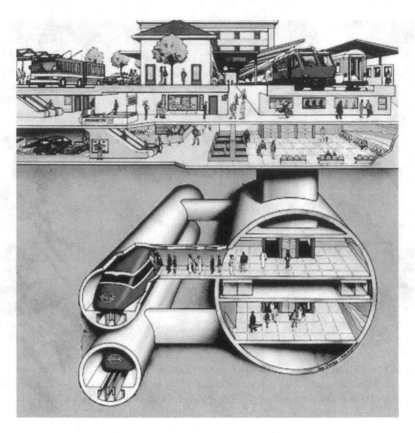

Fig. 8.29 Cross section through the station of Swissmetro. Courtesy of Swissmetro, Geneva, Switzerland.

(a)

(b)

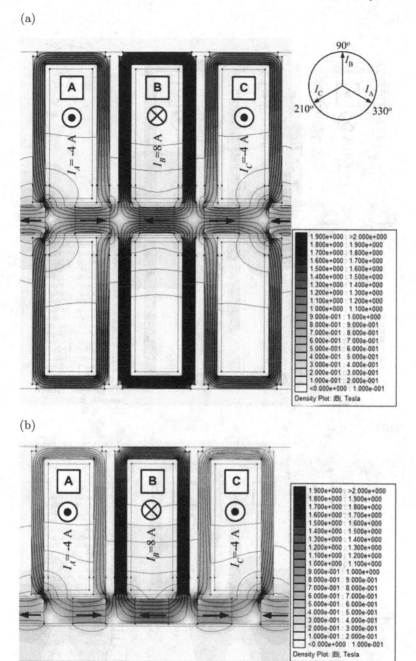

Fig. 4.11. (see color insert.) Magnetic field distribution in longitudinal sections of 3-phase PM LSMs at no-load ($\delta = 0$): (a) flat motor, (b) tubular motor.

(a)

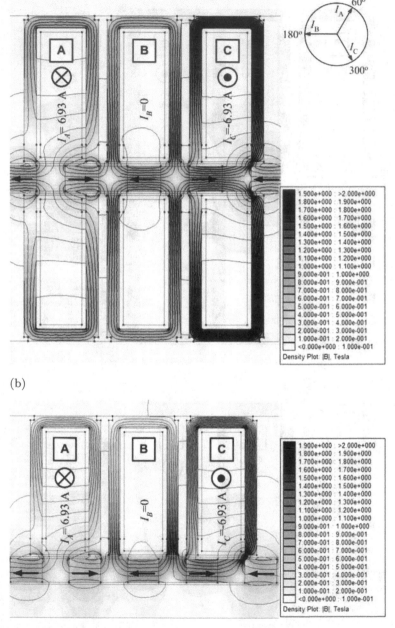

(b)

Fig. 4.12. Magnetic field distribution in longitudinal sections of 3-phase PM LSMs at full load ($\delta = 90°$): (a) flat motor, (b) tubular motor.

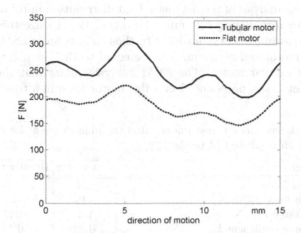

Fig. 4.13. Maximum thrust versus reaction rail position for tubular and flat three-phase PM LSMs.

4.4.5 2D Electromagnetic Field Analysis

The calculated magnetic flux lines in 3-phase PM LSMs are presented in Figs 4.11 to 4.12. Rare-earth sintered NdFeB grade 35 PMs with remanent magnetic flux density $B_r = 1.25$ T and coercivity $H_c = -950$ kA/m at $20°$C have been employed. The relative recoil magnetic permeability is $\mu_{rrec} = 1.048$. Both flat and tubular motors have been considered. The 2D field distribution has been obtained for two load angles: $\delta = 0$ as shown in Fig. 4.11 and $\delta = 90°$ as shown in Fig. 4.12. The maximum armature current is $I_a = 8$ A (Table 4.2). Various shades of grey color denote magnetic flux density values. For the field distribution, the position of three-phase balanced currents is shown in the right top corner. The value of the current at the calculated time instant can be obtained as a projection of the current phasor onto the vertical axis in the complex plane. Under the same supply and load, the magnetic flux density distributions are similar for both flat and tubular motors. However, for the flat motor (Figs. 4.11a and 4.12a), the magnetic flux density distribution in the stator segments is more homogenous as compared with the tubular motor (Figs. 4.11 and 4.12b).

In the tubular LSM, the highest magnetic flux density, as expected, is observed in the region of armature teeth close to the air gap. The lowest magnetic flux density is seen close to the outer surface of the armature core. This is because the cross section of the magnetic flux increases with the radius. The magnetic flux that links the coil turns depends on the input voltage, which causes an increase in the armature current. The magnetic flux is a nonlinear function of the armature current. Consequently, the magnetic saturation depends on the armature current I_a. For the armature current $I_a = 8$ A, some portions of the magnetic circuit become highly saturated (Figs. 4.11 and 4.12).

Comparative analysis of the magnetic field distribution in tubular and flat LSMs allows for formulating recommendations of how to size the motor. To obtain similar parameters of a flat LSM to those of a tubular LSM, the width l_M of the flat motor can be assumed to be equal to the circumference of the air gap of the tubular motor. The FEM analysis shows that the magnetic saturation effect is less pronounced in a flat motor than in a tubular motor.

Table 4.2. Electromagnetic thrust under rated (nominal) current for the analyzed flat and tubular three-phase PM LSMs

Parameter	Flat motor	Tubular motor
Maximum force F_{max}, N	221	305
Minimum force F_{min}, N	146	198
Average force F_{av}, N	180	244
Force ripple coefficient k_r	0.42	0.44
Maximum detent (cogging) force F_{cmax}, N	20	45
Armature current I_a, A	8	8

4.4.6 Calculation of Integral Parameters

The thrust F is the most important parameter of linear motors. In Figs. 4.13 and 4.14, the thrust is plotted against the position of the reaction rail under rated (nominal) armature current.

Forces F_{max}, F_{min}, and F_{av} denote maximum, minimum, and average value of the trust at nominal load angle (Section 4.4) under nominal armature current I_a. The force F_{cmax} is the maximum value of the detent force i.e., the force due to interaction of PM and armature ferromagnetic teeth (salient poles) at zero current state. The force ripple coefficient k_r has been calculated using the formula

$$k_r = \frac{F_{max} - F_{min}}{F_{av}} \tag{4.79}$$

Both the useful electromagnetic thrust and force ripple coefficient k_r are lower in the case of the flat motor.

Although, the results of the 3D analysis are similar to those obtained from the 2D analysis, in both cases of flat and tubular construction, the forces obtained from the 3D simulation are lower. In PM motors, the detent (cogging) thrust affects the useful thrust. The peaks of the thrust are slightly greater in the case of the tubular motor (Figs 4.15 and 4.16). The results of the 2D and 3D FEM analysis clearly show this difference. In fact, the paths of the magnetic flux in a flat motor are closed through the front and back parts of the motor. To minimize the computation time, the 3D analysis has been performed with 1-mm step, and the 2D analysis with 0.25-mm step. Thus, the waveforms obtained from the 2D FEM are smoother (Fig. 4.16).

(a)

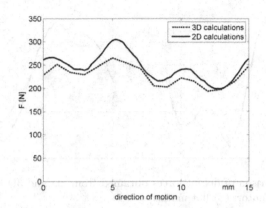

(b)

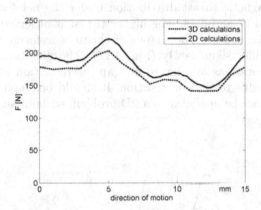

Fig. 4.14. Comparison of thrust obtained from 2D and 3D FEM computations: (a) tubular motor, (b) flat motor.

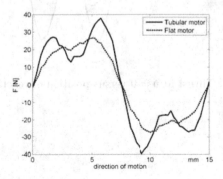

Fig. 4.15. Detent force for the tubular and flat PM LSMs.

(a)   (b)

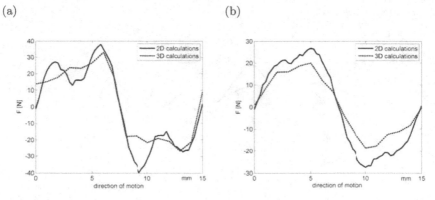

Fig. 4.16. Comparison of detent forces obtained from 2D and 3D FEM computations: (a) tubular motor, (b) flat motor.

The electromagnetic thrust distribution under the net load angle $\delta = 0$ is shown in Figs 4.17 and 4.18. The higher thrust is observed in the case of three-phase tubular motor. The shapes of thrust waveforms of the flat and tubular motors differ significantly (Fig. 4.17). The differences between the 2D and 3D computations are due to air gap discretization and, in the case of the flat motor, due to the construction. It should be emphasized that the flat construction can be analyzed as a 2D problem without significant loss of accuracy.

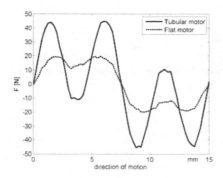

Fig. 4.17. Thrust at $\delta = 0$ versus position of reaction rail.

(a)   (b)

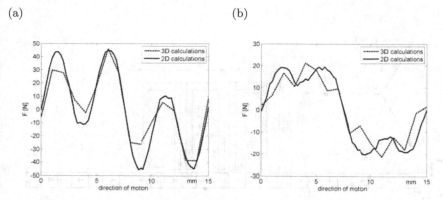

Fig. 4.18. Force at $\delta = 0$ obtained from 3D FEM computations: (a) tubular motor, (b) flat motor.

Examples

Example 4.1

Two three-phase PM LSMs have been studied. The first one is a tubular linear motor (Fig. 4.6) and the second one is a flat motor (Fig. 4.5). The reaction rail (moving part) has been assembled with ring-shaped NdFEB magnets ($2h_M = 8$ mm) and mild steel rings ($w_p = 7$ mm). The dimensions of the tubular LSM are shown in Figs 4.7 and 4.19. The flat LSM has the same cross-sectional dimensions, while its width $l_M = 48.7$ mm. The three coils, each with $N = 280$ turns, are excited with 3-phase sinusoidal armature current the amplitude of which is $I_m = 8$ A.

The objective is to compare the magnetic field distributions excited by the imaginary components of the current system depicted in Fig. 4.19, when the reaction rail is in an aligned position. Also, the modulus of the flux density in the middle of the air gap of both machines should be compared (along section AA' shown in Fig. 4.19). The comparison should be performed for the reaction rail in the aligned position (Fig. 4.19) and for the armature alone, without the reaction rail. The aligned position of the reaction rail is the initial position ($z = 0$) as well. The positive z coordinate is assumed to be in the right direction. Additionally, the inductance of the coil C should be found. For the current system depicted as in Fig. 4.19, the load angle takes its maximum value $\delta = 90°$.

Solution

This problem has been solved using the 2D FEM freeware program written by D. Meeker [144].

First, the final element mesh is created. The portion of the motor geometry being analyzed is shown in Fig. 4.20. An appropriate discretization is the

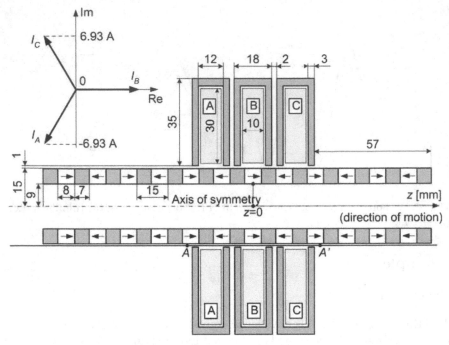

Fig. 4.19. Dimensions of the analyzed three-phase tubular LSM and current phasor diagram. The reaction rail is in aligned position.

second important step in electromagnetic field modeling. The triangular mesh has been implemented (Fig. 4.21). To obtain high accuracy of calculation of the electromagnetic force, the air gap should be discretized very precisely (the largest segment of elements should not exceed 0.3 mm in length). The force acting on the reaction rail is calculated using Maxwell's stress tensor. In this case, the reaction rail should be surrounded with a contour line that is placed 0.5 mm away from the reaction rail [144]. At the outside boundaries (not visible in the figures), the Dirichlet boundary condition $\mathbf{A} = 0$ has been applied. The B-H curve of the mild steel is shown in Fig. 4.8a (solid line). The sintered NdFeB PMs have the remanent magnetic flux density $B_r = 1.25$ T and coercivity $H_c = 950$ kA/m at room temperature $20°$. The phase currents are:

$$i_{aA}(z) = I_m \sin\left(\frac{z}{\tau}180° + 150° + \delta\right)$$
$$i_{aB}(z) = I_m \sin\left(\frac{z}{\tau}180° - 90° + \delta\right) \tag{4.80}$$
$$i_{aC}(z) = I_m \sin\left(\frac{z}{\tau}180° + 30° + \delta\right)$$

(a)

(b)

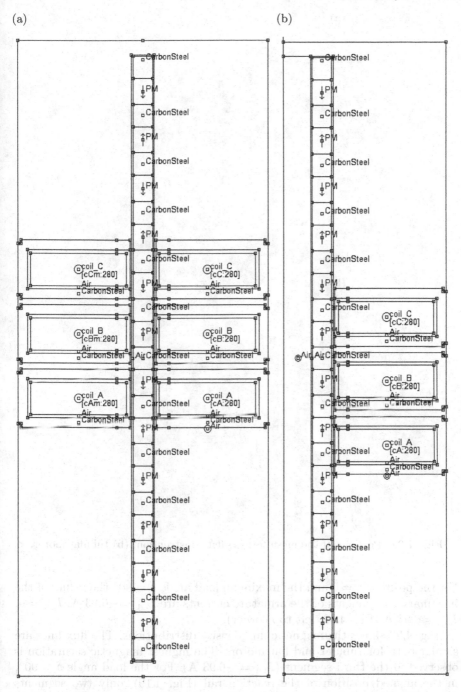

Fig. 4.20. Outline of the geometry of motors for FEM modeling: (a) flat motor, (b) tubular motor.

(a)

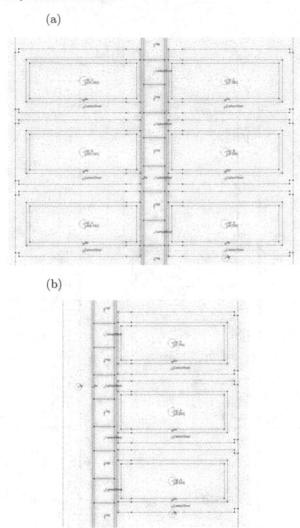

(b)

Fig. 4.21. Portions of the mesh for: (a) flat construction, (b) tubular motor.

For the position $z = 0$ and the maximum load angle $\delta = 90°$ the values of the imaginary components of the armature currents are $I_{aA} = -6.93$ A, $I_{aB} = 0$, $I_{aC} = 6.93$ A (Fig. 4.19, left top corner).

Fig. 4.22 shows the magnetic flux density distributions. The flux lines are similar both for tubular and flat motors. The highest magnetic saturation is observed in the third segment ($I_{aA} = -6.93$ A). For the load angle $\delta = 90°$, in the aligned position of the reaction rail (Fig. 4.19), only two segments (phase A and C) generate the thrust. The middle segment (phase B) does not contribute to the resultant thrust.

(a)

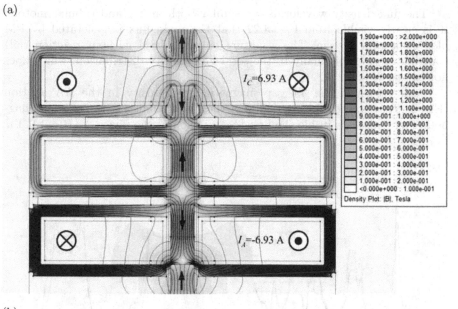

(b)

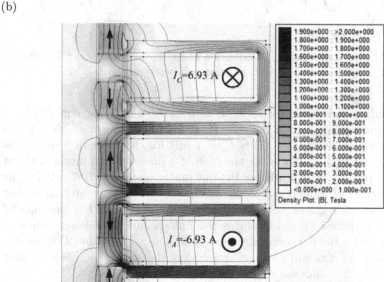

Fig. 4.22. (see color insert.) Magnetic flux density maps in the case of excitation by the imaginary components of armature currents: (a) flat motor, (b) tubular motor.

The flux density waveforms are similar both in flat and tubular motors. It is clearly visible from Fig. 4.22 that the main flux is generated by PM filed excitation system. The armature current (LSM without reaction rail) contributes only minimally to the total flux density distribution, which does not exceed $B = 0.2$ T.

Peak values of the air gap magnetic flux density in the AA' section (Fig. 4.19) are plotted in Fig. 4.23. For the tubular motor, the maximum magnetic flux density is inside the first tooth region of the phase A ($B = 1.9$ T).

(a)   (b)

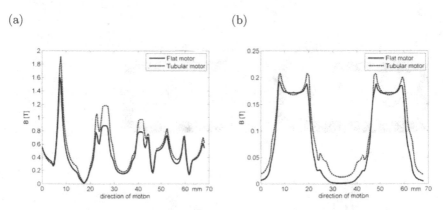

Fig. 4.23. Magnetic flux density distribution along the z-coordinate: (a) in the air gap, (b) near the armature core with reaction rail being removed.

The winding inductance of the coil C is calculated using eqn (4.17) or (4.18) given in Section 4.1. In the case of an assembled motor (with reaction rail), the static inductance (4.17) is $L_s = 7$ mH for the flat motor and $L_s = 12.7$ mH for the tubular motor. In the case of a disassembled motor without the reaction rail, $L_s = 10.5$ mH for the flat motor and $L_s = 16.7$ mH for the tubular motor. The inductance of the assembled motor is greater than that of the disassembled motor because of greater linkage flux. The presence of magnetic flux excited by PMs reduces the linkage flux. The lower values of the inductance for the flat motor are due to simplification of the 3D geometry in the 2D analysis. The end turns (overhangs) of the armature coils have been neglected in the 2D analysis.

The electromagnetic thrust (force) is the most important parameter of a linear motor. To calculate the thrust using Maxwell's stress tensor, the contour of the integration line should be as close as possible to the surface of the reaction rail. In this case, the integration line was at the distance of 0.5 mm from the surface of the reaction rail. The computed thrust is $F = 195$ N for the flat LSM and $F = 268$ N for the tubular LSM.

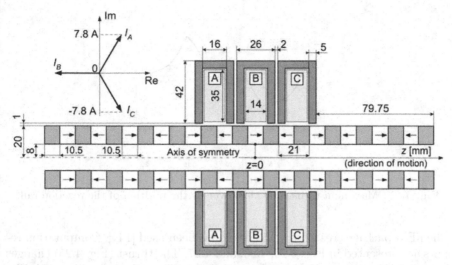

Fig. 4.24. Dimensions of the analyzed three-phase tubular PM LSM and current phasor diagram. The system of the armature currents for $z = 0$ and $\delta = 90°$ is shown in the left top corner.

Example 4.2

Find the magnetic field distribution and its integral parameters for a three-phase tubular LSM with dimensions given in Fig. 4.24 (aligned initial position of the reaction rail). Use the FEM approach. The parameters of the materials including B-H curves for the armature core and reaction rail are the same as in *Example 4.1* (Fig. 4.8a, solid line). The mesh and boundary conditions are assumed to be similar to those in *Example 4.1*. The wire diameter of the armature coils is 1.5 mm and number of turns $N = 190$. The coils are fed with sinusoidal current the peak value of which is $I_m = 9$ A. The load angle has been set as $\delta = 90°$. The coil inductances and thrust should be calculated versus the position of the reaction rail in the interval $0 \leq z \leq 21$ mm.

Solution

To hold the load angle $\delta = 90°$ for the whole range of reaction rail positions, the phase currents must be functions of position of the reaction rail. Three-phase currents are expressed as

$$i_{aA}(z) = I_m \sin\left(\frac{z}{\tau}180° - 30° + \delta\right)$$

$$i_{aB}(z) = I_m \sin\left(\frac{z}{\tau}180° + 90° + \delta\right) \qquad (4.81)$$

$$i_{aC}(z) = I_m \sin\left(\frac{z}{\tau}180° + 210° + \delta\right)$$

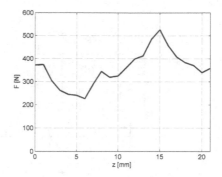

Fig. 4.25. Maximum thrust at $\delta = 90°$ versus the position of the reaction rail.

The FEM package created by D. Meeker has been used [144]. Computation results are presented in Figs 4.25, 4.26, and 4.27. The thrust (Fig. 4.25) changes from 220 N up to 520 N, depending on the position of the reaction rail. Fig. 4.25 also shows that the LSM produces high thrust ripple. The objective function in the optimization procedure should be minimization of the thrust ripple to obtain smoother distribution of the thrust versus the position of the reaction rail.

The static inductance L_s fluctuates with the position of the reaction rail (Fig. 4.26). This is due to interaction of the PM field and armature winding currents. In some positions of the reaction rail, the flux lines have opposite directions, but they coincide one with the other. The main field excitation originates from the PM system so that it always exists in the air gap. Thus, at some positions of the reaction rail, a coil can be coupled with the magnetic flux, although there is no armature current. In these cases, the static inductance tends to infinity. Sometimes, the PM flux and armature reaction flux neutralize each other, which results in the net value of the static inductance L_s.

The magnetic flux distribution is presented in Fig. 4.27 for two positions of the reaction rail, i.e., $z = 6$ mm and at $z = 15$ mm. At $z = 6$ mm, the thrust achieves its minimum value, and $z = 15$ mm, the thrust achieves its maximum value. In both cases, the highest concentration of the magnetic flux lines is in the armature tooth area, in particular at the corners.

Example 4.3

Calculate the main dimensions of a three-phase tubular PM LSM that produces the maximum thrust $F_{max} = 1000$ N. The outline of the motor is given in Fig. 4.6. Dimensions are marked in Fig. 4.7.

The nominal peak current density in the armature winding is assumed as $J_a = 10$ A/mm^2. The magnets are made from sintered NdFeB with remanent magnetic flux density $B_r = 1.25$ T and coercivity $H_c = 950$ kA/m at room

(a)   (b)

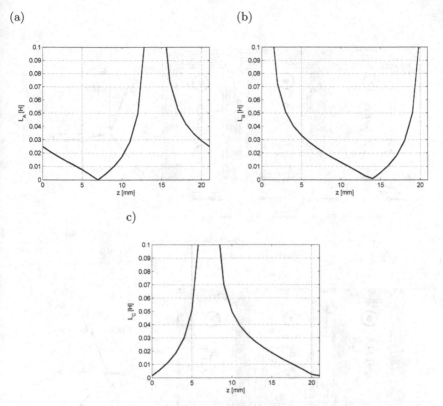

c)

Fig. 4.26. Variation of static inductance of the coils with the position of reaction rail for $\delta = 90°$: (a) coil A, (b) coil B, (c) coil C.

temperature 20°C. All ferromagnetic parts are made from typical mild steel with the B-H curve plotted in Fig. 4.8a, solid line. The desired maximum value of the magnetic flux density in the air gap is around $B_g = 1.5$ T. The wire diameter of the armature winding is $d_w = 2$ mm.

Solution

The magnetic flux density in the air gap is set at $B_g = 1.5$ T. The thrust developed by one segment is estimated on the assumption that the leakage and fringing fluxes are neglected. If the magnetic flux crossing the A-surface of the air gap is known, the electromagnetic force (thrust) can be calculated on the basis of eqn (1.15)[1], i.e.,

---

[1] This equation expresses the attraction force of an electromagnet. However, it can be used for simplified calculation of the thrust of a tubular PM LSM at zero armature current ($I_a = 0$ A) provided that the armature tooth and PM are in misaligned position.

(a)

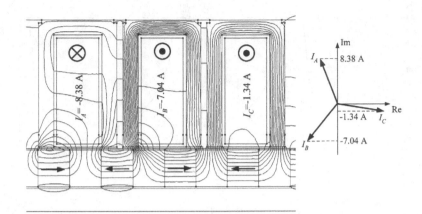

(b)

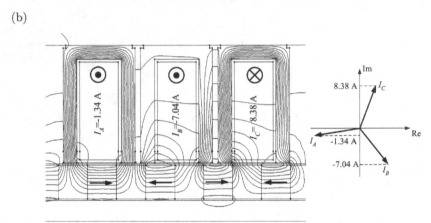

Fig. 4.27. Magnetic field distribution at load angle $\delta = 90°$ for: (a) $z = 6$ mm (minimum thrust); (b) $z = 15$ mm maximum thrust).

$$F = \frac{B_g^2 A}{2\mu_0}$$

where B_g is the magnetic flux density in the air gap, $A = 2(2\pi r w_t)$ is the surface area of air gaps between the armature teeth and the reaction rail (area of two poles North and South), and r is the inner radius of the armature core. Thus, the surface

$$A = \frac{2F\mu_0}{B_g^2} = \frac{2 \times 1000 \times 0.4\pi \times 10^{-6}}{1.5^2} = 1.117 \times 10^{-3} \text{ m}^2$$

Either the width of teeth or their radii are to be known. Putting the inner armature core radius $r = 0.5D_{1in} = 18$ mm and the outer radius of the reaction rail $r_o = 17$ mm, the width of each tooth is

$$w_t = \frac{1}{2}\frac{A}{2\pi r} = \frac{1.117 \times 10^{-3}}{4 \times 3.14 \times 0.018} = 4.941 \text{ mm}$$

The width of the tooth can be rounded to $w_t = 5$ mm. The dimensions of the magnets and ferromagnetic rings should match each other, e.g., the width of the magnet and ferromagnetic ring is twice the width of the armature tooth, i.e.,

$$2h_M = w_p = 2 \cdot w_t = 10 \text{ mm}$$

The inner radius of a PM and ferromagnetic ring should be as small as possible, e.g., $r_i = 4$ mm. Typically, the inner radius should be less than $1/4$ of the outer radius, i.e., $r_i < 0.25r_o$. The inner radius r_i is the same as the outer radius of the nonferromagnetic rod, which keeps all parts of the reaction rail together.

The MMF of the PM depends on its coercivity H_c and height h_M per pole, i.e.,

$$\mathcal{F}_M = 2h_M H_c = 0.01 \times 950\,000 = 9500 \text{ A}$$

Because of the armature reaction, which can weaken the PM excitation field, the armature MMF should be less than half of the magnet MMF. Since the maximum current density and wire diameter ($d_w = 2$ mm) are known, the maximum armature current can be calculated as

$$I_{amax} = J_a \frac{\pi d_w^2}{4} = 10 \times \frac{\pi \times 2.0^2}{4} = 31.4 \text{ A}$$

The number of turns per one coil can be calculated on the basis of the magnet MMF, i.e.,

$$N = \frac{\mathcal{F}_M/2}{I_{amax}} = \frac{4750}{31.4} = 151.27$$

The number of turns per coil is rounded to $N = 151$. To build a tubular PM LSM using separate segments, as shown in Fig. 4.6, the width of the segment "window" w_s has to satisfy the condition

$$2h_M \leq w_s \leq \tau - w_t$$

Thus $10 \text{ mm} \leq w_s \leq 15$ mm. Assuming $w_s = 13$ mm, the width of the coil can be $w_c = 11$ mm. The coil height h_c can be found on the basis of the coil width w_c and number of turns, i.e.,

$$h_c = d_w^2 \frac{N}{w_c} = 2.0^2 \frac{151}{11} \approx 55 \text{ mm}$$

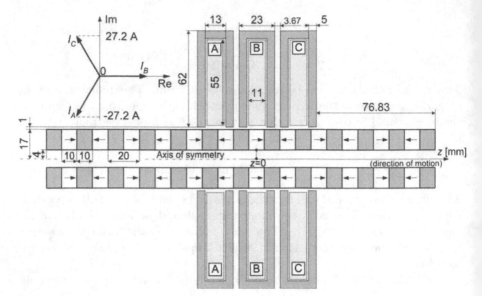

Fig. 4.28. Main dimensions of the designed tubular PM LSM.

The distance d_s between segments is calculated using condition (4.78):

$$d_s = \frac{\tau}{n_{\text{ph}}} - w_{ss} + \tau = \frac{20}{3} - 23 + 20 = 3.67 \text{ mm}$$

All calculated dimensions are shown in Fig. 4.28. The 2D FEM computations give the average force value $F_{av} = 805$ N. The force versus the position of the reaction rail has been calculated in the same way as in *Example 4.1*. The maximum force reaches $F_{max} = 930$ N. Thus, the value obtained from the FEM computations is 7% lower than the target maximum thrust $F_{max} = 1000$ N. The designed tubular PM LSM is characterized by high detent force (Fig. 4.29). Optimization of the LSM construction is needed to minimize the detent force. Possible independent variables in the optimization procedure include the width of armature tooth w_t, distance between stator segments d_s, height of the PM $2h_M$ in the axial direction, width of the ferromagnetic rings w_p, and outer diameter of the reaction rail $2R_o$. Additionally, the shape of the armature tooth can be changed since the concentration of the magnetic flux lines is observed in the tips of teeth (Fig. 4.30). It is also possible to design the PM LSM with two or more armature segments per one phase. In this case, the outside LSM diameter will decrease, and the axial length of the armature system will increase.

Example 4.4

Calculate the main dimensions of a three-phase tubular PMS LSM, which produces the maximum thrust $F_{max} = 1000$ N. Verify the thrust with the

FEM method. Assume the nominal armature current density $J_a = 10$ A/mm^2, wire diameter $d_w = 2$ mm, and the pole pitch $\tau = 50$ mm. Calculate the dimensions of PMs under assumption that the mechanical energy is equal to the energy stored in the PMs. In order to produce the expected thrust, at least 7 PMs and 3 segments of the armature system are needed. The magnets are made of sintered NdFeB with remanent magnetic flux density $B_r = 1.25$ T and coercivity $H_c = 950$ kA/m at 20°C, and the all ferromagnetic parts are made from typical mild steel (Fig. 4.8a, solid line). The expected magnetic energy density of the magnet is $w = 400$ kJ/m^3.

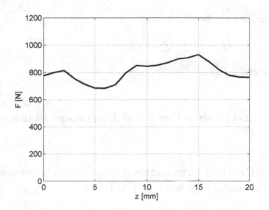

Fig. 4.29. Thrust versus position of the reaction rail at $\delta - 90°$.

Solution

The energy needed to move the reaction rail with the electromagnetic force $F = 1000$ N along the distance $\tau = 50$ mm is

$$W = F\tau = 1000 \times 0.05 = 50 \text{ J}$$

The volume V of all seven magnets is calculated from the energy $W = 50$ J and assumed energy density $w = 400$ kJ/m^3, i.e.,

$$V = \frac{W}{w} = \frac{50}{400000} = 1.25 \times 10^{-4} \text{ m}^3 = 1.25 \times 10^5 \text{ mm}^3$$

The volume of a single magnet V_M of the field excitation system consisting of seven PMs is $V_M = V/7 = 17860$ mm^3.

Now, the dimensions of the PMs must be estimated. Since the pole pitch is $\tau = 50$ mm, and assuming the same dimensions of the magnets and ferromagnetic rings, the width of a single magnet is

$$2h_M = w_p = \frac{\tau}{2} = 25 \text{ mm}$$

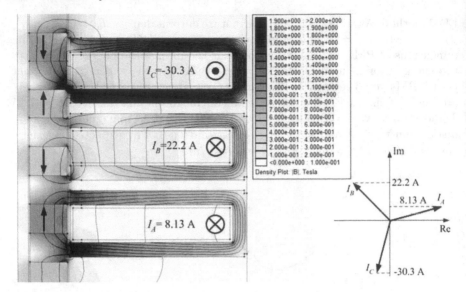

Fig. 4.30. Magnetic field distribution in the longitudinal section of the motor at $z = 15$ mm and $\delta = 90°$.

The inner radius of PMs and ferromagnetic ring should be as small as possible, e.g., $r_i = 4$ mm. The outer diameter of the PM is

$$r_o = \sqrt{\frac{V_M}{\pi(2h_M)} + r_i^2} = \sqrt{\frac{17860}{\pi \times 25} + 4^2} \approx 16 \text{ mm}$$

The width of the armature tooth should be a half of the width of the ferromagnetic ring, i.e.,

$$w_t = \frac{w_p}{2} = \frac{25}{2} = 12.5 \text{ mm}$$

The MMF of the PM per two poles is

$$\mathcal{F}_M = 2h_M H_c = 0.025 \times 950000 = 23\ 750 \text{ A}$$

The maximum current density and wire diameter are known ($d_w = 2$ mm), so that the maximum value of the current is

$$I_{amax} = J_a \frac{\pi d_w^2}{4} = 10\frac{\pi \times 2.0^2}{4} = 31.4 \text{ A}$$

The number of turns per coil is

$$N = \frac{\mathcal{F}_M/2}{I_{amax}} = \frac{23\ 750/2}{31.4} \approx 378$$

To design a tubular PM LSM with separate segments (Fig. 4.7), the width of the "window" w_s has to satisfy the condition

$$2h_M \leq w_s \leq \tau - w_t$$

so that $25 \leq w_s \leq 37.5$ mm. The width w_s of the "window" of each segment can be calculated as the arithmetic mean value of the maximum and minimum width, i.e., $w_s = 0.5(25.0 + 37.5) = 31.25 \approx 32.0$ mm. If the insulation of the coil is 1 mm thick, the width of the coil is $w_c = 30$ mm. Since the distance between wires along the symmetry axis z and in the radial direction is the same, the number of turns per coil can be found as

$$N = \frac{w_c}{d_w} \frac{h_c}{d_w}$$

Thus, the height of the coil

$$h_c = d_w^2 \frac{N}{w_c} = 2.0^2 \times \frac{378}{30} \approx 50 \text{ mm}$$

The distance between segments is

$$d_s = \frac{\tau}{n_{\text{ph}}} - w_{ss} + \tau = \frac{50}{3} - 57 + 50 = 9.67 \text{ mm}$$

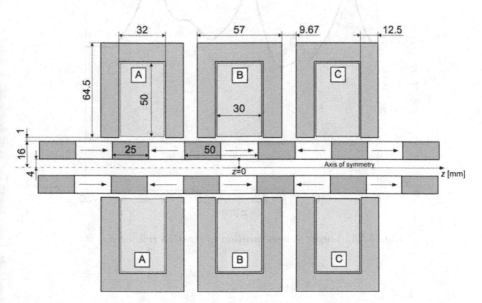

Fig. 4.31. Dimensions of the designed tubular PM LSM.

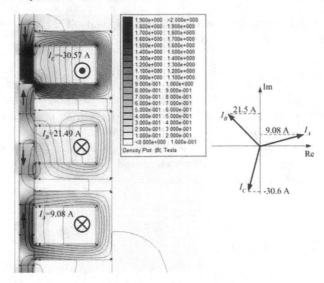

Fig. 4.32. Magnetic flux density distribution in the longitudinal section of the motor (Fig. 4.31) for $z = 38$ mm and load angle $\delta = 90°$.

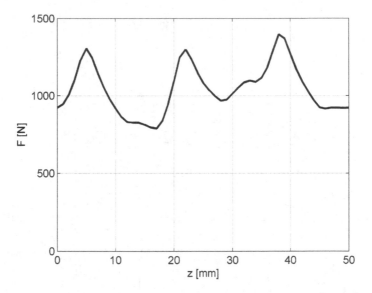

Fig. 4.33. Thrust versus position of reaction rail for $\delta = 90°$.

The dimensions of the motor are shown in Fig. 4.31. For such estimated dimensions, the FEM package can be used to find the electromagnetic field distribution and electromagnetic forces. The magnetic flux distribution as obtained from the 2D FEM is shown in Fig. 4.32. The flux has been obtained under assumption of imaginary values of the three-phase current system shown in Fig. 4.33. Saturation of armature segments exists only in their inner areas, close to the air gap (smaller cross-section area for the magnetic flux).

The thrust versus the position of the reaction rail has been found from the integration of Maxwell stress tensor (Fig. 4.33). The average force $F_{av} = 1043$ N. The maximum value of the force is $F_{max} = 1397$ N. Thus, the value of the maximum force is 40% greater than the target value of the thrust $F_{max} = 1000$ N. The drawback of the construction shown in Fig. 4.31 is high detent force (force ripple). Minimization of the detent force can be done through the optimization process.

Hybrid and Special Linear Permanent Magnet Motors

5.1 Permanent Magnet Hybrid Motors

5.1.1 Finite Element Approach

The magnetic circuits of stepping motors are frequently highly saturated. Therefore, it is often difficult to calculate and analyze motor performance with consistent accuracy by using the classical circuital or field approach. Using the finite element method (FEM) or other numerical methods, more accurate results can be obtained. Since the hybrid linear stepping motor (HLSM) consists of two independent stacks, it is acceptable to model each stack separately. There is one potential difficulty in the modeling of the HLSM. This arises as a result of the tiny air gap, below 0.05 mm. For better accuracy of the calculation, the air gap requires three or four layers of elements, which in turn leads to high aspect ratios.

The fundamentals of the FEM can be found in Chapter 4 and many monographs, e.g., [41, 70, 189, 236]. In the FEM, both the *virtual work method* — eqn (1.13) and *Maxwell stress tensor* — eqn (4.13), can be used for the thrust (tangential force) F_{dx} and normal force F_{dz} calculations. Forces calculated on the basis of the Maxwell stress tensor (4.13) are

$$F_{dx} = \frac{L_i}{\mu_0} \int_l B_x B_z \, dl \qquad F_{dz} = \frac{L_i}{2\mu_0} \int_l (B_z^2 - B_x^2) \, dl \qquad (5.1)$$

where L_i is the width of the HLSM stack, B_x is the tangential component of the magnetic flux density, B_z is the normal component of the magnetic flux density, and l is the integration path. In the method of virtual work, x is the horizontal displacement between forcer and platen, z is the vertical displacement, and W is the energy stored in the magnetic field — eqn (1.13).

The classical virtual work method needs two solutions, and the choice of suitable displacement has direct influence on the calculation accuracy. The following one solution approach based on Coulomb and Meunier's method [46] has been used and implemented here for calculating the tangential force

$$F_{dx} = -\sum_e \left(\frac{\mathbf{B}^T}{\mu_0} \frac{\partial \mathbf{B}}{\partial x} + \frac{\mathbf{B}^T}{\mathbf{B}} 2\mu_0 |\mathbf{Q}|^{-1} \frac{\partial |\mathbf{Q}|}{\partial x} \right) V_e \qquad (5.2)$$

where $\mathbf{B}$ is the matrix of magnetic flux density in the air gap region, $\mathbf{B}^T$ is the transpose matrix of $\mathbf{B}$, $|\mathbf{Q}|$ is the determinant of the Jacobian matrix, and V_e is the volume of an element e. For linear triangular elements, the above equation can further be simplified to the following form:

$$F_{dx} = \sum_e \sum_i \frac{1}{4\mu_0} [(z_2 - z_3)\Delta_1^2 + 2(x_3 - x_2)\Delta_1\Delta_2 - (z_2 - z_3)\Delta_2^2] \qquad (5.3)$$

Similarly, the dual formulation for calculating the normal force is obtained as

$$F_{dz} = \sum_e \sum_i \frac{1}{4\mu_0} [(x_3 - x_2)\Delta_1^2 + 2(z_3 - z_2)\Delta_1\Delta_2 - (x_3 - x_2)\Delta_2^2] \qquad (5.4)$$

In eqns (5.3) and (5.4), x and z are rectangular coordinates of the 2D model, e and i are the numbers of virtually distorted elements and virtually moved nodes within an element, respectively, and subscripts 1, 2 and 3, correspond to the nodes of a triangular element. The parameters Δ_1 and Δ_2 are defined as

$$\Delta_1 = \frac{A_1(x_3 - x_2) + A_2(x_1 - x_3) + A_3(x_2 - x_1)}{\mathbf{Q}} \qquad (5.5)$$

$$\Delta_2 = \frac{A_1(z_3 - z_2) + A_2(z_1 - z_3) + A_3(z_2 - z_1)}{\mathbf{Q}} \qquad (5.6)$$

where A_1 to A_3 are magnitudes of the magnetic vector potential corresponding to each node of a triangular element. The Jacobian matrix becomes

$$\mathbf{Q} = \begin{bmatrix} x_1 - x_3 & z_1 - z_3 \\ x_2 - x_3 & z_2 - z_3 \end{bmatrix} \qquad (5.7)$$

5.1.2 Reluctance Network Approach

In recent years, much research has been done on the modeling of electromechanical energy conversion devices by using the *reluctance network approach* (RNA). Part of this research relates to stepping motors [107, 108, 138, 174]. The RNA is simpler than the FEM and does not require a long computation time [225].

As shown in Fig. 5.1, the fluxes of individual poles are dependent on the PM MMF, winding current, and reluctances. The PM flux Φ_M circulates in

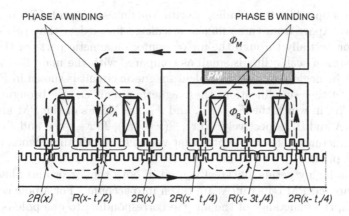

Fig. 5.1. The outline of the magnetic circuit of an HLSM.

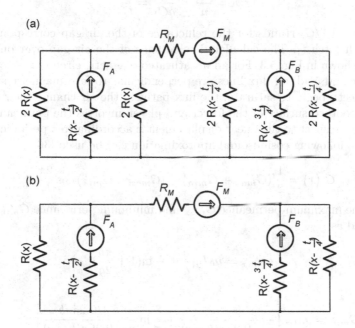

Fig. 5.2. Equivalent magnetic circuit of a two-phase HLSM: (a) magnetic circuit corresponding to Fig. 5.1; (b) simplified magnetic circuit with equal numbers of teeth per pole.

the main loop while the winding excitation fluxes Φ_A and Φ_B create local flux loops. Apart from these fluxes, a leakage flux exists that takes a path entirely or partially through the air or nonferromagnetic parts of the forcer. The amount of such a flux is small as compared with the main flux and can, therefore, be neglected. The equivalent magnetic circuit is shown in Fig. 5.2a.

After further simplification, the magnetic circuit can be brought to that in Fig. 5.2b, in which the $\mathcal{F}_M$, $\mathcal{F}_A$, and $\mathcal{F}_B$ are MMFs of the PM and phase windings A and B, respectively; $\Re(x)$, $\Re(x - \frac{1}{2}t_1)$, $\Re(x - \frac{3}{4}t_1)$, and $\Re(x - \frac{1}{4}t_1)$ are the reluctances of a single pole that vary with tooth alignments, and t_1 is the tooth pitch.

Since a highly permeable steel is used in both the forcer and platen, only the air gap and PM reluctances are taken into account. For a pole consisting of n teeth, the reluctance of the air gap corresponding to one pole is

$$\Re = \frac{\Re_t}{n} = \frac{1}{nG_t} = \frac{1}{G} \tag{5.8}$$

where $\Re_t = 1/G_t$ stands for the reluctance of the air gap corresponding to one tooth pitch t_1. The calculated reluctance of the air gap over one tooth pitch is shown in Fig. 5.3. For an unsaturated magnetic circuit ($\mu \to \infty$) as in Figs 5.4c and 5.4d, all flux lines are perpendicular to ferromagnetic surfaces. As the teeth begin to saturate, the flux paths in the air change their shapes.

The toothed surface of the forcer and platen involves the permeance variation with respect to the linear displacement x according to a periodical function. The following cosinusoidal approximation can be used [36]:

$$G_t(x) = \frac{1}{2}[(G_{max} + G_{min}) + (G_{max} - G_{min}) \cos \frac{2\pi}{t_1} x] \tag{5.9}$$

where the maximum permeance G_{max} and minimum permeance G_{min} can be expressed as

$$G_{max} = \mu_0 L_i [\frac{c}{g} + \frac{2}{\pi} \ln(1 + \frac{\pi b}{2g})] \tag{5.10}$$

$$G_{min} = \mu_0 L_i \left[\frac{b - c}{g + 0.25\pi(b - c)} + \frac{8}{\pi} \ln \frac{g + 0.25\pi b}{g + 0.25\pi(b - c)} \right] \tag{5.11}$$

and c and b are tooth and slot width, respectively. The approximation can further help in finding the derivative of reluctance with regard to the displacement, i.e.,

$$\frac{\partial G_t}{\partial x} = -\frac{\pi}{t_1}(G_{max} - G_{min}) \sin \frac{2\pi}{t_1} x \tag{5.12}$$

The calculated permeance of the air gap over one tooth pitch t_1 and its cosinusoidal approximation [36] are plotted in Fig. 5.5.

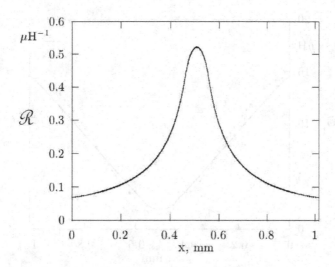

Fig. 5.3. Air gap reluctance distribution over one tooth pitch.

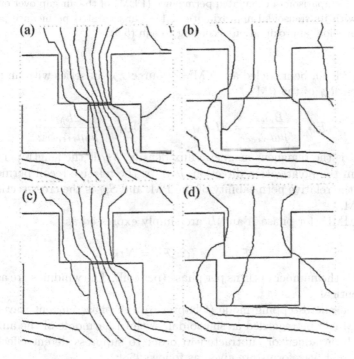

Fig. 5.4. Flux patterns: (a) partially aligned teeth (saturated magnetic circuit), (b) complete misalignment of teeth (saturated magnetic circuit), (c) partially aligned teeth (unsaturated magnetic circuit), (d) complete misalignment of teeth (unsaturated magnetic circuit).

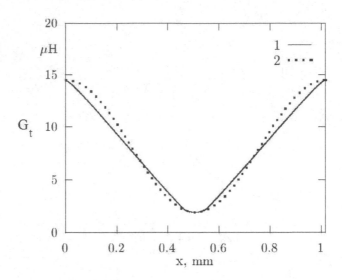

Fig. 5.5. Comparison of calculated permeance (FEM) of the air gap over one tooth pitch t_1 with its sinusoidal approximation. 1 — unsaturated permeance per pole, 2 — cosinusoidal approximation according to eqn (5.9).

The PM can be modeled as an MMF source $\mathcal{F}_M$ in series with an internal reluctance $\Re_M$ of the PM

$$\mathcal{F}_M = \frac{B_r h_M}{\mu_0 \mu_{rrec}} = H_c h_M; \qquad \Re_M = \frac{2h_M}{\mu_0 \mu_{rrec} S_M} \qquad (5.13)$$

where B_r is the remanent magnetic flux density, h_M is the length per pole of the PM in the polarization direction, μ_{rrec} is the relative recoil permeability equal to the relative permeability of the PM, and S_M is the cross-section area of the PM.

The MMFs for phase A and B are simply expressed as

$$\mathcal{F}_A = Ni_A, \quad \mathcal{F}_B = Ni_B \qquad (5.14)$$

where N is the number of turns per phase (per coil). The windings are assumed to be identical.

For the microstepping mode (Chapter 6), the phase current waveform of the HLSM can be regarded as sinusoidal. A third harmonic of the amplitude I_{m3} has been added or subtracted in order to suppress detent effects. The phase current waveforms are given as follows [58]:

$$i_A = I_{m1} \cos(\frac{2\pi x}{t_1} - \phi) \pm I_{m3} \cos(\frac{6\pi}{t_1} - 3\phi) \qquad (5.15)$$

$$i_B = I_{m1} \sin(\frac{2\pi x}{t_1} - \phi) \pm I_{m3} \sin(\frac{6\pi}{t_1} - 3\phi) \qquad (5.16)$$

where ϕ is the phase angle that depends on the load. The tangential force per pole is

$$F_{dxp} = \frac{1}{2}\Phi^2 \frac{\partial \mathfrak{R}}{\partial x} = -\frac{1}{2n}\Phi^2 \left[\frac{1}{G_t^2}\frac{\partial G_t}{\partial x}\right] \qquad (5.17)$$

where Φ is the magnetic flux through the pole, and $\mathfrak{R}$ is the air gap reluctance per pole according to eqn (5.8). Thus, for a $2p$ pole HLSM, the overall available tangential force is

$$F_{dx} = 2pF_{dxp} = -\frac{p}{n}\Phi^2 \left[\frac{1}{G_t^2}\frac{\partial G_t}{\partial x}\right] \qquad (5.18)$$

The normal force can be written in the form of the derivative of coenergy W' with respect to the air gap $z = g$ i.e.,

$$F_{dz} = \frac{\partial W'}{\partial z} = -\frac{p}{n}\Phi^2 \left[\frac{1}{G_t^2}\frac{\partial G_t}{\partial z}\right] \qquad (5.19)$$

and the following simplification can be made [58]:

$$\frac{\partial \mathfrak{R}}{\partial z} = -\frac{1}{nG_t^2}\frac{\partial G}{\partial z} = \frac{\mathfrak{R}_{max} - \mathfrak{R}_{min}}{g_{max} - g_{min}} \qquad (5.20)$$

where $\mathfrak{R}_{max} = 1/G_{max}$, and $\mathfrak{R}_{min} = 1/G_{min}$.

To calculate either the tangential or normal force, it is always necessary to find the magnetic flux Φ on the basis of the equivalent magnetic circuit (Fig. 5.2).

5.1.3 Experimental Investigation

An experimental investigation into a small electrical machine requires high accuracy measurements. All the measured parameters need to be transformed into measurable electric signals. Considerable efforts need to be normally made to eliminate all sources of electromagnetic noise, which would cause electromagnetic interference (EMI) problem.

The fundamental steady-state characteristics are (1) the *force* versus *displacement* characteristic, which gives the relationship between the tangential force and displacement from the equilibrium position; (2) the *force* versus *current* characteristic, which shows how the maximum static force (holding force) increases with the peak excitation current. Unlike in a variable reluctance motor, the static force appears even at current free state (*detent force,* also called *cogging force*).

The instantaneous force is recorded by measuring the output force when the motor is driven in two-phase on the excitation scheme and reaches its

Table 5.1. Design data of the tested L20 HLSM manufactured by Parker Hannifin Corporation, Compumotor Division, Rohnert Park, CA, USA [40]

Specification data	Value
Number of coils	4
Number of turns per coil	57
Wire diameter	0.452 mm
Tooth width	0.4572 mm
Slot width	0.5588 mm
Length of forcer	117.475 mm
Mass of forcer	0.8 kg
Material of forcer	Laminated steel 0.35 mm thick
Platen width	49.53 mm
Platen tooth pitch	1.016 mm
Air gap (mechanical clearance)	0.0127 mm
Peak phase current	2.7 A
PM material	NdFeB ($B_r = 1.23$ T)
PM height	2.54 mm
PM face area	273.79 mm^2
Accuracy (worst case)	±0.09 mm
Repeatability	±0.0025 mm

steady-state operation. It is recommended that different excitation current profiles be applied at different resolution settings.

The stepping resolution of the HLSM has no significant influence on the amplitude of both the tangential and normal ripple forces. Both harmonic-added and harmonic-subtracted schemes work equally well in reducing the amplitude of the force ripple. However, in the case of a normal force, there is better improvement when subtracting the 3rd harmonic from the current profile than when adding it. It has also been found that the amplitude of the tangential force ripple increases with the increase in phase current. The relation between them is not so linear as reported [138]. Higher-order harmonics cannot be neglected when the magnetic circuit becomes highly saturated.

Transient performance measurements are focused on the acceleration (startup) and deceleration (braking). The startup tests can simply be done by supplying power to the HLSM and recording the motor's acceleration. The time interval from the instant at which the power is switched on to the instant at which the HLSM reaches steady-state speed is called the *startup setting* time. For a typical HLSM, it is about 1 s to reach its steady-state speed at no load. As expected, somewhat longer time will be needed when certain load is applied. The braking time is also inversely proportional to the attached mass.

Case Study 5.1. L20 HLSM

The L20 HLSM manufactured by Parker Hannifin Corporation, Compumotor Division [40] has been tested experimentally. The specification data of the L20 HLSM are given in Table 5.1. Then, its characteristics obtained from measurements have been compared with the FEM and RNA results.

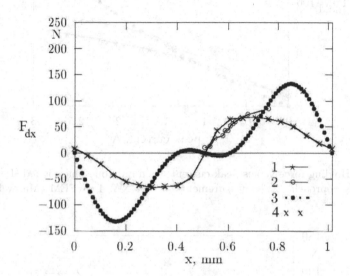

Fig. 5.6. Tangential force versus displacement (when one phase of the motor is fed with 2.7 A current): 1 — FEM, 2 — measurements, 3 — RNA, 4 — Maxwell stress tensor.

The following static characteristics have been considered: (1) static force versus forcer position, and (2) holding force versus peak phase current when only one phase of the HLSM is fed with peak phase current. The results obtained from the FEM, RNA, and measurements are shown in Figs 5.6 and 5.7. It can be seen that the maximum holding force obtained from the FEM, measurements, and RNA are 70 N, 85 N, and 120 N, respectively (Fig. 5.6). The FEM (Coulomb's approach) results correlate well with the experimental results in the case of static characteristics. The RNA tends to overestimate the force since simplifications have been made to calculate the force.

The instantaneous characteristics of the HLSM refer to the output tangential force versus motor position when the HLSM has been powered up and is reaching its steady state. A set of different excitation current waveforms have been used both in testing and computations. They are pure-sine waves and quasi-sinusoidal waves with 4% and 10% of the 3rd harmonics added, respectively.

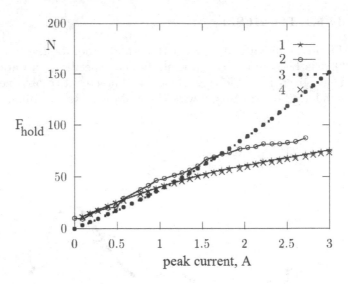

Fig. 5.7. Holding force versus peak current when only one phase is fed. 1 — FEM (Coulomb's approach), 2 — measurements, 3 — RNA, 4 — FEM (Maxwell stress tensor).

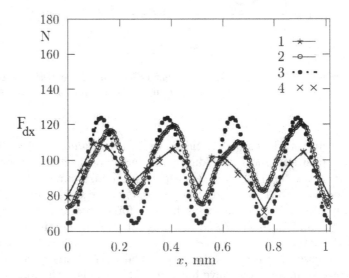

Fig. 5.8. Instantaneous tangential forces versus displacement (when the phase A of the HLSM is driven with pure sine wave and phase B with pure cosine wave). 1 — FEM (Coulomb's approach), 2 — measurements, 3 — RNA, 4 — FEM (Maxwell stress tensor).

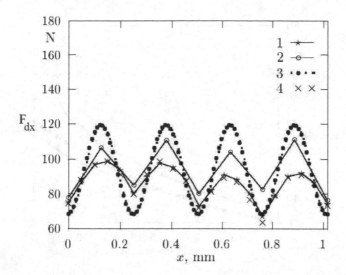

Fig. 5.9. Instantaneous tangential force versus displacement (when 10% of the 3rd harmonic has been injected into the phase current). 1 — FEM (Coulomb's approach), 2 — measurements, 3 — RNA, 4 — FEM (Maxwell stress tensor).

The results obtained from both the FEM and RNA are then compared with measurements as shown in Figs 5.8 and 5.9. Since the HLSM operated at a constant speed of 0.0508 m/s, the force-time curves could be easily transformed to force-displacement curves. In general, the RNA tends to overestimate the force versus displacement as compared with the measured values, while the virtual work method gives more accurate (although underestimated) results. However, the RNA is a very efficient approach from the computation time point of view.

Fig. 5.10 shows the amplitude of tangential force ripple plotted against the peak current. This has been obtained by calculating the maximum force ripple for each peak value of the excitation current.

Owing to the existence of the PM, there is a strong normal force between the forcer and platen even when the HLSM is unenergized (Fig. 5.11). It can also be seen in Fig. 5.11 that the normal force is very sensitive to the small change in the air gap. The FEM results show that the normal force (Fig. 5.12) almost doubled when the HLSM is energized (peak phase current 2.7 A). In this case, the RNA gives results fairly close to the FEM.

An accurate prediction of forces is necessary not only for motor design purposes but also for predicting the performance of a HLSM drive system. Both the FEM and RNA give the results very close to the test results. The accuracy of the FEM depends much on the discretization of the air gap region while the accuracy of the RNA depends on the evaluation of reluctances. The existence of PMs and toothed structures is the main factor in generating ripple

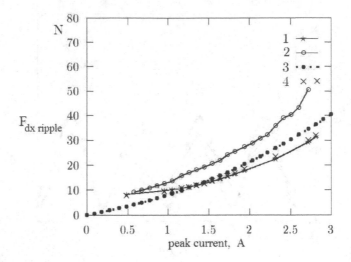

Fig. 5.10. Tangential force ripple amplitude as a function of peak current: 1 — FEM (Coulomb's approach), 2 — measurements, 3 — RNA, 4 — FEM (Maxwell stress tensor).

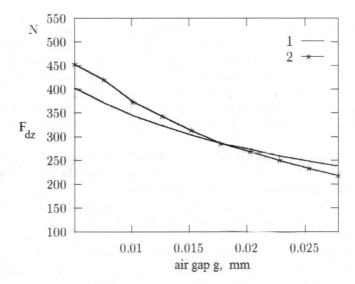

Fig. 5.11. Calculated normal force between forcer and platen versus air gap length at current free state: 1 — FEM (Coulomb's approach), 2 — RNA.

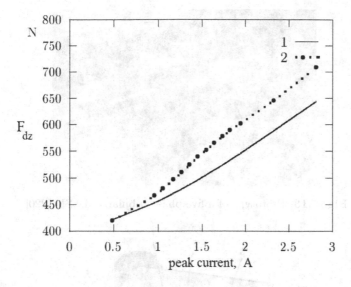

Fig. 5.12. Normal force versus peak current: 1 — FEM (Coulomb's approach), 2 — RNA.

forces. The reduction of ripple forces is necessary to obtain a smooth operation of the motor and minimize audible noise. The force ripple can be suppressed by modifying the input current waveforms. HLSMs with adjustable current profiles offer new techniques of motion control.

5.2 Five-Phase Permanent Magnet Linear Synchronous Motors

Although the most popular are three-phase PM LSMs, their five-phase counterparts show some fault tolerance and reduced detent force [211, 213, 215, 220]. Minimization of the detent force reduces the level of vibration, which is very important in precision mechanisms used in a variety of mechatronics systems. In this section, investigations into five-phase tubular PM LSMs (Fig. 5.13) carried out at the Technical University of Opole, Poland, have been presented [207, 208, 209, 210, 213, 230, P214].

Tubular five-phase PMS LSMs can be designed as linear machines with modular armature units and reaction rails assembled with ring-shaped PMs and soft ferromagnetic materials [214, 216, 217]. Modular armature units allows for building a series of five-phase LSMs with different ratings. The longitudinal section of a five-phase tubular PM LSM consisting of one armature module is shown in Fig. 5.16.

Fig. 5.13. Prototype of a five-phase tubular PM LSM [220].

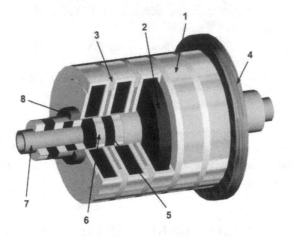

Fig. 5.14. Cutaway view of five-phase tubular PM LSM. 1 — armature segment, 2 — armature coil, 3 — nonferromagnetic ring, 4 — armature cover, 5 — PM, 6 — ferromagnetic ring, 7 — nonferromagnetic tube, 8 — linear slide bearing.

5.2.1 Geometry

The cutaway view of a five-phase tubular PM LSM is shown in Fig. 5.14. The armature coils are magnetically separated. More armature modules can be added to achieve the desired thrust. An axonometric view of a similar flat five-phase PM LSM is shown in Fig. 5.15.

The longitudinal section of a five-phase tubular PM LSM is presented in Fig. 5.16. The reaction rail is assembled of NdFeB 35 PM and mild steel rings. The B-H curve for mild steel is given in Fig. 4.8a [219]. The longitudinal section of a five-phase flat PM LSM is shown in Fig. 5.17.

Flat and tubular five-phase LSMs with the same dimensions have been used for comparative analysis (Figs 5.16 and 5.17). The thrust of the tubular

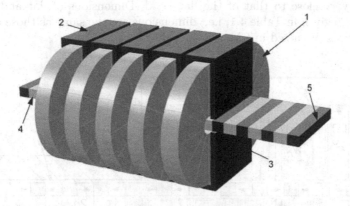

Fig. 5.15. Axonometric view of the flat motor construction. 1 — armature coil, 2 — ferromagnetic core, 3 — tooth of the armature segment, 4 — PM, 5 — flat ferromagnetic bar.

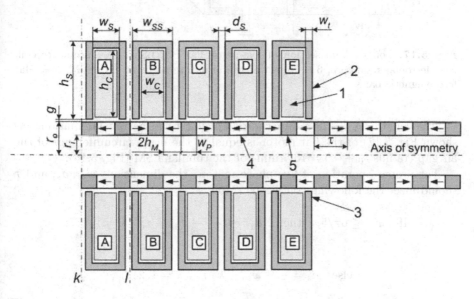

Fig. 5.16. Longitudinal section of five-phase tubular PM LSM. 1 — armature coil, 2 — ferromagnetic core, 3 — tooth of the armature segment, 4 — PM, 5 — ferromagnetic ring.

LSM is very close to that of the flat LSM. Dimensions of flat and tubular motors are given in Table 4.1, i.e., dimensions are the same as those of three-phase LSMs discussed in Chapter 4.

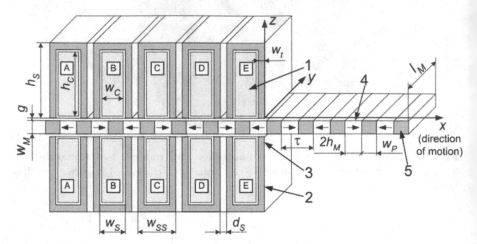

Fig. 5.17. Longitudinal section of five-phase flat PM LSM. 1 — armature coil, 2 — ferromagnetic core, 3 — tooth of the armature segment, 4 — PM, 5 — flat ferromagnetic bar.

The length $l_M = 48.7$ mm (perpendicular to armature laminations, as shown Fig. 5.17) of the flat motor is equal to the mean circumference of the air gap (radius $r_0 + 0.5 = 15.5$ mm) of the tubular LSM (Fig. 5.16).

For proper operation of five-phase motors, the dimensions w_{ss}, d_s, and τ should meet the following conditions [230]:

$$\text{If} \quad w_{ss} \leq 6\tau/5, \quad \text{then} \quad d_s = k \cdot \frac{\tau}{5} - w_{ss} + \tau, \quad k = 1, 2 \ldots$$

$$\text{else} \quad d_s = \frac{\tau}{5} - w_{ss} + k \cdot \tau, \quad k = 1, 2 \ldots$$

where k is the smallest integer for which $d_s > 0$. For main dimensions of five-phase motors as given in Table 4.1, the distance between armature modules amounts to $d_s = 3$ mm.

5.2.2 2D Discretization and Mesh Generation

In the 2D FEM analysis of five-phase LSMs, PDE equations have been solved using the Dirichlet boundary conditions (Fig. 5.18). Only a half of tubular LSM with axial symmetry (Fig. 5.16) can be considered. In FEM modeling,

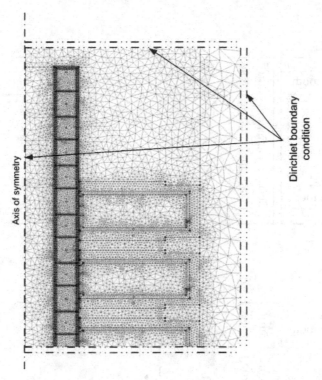

Fig. 5.18. The analyzed area with the Dirichlet boundary conditions.

the fine mesh should be dense enough in order to obtain high accuracy of computations (Fig. 5.19).

The most important region in linear motors, as in any other electrical machines, is the air gap g (mechanical clearance) between the armature and reaction rail. This region is limited by lines m and n in Fig. 5.19. A fine mesh must be applied to the discretization of the air gap. There are subdomains between lines k and l in Fig. 5.16, which include one module of the armature. The PMs, mild steel rings, and windings have been discretized with a mesh of different density.

Considering the air gap, there are two and four 3-node elements in the radial direction for coarse and medium-quality meshes. For the finest mesh, seven 6-node elements in the radial direction have been employed. It is necessary to emphasize that the air gap should be divided at least into two rows of elements. This is visible in Fig. 5.19a, between lines m and n.

After solution of PDEs, the so called differential parameters, e.g., flux density or integral parameters, such as forces of the magnetic field are obtained. The electromagnetic force (thrust) is the most important integral parameter. In Fig. 5.20, the electromagnetic thrust versus the distance z of the moving reaction rail is shown. Results for coarse, medium, and fine meshes have been

(a)

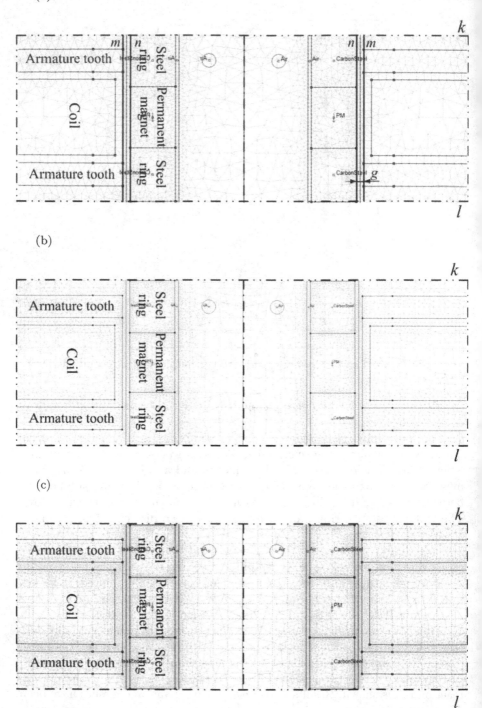

(b)

(c)

Fig. 5.19. Discretization of a tubular PM LSM using meshes of different density:
(a) coarse, (b) medium, (c) fine.

compared. Thrusts developed by the flat and tubular five-phase PM LSMs are presented in Table 5.2.

Table 5.2. Electromagnetic thrust under rated (nominal) current for the analyzed prototypes of flat and tubular five-phase PM LSMs

Parameter	Tubular motor	Flat motor
Maximum force F_{max}, N	447	333
Minimum force F_{min}, N	363	264
Average force F_{av}, N	401	297
Force ripple coefficient k_r	0.21	0.23
Maximum detent (cogging) force F_{cmax}, N	35	32
Armature rms current I_a, A	8	8

The coefficient of thrust ripple k_r is defined by eqn (4.79). The errors in the average force value and thrust ripple coefficient for different mesh densities have been estimated as

$$\delta F_{av} = \frac{F_{av}^{fine} - F_{av}}{F_{av}^{fine}} \tag{5.21}$$

$$\delta k_r = \frac{\left| k_r^{fine} - k_r \right|}{k_r^{fine}} \tag{5.22}$$

Calculation results of the thrust, coefficient of thrust ripple, and errors are given in Table 5.3.

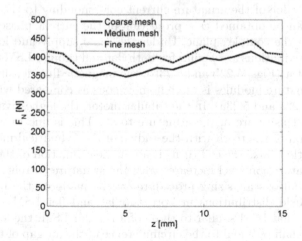

Fig. 5.20. Comparison of force calculation results for different mesh densities.

Table 5.3. Comparison of thrust, coefficient of thrust ripple and errors for different mesh densities (Fig. 5.19)

Mesh	Number of elements	F_{max} N	F_{min} N	F_{av} N	δF_{av} %	k_r —	δk_r %	CPU time s
Coarse	10234	415	347	377	6.5	0.178	8.7	5
Medium	28973	444	365	402	0.2	0.198	1.5	25
Fine	67617	445	366	403	—	0.195	—	51

Table 5.3 shows that further increase in the number of elements of the air gap does not change the average force F_{av}. Thus, overdiscretization causes longer calculation time without significant improvement in most of the parameters.

In Fig. 5.21, the fine meshes of tubular and flat motors are drawn. For both cases, the air gap region mesh density is higher than the mesh density of other regions. A dense mesh in the air gap region is necessary to obtain more accurate computation results of electromagnetic forces. Maxwell's stress tensor should be integrated near the ferromagnetic edges.

5.2.3 2D Electromagnetic Field Analysis

Figs 5.22 and 5.23 show the calculated 2D magnetic field distribution, in particular, magnetic flux lines. The 2D field distribution has been obtained for two load angles: $\delta = 0$ (Fig. 5.22) and maximum angle $\delta = 90^\circ$ (Fig. 5.23). The motor draws maximum armature current $I_a = 8$ A (Table 5.2). The various gray shades visualize different values of the magnetic flux density. In the top right corners of Figs 5.22b and 5.23b, the five-phase system of currents is shown. The value of the armature current corresponding to the calculated time instant can be obtained as a projection of the current phasor onto the vertical axis in the complex plane. Under the same supply and loading, the flux density distributions are similar for both flat and tubular LSMs. However, for the flat motor (Figs. 5.22b and 5.23b), the magnetic flux density distribution in the armature modules is more homogenous as compared with tubular motor (Figs. 5.22a and 5.23a). In the tubular motor, the highest values of the magnetic flux density are in the armature teeth. This is due to variation of the cross section of the teeth with the radius of the cylindrical module. The magnetic flux that links the coil turns is a nonlinear function of the armature current. The saturation level increases with the armature current I_a.

Recommendations for sizing procedure can be made on the basis of the 2D magnetic field distributions in both tubular and flat LSMs. To obtain parameters of a flat LSM similar to those of a tubular LSM, the length l_M of the flat motor shall be equal to the circumference of the air gap of the tubular motor. The magnetic saturation effects are less evident in the flat motor.

(a)

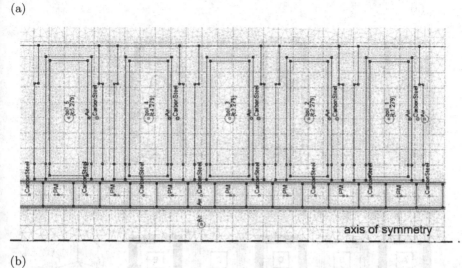

(b)

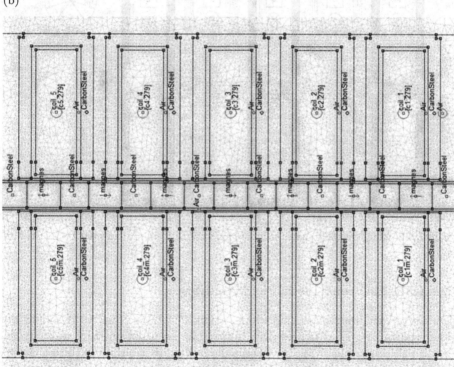

Fig. 5.21. Discretization mesh of five-phase PM LSMs: (a) tubular motor, (b) flat motor.

(a)

(b)

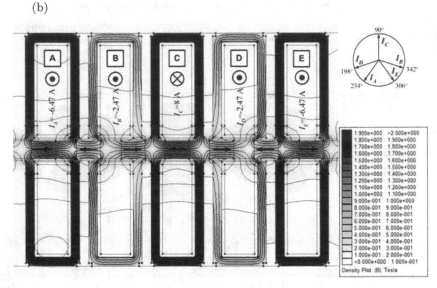

Fig. 5.22. Magnetic field distribution at load angle $\delta = 0$: (a) tubular LSM, (b) flat LSM.

5.2.4 3D Electromagnetic Field Analysis

For better accuracy of computations, a flat PM LSM (Fig. 5.15) should be analyzed in 3D. The 3D field distribution has been obtained for the current system shown in Fig. 5.23. The 3D magnetic flux distribution in the armature and reaction rail of the five-phase PM LSM is presented in Fig. 5.24. The highest concentration of flux lines is observed in the first and the fourth segment of the armature (on the left side of the motor). In PM, the magnetic field is rather homogenous. Inside the ferromagnetic pieces of bars located between PMs, the highest magnetic flux density is observed in the corners.

(a)

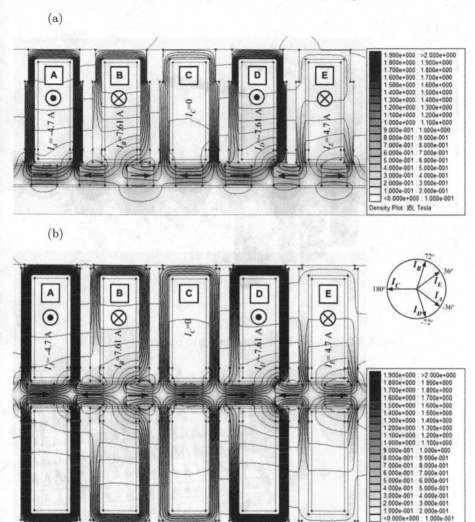

(b)

Fig. 5.23. Magnetic field distribution at maximum load angle $\delta = 90°$: (a) tubular LSM, (b) flat LSM.

The field integral parameters have also been found from the 3D analysis. The most important is thrust. According to the 3D analysis, the electromagnetic thrust $F = 282$ N, while according to the 2D analysis, the thrust $F = 304$ N. The width $l_M = 48.7$ mm is sufficient enough to perform only the 2D analysis. However, if this dimension is smaller, 3D analysis should be executed.

The distribution of the magnetic flux density modulus has been plotted in Fig. 5.25. The field distribution in the portion of full region in the xy

Fig. 5.24. 3D field distribution in a five-phase flat PM LSM.

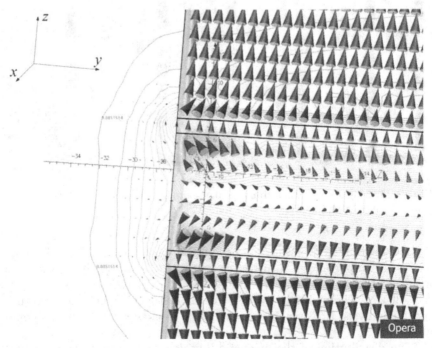

Fig. 5.25. Magnetic flux density distribution in the air gap.

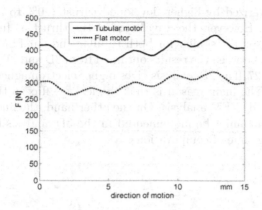

Fig. 5.26. Electromagnetic thrust developed by five-phase LSM versus position of the reaction rail at $\delta = 90°$.

(a)   (b)

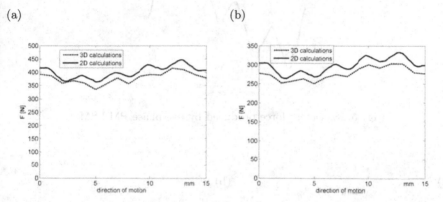

Fig. 5.27. Comparison of the results obtained from 2D and 3D FEM analysis for: (a) tubular LSM, (b) flat LSM.

plane inside the ferromagnetic casing of coil A has been chosen (Fig. 5.17). Arrows represent the flux density vectors. The 3D behavior of the magnetic field distribution is clearly visible. The electromagnetic thrust arises from the magnetic energy stored in the air gap (Fig. 5.25). The field analysis allows for prediction of the magnetic flux polarity in all parts of the reaction rail.

5.2.5 Electromagnetic Thrust and Thrust Ripple

The most important integral parameter, i.e., the thrust plotted against the position δ of the reaction rail at armature current corresponding to $\delta = 90°$, is shown in Figs 5.26 and 5.27. The thrust ripple produced by the five-phase LSM is much lower than that in the case of the three-phase motor (Chapter 4). In general, tubular LSMs with the same phase number m_1 as flat

LSMs are characterized by higher developed thrust (20% to 30%) than flat LSMs (Fig. 5.26). However, the waveforms of the thrust as functions of the reaction rail position are similar for both tubular and flat LSMs.

The difference between the results obtained from 3D and 2D FEM analysis is shown in Fig. 5.27. The 2D analysis gives higher electromagnetic thrust than the 3D analysis. The main reason is that the leakage flux in the end regions is included in the 3D FEM analysis. On the other hand, the fine mesh used in the 2D simulation cannot be implemented in the 3D analysis because of the unacceptably long time of computations.

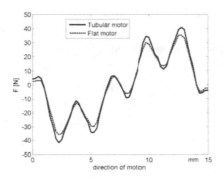

Fig. 5.28. Detent force produced by five-phase PM LSM.

(a)   (b)

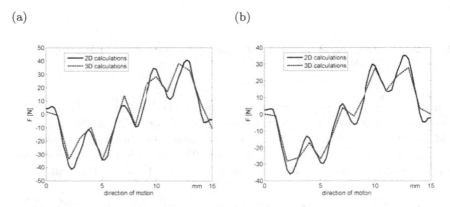

Fig. 5.29. Comparison of detent thrust calculated on the basis of 2D and 3D FEM analysis for (a) tubular LSM, (b) flat LSM.

To obtain the detent force, the FEM analysis of zero armature current $I_a = 0$ has been performed. The average and maximum force increases with the PM and ferromagnetic ring thickness. For dimensions shown in Table 4.1, the detent force for flat and tubular LSM is nearly the same (Fig. 5.28). The highest value of the detent force is for $z = 2.5$ mm and $z = 13$ mm, i.e., nearly 40 N (Figs. 5.28 and 5.29), which is approximately equal to 10% of the thrust. Although, the detent force is significant, the tubular construction is more economical due its higher developed thrust.

The detent force in turn affects the thrust. The number of maxima and minima in detent force waveforms depends on the number of phases (Figs. 4.15 and 5.28). The trust obtained from the 2D FEM analysis (Fig. 5.30) differs from that obtained from the 3D analysis (Fig. 5.31). It is especially visible in the case of the tubular LSM (Fig. 5.31a). This is due to the coarse discretization of the air gap (mechanical clearance) using tetrahedral elements. Considering also CPU time, the user of the 3D FEM software is forced to assume an insufficient number of elements of the mesh.

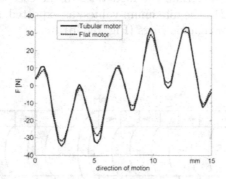

Fig. 5.30. Electromagnetic force of five-phase LSM versus position of the reaction rail at $\delta = 0$.

(a)   (b)

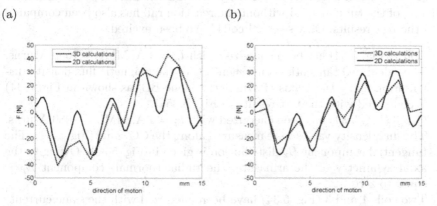

Fig. 5.31. Electromagnetic force calculated on the basis of the 3D FEM at $\delta = 0$ for (a) tubular PM LSM, (b) flat PM LSM.

5.2.6 Experimental Verification

Magnetic Flux Density Distribution

The magnetic flux density distribution at the active surfaces of the armature core and reaction rail has been verified experimentally. Since it is difficult to measure the flux density in the air gap, the magnetic fields excited by the reaction rail (PMs) and the armature (input currents) have been measured separately. The field distribution has been measured along the z-axis.

To verify the distribution of the magnetic flux density excited by PMs, the segment AA' of the reaction rail without the armature has been considered (Fig. 5.32). The calculated components B_z and B_r have been compared with the test results (Fig. 5.33). In the case of the distribution of tangential component B_z, the end effects are not significant (Fig. 5.33a). The first maximum of B_z is about 12% lower than the next one. In the case of the distribution of radial component B_r, the maximum values, as expected, are observed near the edges of the ferromagnetic rings (Fig. 5.33b).

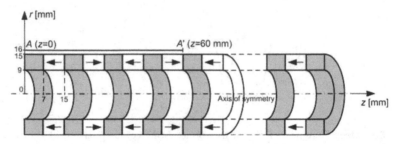

Fig. 5.32. Zone AA' of the reaction rail of tubular PM LSM.

The calculated magnetic flux density distribution excited by the armature winding of the tubular LSM without the reaction rail has also been compared with the test results. Only selected coils have been excited.

(a) Coil 1 (Fig. 5.34) has been energized with $I_a = 4$ A, MMF = 1080 Aturns. The calculated tangential component B_z of the magnetic flux density distribution along the z-axis of symmetry (zone BB' as shown in Fig. 5.34) is compared with the test results in Fig. 5.35a).

(b) Coil 3 (Fig. 5.34) has been energized with $I_a = 4$ A, MMF = 1080 Aturns. The flux density values were measured along the CC' zone (Fig. 5.34). The tangential component B_z distribution is given in Fig. 5.35b. Owing to the axial symmetry of the armature, the radial (normal) component B_r is equal to zero.

(c) Two coils 1 and 3 (Fig. 5.34) have been energized with the same currents $I_a = 4$ A. Distributions of the magnetic flux density components B_r and B_z are plotted in Fig. 5.36. However, the maximum values of tangential

(a)   (b)

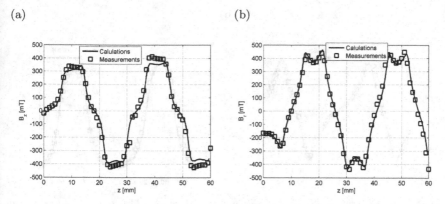

Fig. 5.33. Magnetic flux density components along the AA' zone of the reaction rail: (a) tangential component B_z; (b) radial (normal) component B_r.

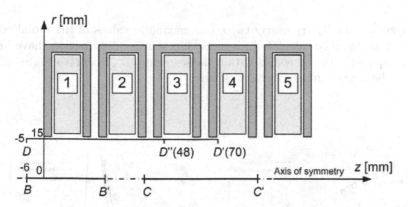

Fig. 5.34. Auxiliary sketch showing location of measurement zones.

(a)   (b)

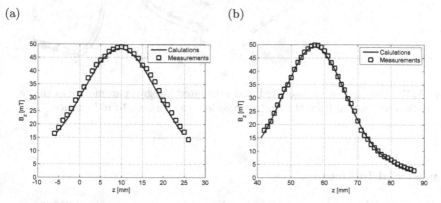

Fig. 5.35. Distribution of tangential component B_z of the flux density at $I_a = 4$ A: (a) along BB' zone with coil 1 being energized; b) along CC' zone with coil 3 being energized.

(a)   (b)

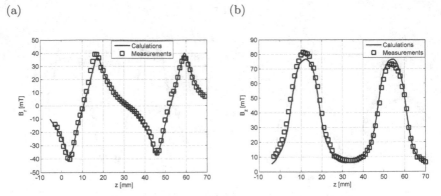

Fig. 5.36. Distribution of magnetic flux density along DD' zone at $I_a = 4$ A in coils 1 and 3: (a) radial component B_r, (b) tangential component B_z.

components B_z are nearly twice the maximum values of the radial component B_r. The calculated magnetic flux density distributions have been compared with test results within the segment DD', which is situated close to the active surface of the armature.

(a)   (b)

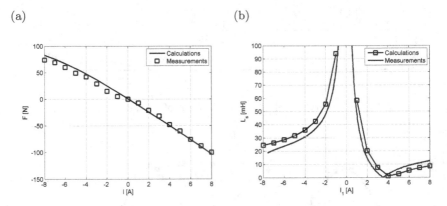

Fig. 5.37. Comparison of measured and calculated integral parameters in the case of reaction rail moved from the central position: (a) thrust, (b) inductance.

Thrust and Inductance

In Fig. 5.37a the electromagnetic thrust is plotted against the armature current when the coil 1 is energized. Fig. 5.37b shows the inductance of coil 1 versus current. Calculations have been compared with test results. The reaction rail has not been positioned in the center of the armature system. In the central position, the armature and reaction rail symmetry axes normal to the active surfaces coincide. Assuming symmetry axis at $z = 0$, the reaction rail symmetry axis normal to its active surface has been moved to $z = 7.5$ mm.

5.3 Tubular Linear Reluctance Motors

A tubular linear reluctance motor (LRM) belongs to the group of cyclic actuators with oscillating movement of its movable parts (reaction rails) [27], [160]. So far, reaction rails have been assembled of solid ferromagnetic rods or cylinders [27, 117].

In this section, a laminated silicon steel reaction rail has been proposed. The effects of nonlinearity and laminations on the magnetic field distribution and integral parameters of tubular LSMs have been investigated.

In the case of coreless armature winding (open magnetic system) of a tubular LRM, the traveling magnetic field is generated in the z-direction (Fig. 5.38) by a system of coils. The tubular armature consists of solenoid-type concentrated-parameter coils that constitute a three-phase winding.

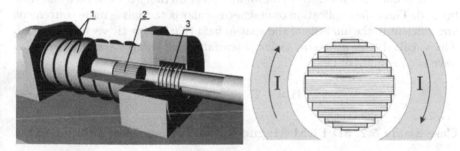

Fig. 5.38. Tubular LRM with coreless armature system and the reaction rail. 1 — armature coils, 2 — laminated stack, 3 — spring.

The armature with ferromagnetic core (closed magnetic system) has also been considered. A simplified computer-generated 3D image of a tubular LRM with the armature ferromagnetic core is shown in Fig. 5.39. The external part of the armature core consists of a laminated ferromagnetic cylinder and end disks. The thickness of the cylinder and end disks is d_k. In the analysis of the electromagnetic field, the air gaps between the cylinder and end disks have been neglected.

Poisson's equation (4.4) for the tubular LRM can be written in the following form: [208, 209, 210, 213, 215, 249]

$$\nabla \times \left(\nabla \times \mathbf{A}\, \frac{1}{\mu(B)}\right) = \mathbf{J} + \sigma\left(\mathbf{v} \times \nabla \times \mathbf{A} - \frac{\partial \mathbf{A}}{\partial t}\right) \qquad (5.23)$$

(a)   (b)

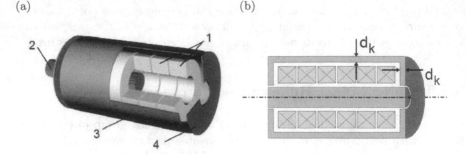

Fig. 5.39. Tubular LRM with ferromagnetic armature core (a) 3D computer created image, (b) longitudinal section. 1 — coils, 2 — moving rod (runner), 3 — ferromagnetic housing, 4 — ferromagnetic end disks.

in which **A** is the magnetic vector potential, **v** is the velocity of the magnetic field excited by the current density **J**, and σ is the electric conductivity. For slow time-varying magnetic fields (slowly moving raction rail), the second term $\sigma(\mathbf{v} \times \nabla \times \mathbf{A} - \partial\mathbf{A}/\partial t)$ on the right-hand side of enq (5.23) vanishes.

To include the anisotropy of laminated reaction rail, 3D FEM analysis must be used. Thus, the application of magnetic scalar potentials is more convenient and efficient in the analysis of the spatial field than magnetic vector potential. The so-called *total magnetic scalar potential* ψ and *reduced scalar potential* ϕ have been employed, i.e.,

$$\mathbf{H} = -\nabla\psi; \quad \nabla^2\phi = 0 \qquad (5.24)$$

Case Study 5.2. 3D FEM Magnetic Field Analysis in tubular LRM

The 3D FEM analysis have been performed using the Opera-3d commercial package [25, 162]. Steady-state characteristics and electromagnetic parameters of LRMs have been found using the FEMM 4.0 package [144].

In the analyzed LRM, the length of the armature winding is 160 mm. The outside and inside diameters of coils are $d_o = 85$ mm and $d_i = 39$ mm, respectively. The reaction rail is stacked of silicon steel sheets and inserted in a nonferromagnetic tube with its inside diameter $d = 29$ mm. Thus, the nonferromagnetic air gap between the armature wires and the ferromagnetic core of the reaction rail is $g = 5$ mm. The mass of the reaction rail is 1.05 kg, and its length is 190 mm. The air gap between the armature end disks and the reaction rail is 2 mm.

The number of turns per phase is $N = 2400$ both for the coreless armature and armature with ferromagnetic core. The reaction rail and armature core are made of EP470-50A silicon steel [205, 206]. Since the magnetic circuit is laminated both axially and radially (Fig. 5.38), the electromagnetic field analysis must include anisotropy.

Magnetic Field in Laminated Reaction Rail

It has been assumed [205, 206] that the magnetization curve B-H of the reaction rail in the direction of motion (z-axis) is typical, such as that given in catalogues of electrical sheet steels. The B-H curve in the direction perpendicular to the plane of laminations (Fig. 5.38b) is extremely different from that in the z-direction. Manufacturers of electrical sheet steels normally do not provide this B-H curve. However, if the thickness of laminations and difference between magnetic permeabilities in both directions are known, the magnetization curve in the direction perpendicular to the plane of laminations can be reconstructed, e.g., [207].

The 3D FEM mesh contains over 280 000 elements. The anisotropy effect is especially visible in the nonsaturated core (Fig. 5.40a), when the armature current is small, e.g., $I_a = 1.0$ A, and the reaction rail is in the central position. For comparison, the computation results for the isotropic core (solid steel) of reaction rail are presented in Fig. 5.40b. The "step" corners of the anisotropic stack are more saturated than those of the isotropic stack, even for small armature current $I_a = 1.0$ A (Fig. 5.40b). Increase in the armature current to $I_a = 8.0$ A causes the reaction rail to saturate (Fig. 5.41). However, at each end of the reaction rail, the inner area of the cross section remains unsaturated.

Magnetic Field in Assembled LRM

A tubular LRM with armature furnished with ferromagnetic core is shown in Fig. 5.39. The armature winding is enclosed with a laminated silicon steel core. To simplify the analysis, eddy currents in the laminated armature core have been neglected. Calculations have been executed for quantitative changes in the armature current and the thickness d_k of the stator cylindrical core (Fig. 5.39) and for various axial positions of the reaction rail.

The ferromagnetic cylindrical core with its thickness $d_k = 1.5$ mm (Fig. 5.42b) is rather a magnetic shield for armature winding than a part of the magnetic circuit. For small armature current $I_a = 1.0$ A, the average magnetic flux density value in the reaction rail is 0.75 T in the case of coreless armature winding (Fig. 5.42a) and 0.95 T in the case of armature winding with ferromagnetic core (Fig. 5.42b).

The shielding effect is also visible in Fig. 5.43 where the armature winding is energized with high armature current $I_a = 8$ A. The thin cylindrical ferromagnetic core (external part of the armature magnetic circuit) becomes highly saturated. Because of high-level of saturation of the armature core (Fig. 5.43), the magnetic flux distribution in the reaction rail is almost the same for coreless armature winding and armature winding with ferromagnetic core. The average magnetic flux density value in the reaction rail is nearly 1.9 T. However, in the case of the armature with ferromagnetic core, the magnetic flux is forced to concentrate in the edges of the armature core (Fig. 5.43b).

(a)

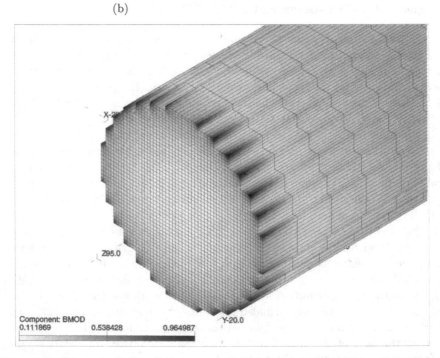

(b)

Fig. 5.40. 3D magnetic flux density map for (a) anisotropic reaction rail stacked of laminations, (b) isotropic reaction rail made of solid steel.

(a)

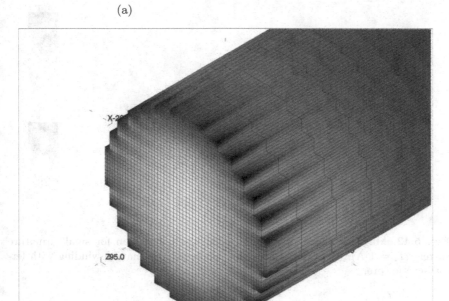

(b)

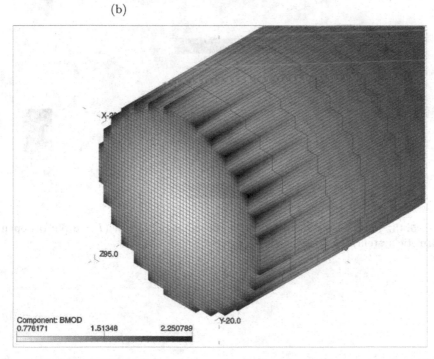

Fig. 5.41. Map of the modulus of the magnetic flux density for saturated reaction rail: (a) anisotropic; (b) isotropic.

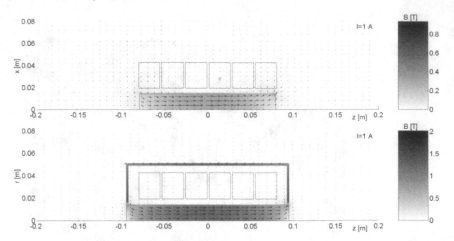

Fig. 5.42. Magnetic flux distribution in longitudinal section for small armature current ($I_a = 1$ A): (a) coreless armature winding; (b) armature winding with ferromagnetic core.

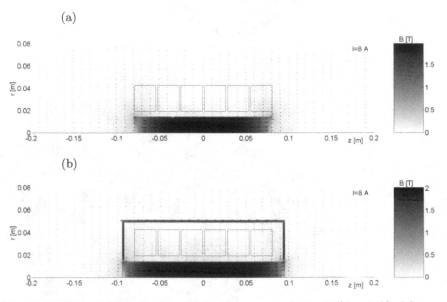

Fig. 5.43. Flux density distribution for the saturated circuit ($I = 8$ A): (a) open magnetic system; (b) closed magnetic system.

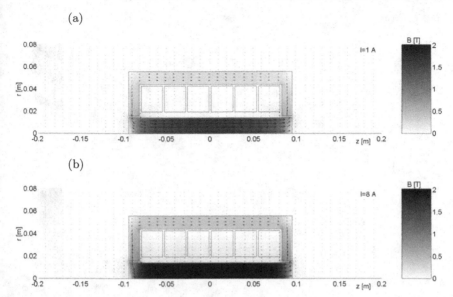

Fig. 5.44. Magnetic flux density distribution in the case of thick cylindrical part of the armature magnetic circuit at (a) $I_a = 1.0$ A, (b) $I_a = 8.0$ A.

When the external cylindrical part of the armature magnetic circuit is relatively thick ($d_k = 12$ mm), the magnetic flux distribution for coreless armature system and armature winding with ferromagnetic core considerably differ from each other. It is especially visible for high value of the armature current $I_a = 8$ A (Fig. 5.44b). The magnetic flux density in the armature winding area is greater than that in the coreless armature system (Fig. 5.43a). Even for a small value $I_a = 1.0$ A of the armature current, the maximum values of the flux density in the armature system with ferromagnetic core are nearly twice greater, i.e., $B = 1.0$ to 1.5 T (Fig. 5.44a) than those in the case of coreless armature winding (Fig. 5.42a).

More interesting is the behavior of the magnetic flux when the reaction rail is not in the armature central position. For example, if at $I_a = 8.0$ A half of the reaction rail is outside the armature system, the smaller portion of the reaction rail covered by the armature magnetic circuit becomes highly saturated (Fig. 5.45a). The reaction rail is more saturated in the case of the armature with ferromagnetic core (Fig. 5.45b) than in the case of coreless armature.

Experimental Verification

The distribution of B_z and B_r components along the z-axis obtained from calculations and measurements are plotted in Figs. 5.46 and 5.47. Measurements have been performed only for the tubular LRM with coreless armature

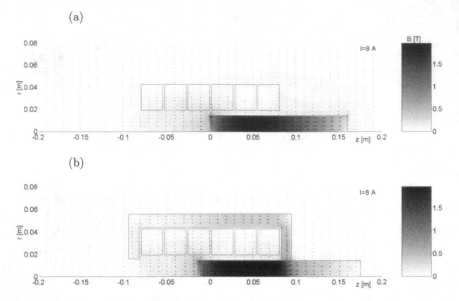

Fig. 5.45. Magnetic flux distribution at $I_a = 8$ A and reaction rail not positioned in the center of the armature system: (a) coreless armature winding; (b) armature winding with ferromagnetic core.

system for the reaction rail positioned at $z = 80$ mm with respect to the central position. A good agreement between calculation and test results confirms the correctness of computer simulations.

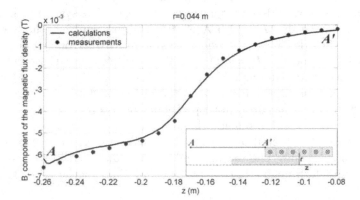

Fig. 5.46. Distribution of radial component B_r of magnetic flux density along the zone AA'.

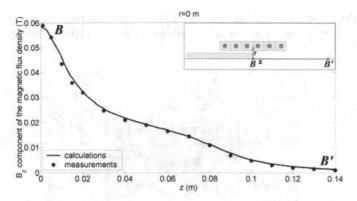

Fig. 5.47. Distribution of tangential component B_z of magnetic flux density along the zone BB'.

5.4 Linear Oscillatory Actuators

The linear oscillatory actuator (LOA) of the reluctance type belongs to the group of short-stroke actuators [68, 160]. Their construction is very simple(Fig. 5.48): the stationary part with coils is fed from an a.c. source, while the movable part constitutes a ferromagnetic bar. LOAs convert electrical energy into sustained short-stroke reciprocating motion [252]. Applications include, but are not limited to, reciprocating compressors, vibrators, pumps, shuttles of weaving looms, electric hammers, etc.

In an LOA, the magnetic energy depends on the position of the moving bar. The energy takes maximum value when the moving bar is in the center of the stationary part ($z = 0$), as shown in Fig. 5.48. As the bar moves left or right ($z \neq 0$), the magnetic energy decreases. A change in energy produces the electromagnetic force acting on the moving bar.

The RLC circuit consisting of the stationary coil and series capacitor is in resonance when the center of the moving ferromagnetic bar approaches the edge of the coil. At this instant, the current in the coil abruptly increases and high electromagnetic force pulls the bar into the center of the coil and further toward the opposite edge of the coil. The resonance occurs again and the cycle is repeated.

LOAs have been widely studied using the circuital approach, e.g., [31, 141, 160], and the FEM, e.g., [252]. This section is devoted to electromagnetic field analysis, integral parameters, and calculations of dynamic performance characteristics of LOAs.

5.4.1 2D Electromagnetic Field Analysis

The electromagnetic field analysis in a LOA (Fig. 5.48) can be brought to the analysis of a magnetically open problem [231, 204, 250]. In general, the problem should be analyzed using a method that does not require any limitation of the calculated area, e.g. boundary integral method (BIM) [204, 250].

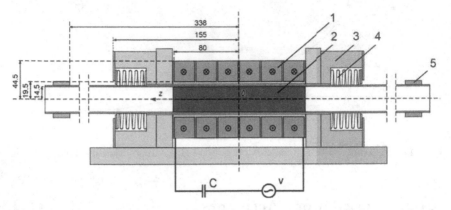

Fig. 5.48. Longitudinal section of a LOA. 1 — stationary coil, 2 — moving bar, 3 — housing, 4 — spring, 5 — limiter.

Since the magnetic circuit is open (Fig. 5.48), the magnetic flux density in each portion of the magnetic circuit is relatively low. When the mechanical frequency of the moving bar is low (2 to 3 Hz), the displacement currents and eddy-current losses in conductive parts of an LOA can be neglected. With these simplifications, the electromagnetic field is governed by eqn (5.23), in which $\sigma(\mathbf{v} \times \nabla \times \mathbf{A} - \partial \mathbf{A}/\partial t) = 0$

For axial symmetry of the LOA, the electromagnetic field can be described by the so-called *reduced scalar potential* ϕ, which depends on the angular component A_φ of the magnetic vector potential $\mathbf{A}$ and the radius r (the distance between source and field point) i.e., $\phi = A_\varphi r$. Thus

$$\frac{\partial}{\partial r}\left(\frac{1}{\mu(B)}\frac{1}{r}\frac{\partial \phi}{\partial r}\right) + \frac{\partial}{\partial z}\left(\frac{1}{\mu(B)}\frac{1}{r}\frac{\partial \phi}{\partial z}\right) = -J \qquad (5.25)$$

See also eqn (5.24). On the basis of the B-H curve, the relative magnetic permeability μ_r of the moving bar has been determined. The tangential B_z and radial B_r components of the magnetic flux density $\mathbf{B}$ have been calculated as

$$B_z = \frac{1}{r}\frac{\partial \phi}{\partial r}; \quad B_r = -\frac{1}{r}\frac{\partial \phi}{\partial z} \qquad (5.26)$$

To obtain the reliable values of the magnetic flux density in the vicinity of the stationary part, the dimensions of the investigated region must be at least four times greater than the longest dimension of the LOA [204].

5.4.2 Magnetic Flux, Force, and Inductance

The inductance of the stationary coil can be found either from the total energy of the magnetic field stored in the coil or from the magnetic flux Ψ linked with the coil. The magnetic energy is expressed by eqn (4.33), and the linkage flux by eqn (4.16). When the system is unsaturated, the average total inductance of the stationary coil is expressed by eqn (4.30), and when the system comes into saturation, it is convenient to use eqn (4.18).

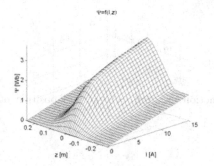

Fig. 5.49. Distribution of magnetic flux linkage as a function of input current and position of moving bar.

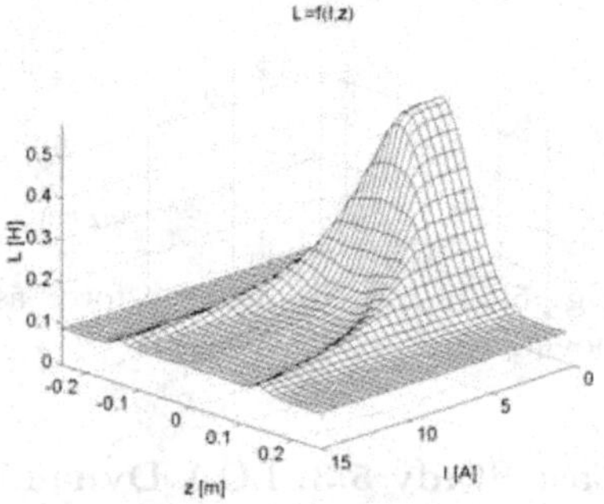

Fig. 5.50. Dynamic inductance of the coil as a function of input current and position of moving bar.

As known, the magnetic flux Ψ linking the coil turns depends on the input current. For the LOA shown in Fig. 5.48 and input current exceeding $I = 5$ A, the function $\Psi = f(I)$ plotted in Fig. 5.49 is strongly nonlinear. For $I \geq 10$ A, the system is saturated, and the flux Ψ minimally changes with the current I.

For the position $z > 0.12$ m of the moving bar, the magnetic flux is practically independent of z. The moving bar is outside the stationary coil, and the coil behaves as an air-cored coil. Thus, the flux linkage is a linear function of the current I.

It is evident from Fig. 5.50 that the inductance L of the stationary coil decreases as the current increases. For the dead zone of the bar (in the middle of the stationary coil), the inductance decreases from $L_{max} = 0.55$ H to $L_{min} = 0.1$ H as the current increases from 1.0 to 15.0 A.

When the current is greater than 2.0 A, the moving bar gets slightly saturated and the inductance drops somewhat. The bottom margin of the inductance (for the current resonance) is nearly 0.1 H.

The electromagnetic force acting on the moving bar as a function of the input current and bar position is plotted in Fig. 5.51. The direction of the electromagnetic force depends on the position of the moving bar, i.e. for $z < 0$, the force is positive, and for $z > 0$, the force in negative.

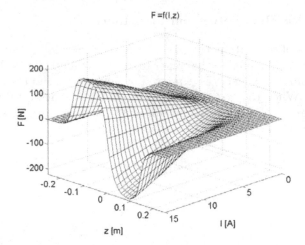

Fig. 5.51. Electromagnetic force as a function of input current and position of moving bar.

Case Study 5.3. LOA Dynamics

An LOA can be desribed by the following equations:

- electric balance equation

$$v(t) = L\,[z, i(t)]\,\frac{\partial i(t)}{\partial t} + \frac{\partial \Psi\,(z, i(t))}{\partial z}\frac{\partial z}{\partial t} + \frac{1}{C}\int i(t)dt + R \cdot i(t) \quad (5.27)$$

- mechanical balance equation

$$m\frac{d^2 z}{dt^2} + D_v\frac{dz}{dt} + f_s = f \quad (5.28)$$

in which $v(t)$ is the input voltage, $L(z, i)$ is the inductance of the stationary coil, $\Psi(z, i)$ is the magnetic flux linked with the stationary coil, C is the series capacitance in the stationary coil circuit, R is the stationary coil resistance, m is the mass of moving bar including load, D_v is the damping (friction) constant, $f = \partial W'/\partial z$ is the electromagnetic force calculated on the basis of magnetic coenergy W', $f_s = k_s(z_0 - z)$ is the spring force, and k_s is the spring constant (stiffness). See also eqns (1.28) amd (1.29).

To study the dynamic behavior of the LOA, it has been assumed that

(a) the moving bar position changes from $z(t) = 0$ to $|z(t)| = 0.188$ m;
(b) the coefficients in eqn (5.28) are $m = 1.05$ kg, $k_{fr} = 2$ Ns/m, $k_s = 2$ kN/m;
(c) electrical parameters in eqn (5.27) are $R = 8.5\ \Omega$, $C = 100\ \mu$F, $z_0 = 0.188$ m.

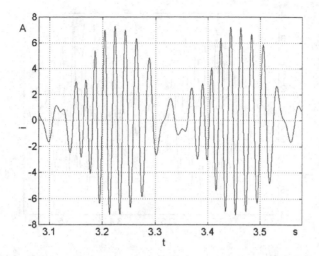

Fig. 5.52. Computed waveform of the current in stationary coil at $v = 50$ V.

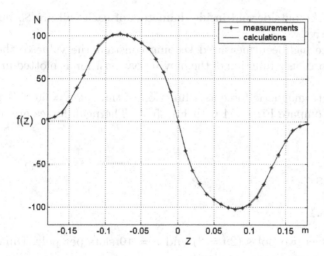

Fig. 5.53. Static force acting on moving bar at $I = 8.0$ A.

A direct solution to eqns (5.27) and (5.28) is not possible without physical simplifications. Eqns (5.27) and (5.28) have been solved using Matlab and toolboxes PDE and Simulink [167]). After computing the integral parameters of the magnetic field, the combined kinetic and electric field equations have been solved simultaneously. The computed results for $v = 50$ V are given in Fig. 5.52.

Experimental verifications of computer simulations have been done for the moving 160-mm long steel bar . The mass of the bar is 1.05 kg. The stationary winding consists of six coils connected in series, each wound with

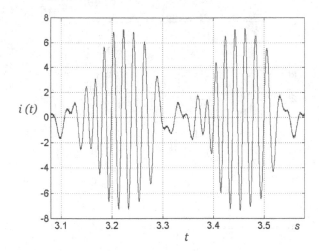

Fig. 5.54. Experimental results for the stator current, $v = 50$ V.

400 turns. The outside and inside diameters of each coil are 85 mm and 39 mm, respectively.

The force has been measured keeping constant the value of the current. The static force as a function of the moving bar position is plotted in Fig. 5.53.

The electromagnetic force as a function of time at $v = 50$ V is plotted in Fig. 5.54. Compare Fig. 5.54 with Fig. 5.52. The peak forces are almost the same.

Examples

Example 5.1

An HLSM has two poles $(2p = 2)$ and $n = 10$ slots per pole. Dimensions of the magnetic circuit are as follows

- slot width $b = 0.56$ mm
- tooth width $c = 0.46$ mm
- air gap $g = 0.014$ mm
- length of the forcer stack (perpendicular to the direction of motion) $L_i = 12$ mm
- width of pole (along the platen) $w_p = 22$ mm

The stacking factor $k_i = 0.97$. Find the variation of the air gap permeance $G_t(x)$ and variation of the tangential force per pole $F_{dx}(B_g, x)$ for the air gap magnetic flux density $B_g = 0.25$ T, 0.5 T, and 0.75 T.

Solution

Pole pitch of the forcer

$$t_1 = b + c = 0.56 + 0.46 = 1.02 \text{ mm}$$

Maximum air gap permeance according to eqn (5.10)

$$G_{max} = 0.4\pi \times 10^{-6} \times 0.012 \left[\frac{0.46}{0.014} + \frac{2}{\pi} \ln \left(1 + \frac{\pi \times 0.56}{2 \times 0.014} \right) \right] = 5.354 \times 10^{-7} \text{ H}$$

Mnimum air gap permeance according to eqn (5.11)

$$G_{max} = 0.4\pi \times 10^{-6} \times 0.012 \left[\frac{0.56 - 0.46}{0.014 + 0.25 \times \pi \times (0.56 - 0.46)} \right.$$

$$\left. + \frac{8}{\pi} \ln \left(\frac{0.014 + 0.25 \times \pi \times 0.56}{0.014 + 0.25 \times \pi \times (0.56 - 0.46)} \right) \right] = 7.735 \times 10^{-8} \text{ H}$$

Variation of air gap permeance with the x-axis is expressed by eqn (5.9) and plotted in Fig. 5.55.

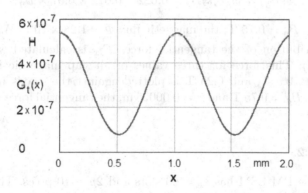

Fig. 5.55. Variation of air gap permeance with the x-axis.

The first derivative of permeance with respect to the x-axis is given by eqn (5.12), i.e.,

$$\frac{\partial G_t}{\partial x} = -\frac{\pi}{0.00102} (5.354 \times 10^{-7} - 7.735 \times 10^{-8}) \sin \left(\frac{2\pi}{0.00102} x \right)$$

$$= -0.001411 \sin(6160x)$$

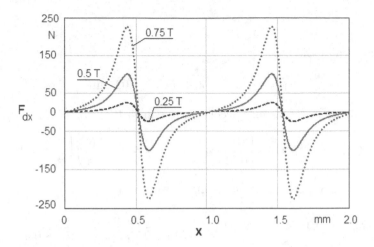

Fig. 5.56. Variation of tangential force F_{dx} with the x-axis at the air gap magnetic flux density $B_g = 0.25$ T, 0.5 T, and 0.25 T.

Putting, e.g., $x = 0.0008$ m, the first derivative $\partial G_t / \partial x = 0.00138$ H/m. The magnetic flux

$$\Phi(B_g) = w_p L_i k_i B_g = 0.022 \times 0.012 \times 0.97 \times B_g$$

Putting, e.g., $B_g = 0.75$ T, the magnetic flux $\Phi = 1.28 \times 10^{-4}$ Wb.

The distribution of the tangential forces F_{dx} is calculated with the aid of eqn (5.18). This force for three values of air gap magnetic flux density $B_g = 0.25$ T, 0.5 T, and 0.25 T is plotted against the x-axis in Fig. 5.56. Putting, e.g., $B_g = 0.75$ T and $x = 0.00025$ m, the tangential force $F_{dx} = 52.95$ N.

Example 5.2

A three-phase PM LSM has $s_1 = 60$ slots and $2p = 10$ poles. The armature winding is Y-connected and fed with $V_{1L} = 400$ V rms (line-to-line). The linear synchronous speed is $v_{s3} = 8$ m/s, the rated (nominal) thrust is $F_x = 1200$ N, and the efficiency × power factor product is $\eta \cos \phi = 0.68$.

Redesign this three-phase LSM into a five-phase motor assuming that (a) the number of slots per pole per phase is the same, (b) the output power is the same, (c) the $\eta \cos \phi$ product is the same, and (d) the phase windings are Y-connected.

Solution

(a) Three-phase motor

The number of slots per pole per phase

$$q_1 = \frac{s_1}{2pm_1} = \frac{60}{10 \times 3} = 2$$

must be kept the same for both three-phase and five-phase motors. The number of slots per pole

$$Q_1 = \frac{s_1}{2p} = \frac{60}{10} = 6$$

Output power

$$P_{out} = v_{s3}F_x = 8.0 \times 1200 = 9600 \text{ W}$$

Phase voltage

$$V_1 = \frac{V_{1L}}{\sqrt{3}} = \frac{400}{\sqrt{3}} = 230.94 \text{ V}$$

Armature current of three-phase motor

$$I_{a3} = \frac{P_{out}}{m_1 V_1 \eta \cos \phi} = \frac{9600}{3 \times 230.94 \times 0.68} = 20.38 \text{ A}$$

(b) Five-phase motor

Armature current of five-phase motor

$$I_{a5} = \frac{P_{out}}{5V_1 \eta \cos \phi} = \frac{9600}{5 \times 230.94 \times 0.68} = 12.23 \text{ A}$$

For five-phase motor the number of poles must be changed because the armature core is the same (60 slots) and, according to assumption (a), the number of slots per pole per phase must be the same ($q_1 = 2$).
    The number of pole pairs of five-phase motor

$$p = \frac{s_1}{2q_1 m_1} = \frac{60}{2 \times 2 \times 5} = 3$$

The number of slots per pole

$$Q_{15} = \frac{60}{6} = 10$$

The pole pitch of five-phase motor increases approximately in proportion to $Q_{15}/Q_{13} = 10/6$, so that the linear speed

$$v_{s5} = v_{s3}\frac{Q_{15}}{Q_{13}} = 8\frac{10}{6} = 13.3 \text{ m/s}$$

The thrust of five-phase motor

$$F_{x5} = \frac{9600}{13.3} = 720 \text{ N}$$

For constant output power (approximately for constant input power), the current of five-phase motor is reduced, the speed increases (assuming $q_1 = 2 = const$), and the thrust decreases.

Suppose that the armature current of both 3-phase and 5-phase motors is the same, i.e., $I_{a5} = I_{a3} = 20.38$ A. The output power of five-phase motor at the same value of $\eta \cos \phi = 0.68$ product is

$$P_{out5} = 5V_1 I_{a5}\eta \cos \phi = 5 \times 230.94 \times 20.38 \times 0.68 = 16\,000 \text{ W}$$

The thrust for the same armature current and increased linear speed

$$F_{x5} = \frac{16\,000}{13.3} = 1200 \text{ N}$$

The thrust for the same pole pitch (linear speed $v_{s5} = v_{s3} = 8$ m/s)

$$F_{x5} = \frac{16\,000}{8.0} = 2000 \text{ N}$$

If the current and pole pitch are the same, the thrust of five-phase motor increases.

Example 5.3

Consider a three-phase tubular LSM as in *Example 4.3*. Assuming the same dimensions of all parts (Fig. 4.28), the same peak current $I_m = 31.4$ A (Fig. 5.57), and the same load angle $\delta = 90°$, find the magnetic field distribution and integral parameters of a five-phase tubular LSM using the FEM approach. Material parameters and boundary conditions are the same as in *Example 4.1*. The thrust should be calculated as a function of the the reaction rail position in the interval $0 \le z \le 20$ mm.

Solution

In the case of five-phase construction, the distance d_s between segments should be recalculated with the aid of eqn (4.78). For suitable distance between segments, i.e.,

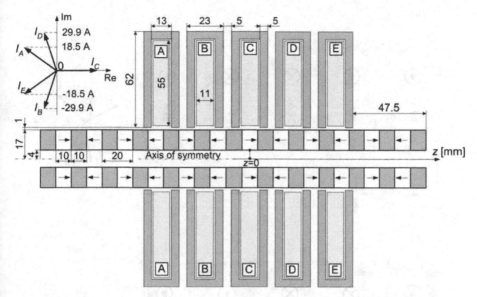

Fig. 5.57. Dimensions of the analyzed five-phase tubular PM LSM and current phasor diagram. The system of the armature currents for $z = 0$ and $\delta = 90°$ is shown in the left top corner.

$$d_s = k\frac{\tau}{n_{ph}} - w_{ss} + \tau = 2\frac{20}{5} - 23 + 20 = 5 \text{ mm}$$

the coefficient k (Section 4.4) have been assumed $k = 2$. To keep the same load angle $\delta = 90°$ for the whole range of reaction rail movement, the current waveforms must be functions of the reaction rail position. Thus, the system of 5-phase currents is expressed as

$$i_{aA}(z) = I_m \sin\left(\frac{z}{\tau}180° + 54° + \delta\right)$$

$$i_{aB}(z) = I_m \sin\left(\frac{z}{\tau}180° - 198° + \delta\right)$$

$$i_{aC}(z) = I_m \sin\left(\frac{z}{\tau}180° - 90° + \delta\right)$$

$$i_{aD}(z) = I_m \sin\left(\frac{z}{\tau}180° + 18° + \delta\right)$$

$$i_{aE}(z) = I_m \sin\left(\frac{z}{\tau}180° - 234° + \delta\right)$$

For dimensions as in Fig. 5.57, the thrust is determined on the basis of the magnetic field analysis, as an integral parameter. The magnetic field distribution is calculated for several positions of the reaction rail. The magnetic field distribution for two positions ($z = 6$ mm and $z = 16$ mm) of the reaction rail is plotted in Fig. 5.58.

(a)

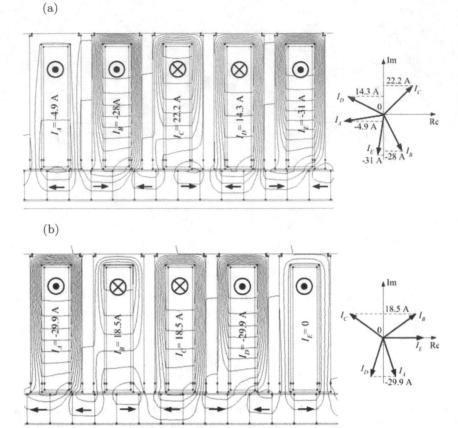

(b)

Fig. 5.58. Magnetic field distribution at load angle $\delta = 90°$ for: (a) $z = 6$ mm (minimum thrust), (b) $z = 16$ mm maximum thrust).

The minimum value of the thrust is for $z = 6$ mm. For $z = 16$ mm the thrust takes the maximum value. In both cases the highest concentration of magnetic flux lines is in the area of armature teeth.

In the case of three-phase LSM, the thrust (Fig. 5.59) changes from 680 N up to 930 N depending on the position of the reaction rail. The thrust developed by the five-phase LSM is approximately 1.67 higher than that developed by the three-phase motor (Fig. 5.59). This is becuase phase currents of both motors have been assumed the same.

The five-phase LSM produces relatively low thrust ripple ($k_r = 0.164$), approximately 50% lower than the thrust ripple of the three-phase motor ($kr = 0.306$).

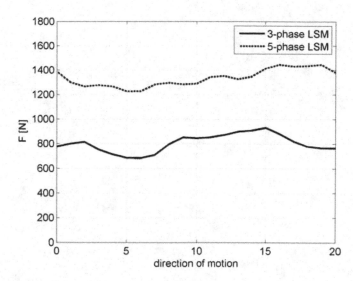

Fig. 5.59. Maximum thrust at $\delta = 90°$ versus the reaction rail position.

6

Motion Control

6.1 Control of AC Motors

Control variables are classified into input variables (input voltage, input frequency), output variables (speed, angular displacement, torque), and internal variables (armature current, magnetic flux). The mathematical model of an a.c. motor is nonlinear, which for small variations of the input voltage, input frequency, and output speed can be linearized.

Scalar control methods are based on changing only the amplitudes of controlled variables. A typical example is to maintain constant torque (magnetic flux) of a.c. motors by keeping constant the V/f ratio. The scalar control can be implemented both in the *open loop* (most of industrial applications) and *closed loop* control systems with speed feedback.

In the *vector oriented control* method (Fig. 6.1), both the amplitudes and phases of the space vectors of variables are changed. Vector control based upon the *field orientation* principle uses the analogy between the a.c. (induction or synchronous) motor and the d.c. commutator motor. The active and reactive currents are decoupled, which, in turn, determine the thrust and magnetic flux, respectively.

Standard controllers have a fixed structure and constant parameters. However, the parameters of electric motors are variable, e.g., winding resistances are temperature dependent, and winding inductances are magnetic saturation dependent. Consequently, the deterioration of the dynamic behavior of the drive can even lead to its instability. In *adaptive controllers*, variable parameters or structure are adjusted to the change in parameters of the drive system.

In a *self-tuning adaptive control*, the controller parameters are tuned to adapt to the drive parameter variation. In a *model reference adaptive control*, the drive response is forced to track the response of a reference model irrespective of the drive parameter variation. This can be achieved by storing the fixed parameter reference model in the computer memory. Some model reference control methods are based on *search strategy*.

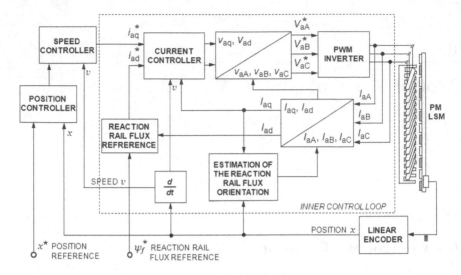

Fig. 6.1. Vector oriented control of an a.c. motor.

The *sliding mode control* is a variable structure control technique similar to the adaptive model reference control. In sliding mode control, the response of the drive system is insensitive to the parameter and load variations. Thus, this method is suitable for servo drives, e.g., machine tools, robots, factory automation systems, etc. The drive system is forced to "slide" along the so-called *predefined trajectory* in the *state space* by a switching control algorithm independently of change in its parameters and load.

The above classical control methods use a mathematical model of the controlled system either in the form of a transfer function or in the form of state space equations. Neural and fuzzy logic control are based on *artificial intelligence* and do not need any mathematical models.

Neural control applies an *artificial neural network* as a controller or emulator of the dynamic system. The artificial neural network is a network of artificial neurons that simulate the nervous system of the human brain. Each neuron in a single layer is connected with all neurons of the neighboring layers with the aid of so-called *synapses*[1]. The neural controller with its associative property memory can create a nonlinear relationship between its input and output values. Since there is no flux estimator, the input–output relationship has to be trained on a sufficient number of samples. The training can be performed either online or offline.

A controller that is based on *fuzzy logic* uses the experience and intuition of a human plant operator. The memory of a fuzzy controller creates fuzzy logic rules, e.g., if the speed of an LSM is *slow*, and the reference speed is

---

[1] In medicine, a synapse is a point at which a nerve charge passes from one basic reaction unit cell to another.

fast, then set the input signal (frequency) is *high*. The main advantage of fuzzy control is that a strictly nonlinear or unknown system can be controlled by linguistic variables. Plants that are difficult to model using conventional parameter identification techniques can be made controllable by implementing *human expert knowledge*.

6.2 EMF and Thrust of PM Synchronous and Brushless Motors

6.2.1 Sine-Wave Motors

A three-phase (multiphase) armature winding with distributed parameters produces sinusoidal or quasi-sinusoidal distribution of the MMF. For a sinusoidal distribution of the air gap magnetic flux density, the first harmonic of the excitation flux can be found on the basis of eqn (3.26). The excitation magnetic flux calculated on the basis of the maximum air gap magnetic flux density B_{mg} is then $\Phi_f \approx \Phi_{f1} = (2/\pi)L_i\tau k_f B_{mg}$, where the *form factor of the excitation field* $k_f = B_{mg1}/B_{mg}$ is according to eqn (3.10). Assuming that the instantaneous value of the EMF induced in a single stator conductor by the first harmonic of the magnetic flux density is $e_{f1} = E_{mf1}\sin(\omega t) = B_{mg1}L_i v_s \sin(\omega t) = 2fB_{mg1}L_i\tau\sin(\omega t)$, the *rms* EMF is $E_{mf1}/\sqrt{2} = \sqrt{2}fB_{mg1}L_i\tau = (1/2)\pi\sqrt{2}f(2/\pi)B_{mg1}L_i\tau$. For two conductors or one turn, $E_{mf1}/\sqrt{2} = \pi\sqrt{2}f(2/\pi)B_{mg1}L_i\tau$. For $N_1 k_{w1}$ turns, where k_{w1} is the winding factor, the *rms* EMF is

$$E_f \approx E_{f1} = \pi\sqrt{2}f N_1 k_{w1}\alpha_i k_f B_{mg}L_i\tau$$

$$= \frac{\pi}{\tau}\frac{1}{\sqrt{2}}N_1 k_{w1}\Phi_f v_s = c_E \Phi_f v_s \tag{6.1}$$

where $2/\pi$ is replaced by α_i and the EMF (armature) constant is

$$c_E = \frac{\pi}{\tau}\frac{1}{\sqrt{2}}N_1 k_{w1} \tag{6.2}$$

For $\Phi_f = const$ a new EMF constant

$$k_E = c_E \Phi_f \tag{6.3}$$

can be used. Assuming a negligible difference between the *d*- and *q*-axis synchronous reactances, i.e., $X_{sd} \approx X_{sq}$, the electromagnetic (air gap) power $P_{elm} \approx m_1 E_f I_{aq} = m_1 E_f I_a \cos\Psi$. Note that in academic textbooks the electromagnetic power is usually calculated neglecting the armature winding resistance R_1 as $P_{elm} \approx P_{in} = m_1 V_1 I_a \cos\phi = m_1 V_1 I_a \cos(\delta + \Psi)$, where $\cos((\delta + \Psi) = E_f \sin\delta/(I_a X_{sd})$. Putting E_f according to eqn (6.1) and $v_s = 2f\tau$, the electromagnetic thrust developed by the LSM is

$$F_{dx} = \frac{P_{elm}}{v_s} = \frac{m_1 E_f I_a}{v_s} \cos\Psi$$

$$= m_1 \frac{\pi}{\tau} \frac{1}{\sqrt{2}} N_1 k_{w1} \Phi_f I_a \cos\Psi = \frac{m_1}{2} c_F \Phi_f I_a \cos\Psi \qquad (6.4)$$

where the *thrust constant* is

$$c_F = 2c_E = \frac{\pi\sqrt{2}}{\tau} N_1 k_{w1} \qquad (6.5)$$

For $\Phi_f = const$ a new thrust constant

$$k_F = c_F \Phi_f \qquad (6.6)$$

simplifies prediction of thrust-current characteristics. The maximum thrust

$$F_{dxmax} = \frac{m_1}{2} c_F \Phi_f I_a = \frac{m_1}{2} k_E I_a \qquad (6.7)$$

is for the angle $\Psi = 0°$, which means that $\delta = \phi$ (Fig. 6.2). In such a case, there is no demagnetizing component Φ_{ad} of the armature reaction flux and the air gap magnetic flux density takes its maximum value. The EMF E_f is high so it can better balance the input voltage V_1, thus minimizing the armature current I_a. When $\Psi \approx 0°$, the low armature current $I_a \approx I_{aq}$ is mainly torque producing. An angle $\Psi = 0°$ results in a decoupling of the rotor flux Φ_f and the armature flux Φ_a.

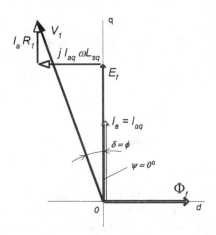

Fig. 6.2. Phasor diagram at $I_{ad} = 0$.

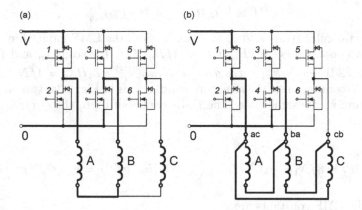

Fig. 6.3. IGBT inverter-fed armature circuits of PM LBMs: (a) Y-connected phase windings, (b) Δ-connected armature windings.

6.2.2 Square-Wave (Trapezoidal) Motors

PM LBMs predominantly have the reaction rail designed with surface PMs and large effective pole-arc coefficient $\alpha_i^{(sq)} = b_p/\tau$. The three-phase armature winding with distributed parameters can be Y, or Δ-connected. For the Y-connected winding, as in Fig. 6.3, two phases conduct the armature current at the same time.

For a.c. synchronous motors [70],

$$\mathbf{V}_1 = \mathbf{E}_f + \mathbf{I}_{ad}(R_1 + jX_{sd}) + \mathbf{I}_{aq}(R_1 + jX_{sq}) \tag{6.8}$$

Rectangular (trapezoidal) waveforms in the armature winding of an inverter-fed LBM correspond to the operation of a d.c. commutator motor. For a d.c. motor $\omega \to 0$, then eqn (6.8) becomes

$$V = E_f + RI_a^{(sq)} \tag{6.9}$$

where R is the sum of two-phase resistances in series (for Y-connected phase windings), and E_f is the sum of two-phase EMFs in series, V is the d.c. input voltage supplying the inverter, and $I_a^{(sq)}$ is the flat-top value of the square-wave current equal to the inverter input current.

For an ideal rectangular distribution of $B_{mg} = const$ in the interval of $0 \le x \le \tau$ or from 0 to 180 electrical degrees

$$\Phi_f^{(sq)} = L_i \int_0^\tau B_{mg} dx = \tau L_i B_{mg}$$

Including the pole shoe width $b_p < \tau$ and a fringing flux, the excitation flux is somewhat smaller:

$$\Phi_f^{(sq)} = b_p L_i B_{mg} = \alpha_i^{(sq)} \tau L_i B_{mg} \tag{6.10}$$

For a rectangular (trapezoidal) wave excitation, the EMF induced in a single turn (two conductors) is $2B_{mg}L_i v = 4fB_{mg}L_i\tau$. Including b_p and fringing flux, the EMF for $N_1 k_{w1}$ turns $e_f = 4fN_1 k_{w1}\alpha_i^{(sq)}\tau L_i B_{mg} = 4fN_1 k_{w1}\Phi_f^{(sq)}$. For the Y-connection of the armature windings, as in Fig. 6.3a, two phases are conducting at the same time. The EMF contributing to the electromagnetic power is

$$E_f = 2e_f = 8fN_1 k_{w1}\alpha_i^{(sq)}\tau L_i B_{mg} = \frac{4}{\tau}N_1 k_{w1}\Phi_f^{(sq)} v = c_{Edc}\Phi_f^{(sq)} v = k_{Edc} v \tag{6.11}$$

where the EMF constants are

$$c_{Edc} = \frac{4}{\tau}N_1 k_{w1} \tag{6.12}$$

and

$$k_{Edc} = c_{Edc}\Phi_f^{(sq)} \tag{6.13}$$

The electromagnetic thrust developed by the LSM is expressed by the equation

$$F_{dx} = \frac{P_g}{v} = \frac{E_f I_a^{(sq)}}{v} = \frac{4}{\tau}N_1 k_{w1}\Phi_f^{(sq)} I_a^{(sq)} = c_{Fdc}\Phi_f^{(sq)} I_a^{(sq)} = k_{Fdc} I_a^{(sq)} \tag{6.14}$$

in which

$$c_{Fdc} = c_{Edc}; \qquad k_{Fdc} = c_{Fdc}\Phi_f^{(sq)} \tag{6.15}$$

and $I_a^{(sq)}$ is the flat-top value of the phase current. Eqn (6.14) indicates that the thrust developed by the PM LBM can be controlled directly by varying the current.

For $v = v_s$ and $\Psi = 0°$ the ratio F_{dx} of a *square-wave motor* to F_{dx} of a *sinewave motor* is

$$\frac{F_{dx}^{(sq)}}{F_{dx}} = \frac{4\sqrt{2}}{\pi m_1}\frac{\Phi_f^{(sq)}}{\Phi_f}\frac{I_a^{(sq)}}{I_a} \tag{6.16}$$

For three-phase motors,

$$\frac{F_{dx}^{(sq)}}{F_{dx}} \approx 0.6\frac{\Phi_f^{(sq)}}{\Phi_f}\frac{I_a^{(sq)}}{I_a} \tag{6.17}$$

6.3 Model of PM Motor in dq Reference Frame

Control algorithms of sinusoidally excited synchronous motors frequently use the d-q model of synchronous machines. The d-q dynamic model is expressed in a *rotating reference frame* that moves at synchronous speed ω. The time-varying parameters are eliminated, and all variables are expressed in *orthogonal* or *mutually decoupled* d and q axes.

A synchronous machine is described by the following set of differential equations:

$$v_{1d} = R_1 i_{ad} + \frac{d\psi_d}{dt} - \omega\psi_q \tag{6.18}$$

$$v_{1q} = R_1 i_{aq} + \frac{d\psi_q}{dt} + \omega\psi_d \tag{6.19}$$

$$v_f = R_f I_f + \frac{d\psi_f}{dt} \tag{6.20}$$

$$0 = R_D i_D + \frac{d\psi_D}{dt} \tag{6.21}$$

$$0 = R_Q i_Q + \frac{d\psi_Q}{dt} \tag{6.22}$$

The linkage fluxes in the above equations are defined as

$$\psi_d = (L_{ad} + L_1)i_{ad} + L_{ad}i_D + \psi_f = L_{sd}i_{ad} + L_{ad}i_D + \psi_f \tag{6.23}$$

$$\psi_q = (L_{aq} + L_1)i_{aq} + L_{aq}i_Q = L_{sq}i_{aq} + L_{aq}i_Q \tag{6.24}$$

$$\psi_f = L_{ad}i_{ad} + (L_{ad} + L_{lf})i_{fd} + L_{ad}i_D \tag{6.25}$$

$$\psi_D = L_{ad}i_{ad} + (L_{ad} + L_D)i_D + \psi_f \tag{6.26}$$

$$\psi_Q = L_{aq}i_{aq} + (L_{aq} + L_Q)i_Q \tag{6.27}$$

where v_{1d} and v_{1q} are d- and q-axis components of terminal voltage, ψ_f is the maximum flux linkage per phase produced by the excitation system; R_1 is the armature winding resistance; L_{ad}, L_{aq} are d- and q-axis components of the armature self-inductance, respectively; $\omega = \pi v_s/\tau$ is the angular frequency of the armature current; τ is the pole pitch; v_s is the linear synchronous velocity; i_{ad}, i_{aq} are the d- and q-axis components of the armature current, respectively; i_D, i_Q are d- and q-axis components of the damper current, respectively. The field winding resistance, which exists only in the case of electromagnetic excitation, is R_f, the field excitation current is I_f, the excitation linkage flux is ψ_f, and the field leakage inductance is L_{lf}. The damper resistance and inductance in the d-axis are R_D and L_D, respectively. The damper resistance and inductance in the q-axis are R_Q and L_Q, respectively. The resultant armature inductances are

$$L_{sd} = L_{ad} + L_1, \qquad L_{sq} = L_{aq} + L_1 \qquad (6.28)$$

where L_{ad} and L_{aq} are self-inductances in the d and q-axis, respectively, and L_1 is the leakage inductance of the armature winding per phase. In a three-phase machine, $L_{ad} = (3/2)L'_{ad}$ and $L_{aq} = (3/2)L'_{aq}$, where L'_{ad} and L'_{aq} are self-inductances of a single-phase machine.

The excitation linkage flux in eqn (6.25) is $(L_{ad} + L_{lf})i_{fd}$, while $L_{ad}i_{ad}$ is the armature reaction flux. In the case of a PM field excitation, the fictitious current (equivalent MMF) is $I_f = H_c h_M$ and $i_{fd} = \sqrt{2}I_f$.

For no damper winding, i.e., $i_D = i_Q = 0$, the voltage equations in the d- and q-axis are

$$v_{1d} = R_1 i_{ad} + \frac{d\psi_d}{dt} - \omega\psi_q = \left(R_1 + \frac{dL_{sd}}{dt}\right) i_{ad} - \omega L_{sq} i_{aq} \qquad (6.29)$$

$$v_{1q} = R_1 i_{aq} + \frac{d\psi_q}{dt} + \omega\psi_d = \left(R_1 + \frac{dL_{sq}}{dt}\right) i_{aq} + \omega L_{sd} i_{ad} + \omega\psi_f \qquad (6.30)$$

The matrix form of voltage equations in terms of inductances L_{sd} and L_{sq} is

$$\begin{bmatrix} v_{1d} \\ v_{1q} \end{bmatrix} = \begin{bmatrix} R_1 + \frac{d}{dt}L_{sd} & -\omega L_{sq} \\ \omega L_{sd} & R_1 + \frac{d}{dt}L_{sq} \end{bmatrix} \begin{bmatrix} i_{ad} \\ i_{aq} \end{bmatrix} + \begin{bmatrix} 0 \\ \omega\psi_f \end{bmatrix} \qquad (6.31)$$

For the steady-state operation $(d/dt)L_{sd}i_{ad} = (d/dt)L_{sq}i_{aq} = 0$, $\mathbf{I}_a = I_{ad} + jI_{aq}$, $\mathbf{V}_1 = V_{1d} + jV_{1q}$, $i_{ad} = \sqrt{2}I_{ad}$, $i_{aq} = \sqrt{2}I_{aq}$, $v_{1d} = \sqrt{2}V_{1d}$, $v_{1q} = \sqrt{2}V_{1q}$, $E_f = \omega L_{fd}I_f/\sqrt{2} = \omega\psi_f/\sqrt{2}$ [60]. The quantities ωL_{sd} and ωL_{sq} are known as the d- and q-axis synchronous reactances, respectively. Eqn (6.31) can be brought to the form (3.32).

The instantaneous power input to the three-phase armature is

$$p_{in} = v_{1A}i_{aA} + v_{1B}i_{aB} + v_{1C}i_{aC} = \frac{3}{2}(v_{1d}i_{ad} + v_{1q}i_{aq}) \qquad (6.32)$$

The power balance equation is obtained from eqns (6.29) and (6.30), i.e.,

$$v_{1d}i_{ad} + v_{1q}i_{aq} = R_1 i_{ad}^2 + \frac{d\psi_d}{dt}i_{ad} + R_1 i_{aq}^2 + \frac{d\psi_q}{dt}i_{aq} + \omega(\psi_d i_{aq} - \psi_q i_{ad}) \quad (6.33)$$

The last term $\omega(\psi_d i_{aq} - \psi_q i_{ad})$ accounts for the electromagnetic power of a single-phase, two-pole synchronous machine. For a three-phase machine

$$p_{elm} = \frac{3}{2}\omega(\psi_d i_{aq} - \psi_q i_{ad}) = \frac{3}{2}\omega[(L_{sd}i_{ad} + \psi_f)i_{aq} - L_{sq}i_{ad}i_{aq}]$$

$$= \frac{3}{2}\omega[\psi_f + (L_{sd} - L_{sq})i_{ad}]i_{aq} \tag{6.34}$$

The electromagnetic thrust of a three-phase LSM is

$$F_{dx} = \frac{p_{elm}}{v_s} = \frac{3}{2}\frac{\pi}{\tau}[\psi_f + (L_{sd} - L_{sq})i_{ad}]i_{aq} \quad N \tag{6.35}$$

where $v_s = 2f\tau = \omega\tau/\pi$. Compare eqn (6.35) with eqn (3.41).

The relationships between i_{ad}, i_{aq} and phase currents i_{aA}, i_{aB}, and i_{aC} are,

$$i_{ad} = \frac{2}{3}\left[i_{aA}\cos\omega t + i_{aB}\cos\left(\omega t - \frac{2\pi}{3}\right) + i_{aC}\cos\left(\omega t + \frac{2\pi}{3}\right)\right] \tag{6.36}$$

$$i_{aq} = -\frac{2}{3}\left[i_{aA}\sin\omega t + i_{aB}\sin\left(\omega t - \frac{2\pi}{3}\right) + i_{aC}\sin\left(\omega t + \frac{2\pi}{3}\right)\right] \tag{6.37}$$

The reverse relations, obtained by simultaneous solution of eqns (6.36) and (6.37) in conjunction with $i_{aA} + i_{aB} + i_{aC} = 0$, are

$$i_{aA} = i_{ad}\cos\omega t - i_{aq}\sin\omega t$$

$$i_{aB} = i_{ad}\cos\left(\omega t - \frac{2\pi}{3}\right) - i_{aq}\sin\left(\omega t - \frac{2\pi}{3}\right) \tag{6.38}$$

$$i_{aC} = i_{ad}\cos\left(\omega t + \frac{2\pi}{3}\right) - i_{aq}\sin\left(\omega t + \frac{2\pi}{3}\right)$$

Including the *zero-sequence current*

$$i_0 = \frac{1}{3}(i_{aA} + i_{aB} + I_{aC}) \tag{6.39}$$

the relationship between $dq0$ and ABC currents can be written in the following matrix form:

$$\begin{bmatrix} i_{ad} \\ i_{aq} \\ i_0 \end{bmatrix} = \frac{2}{3}\begin{bmatrix} \cos(\omega t) & \cos\left(\omega t - \frac{2\pi}{3}\right) & \cos\left(\omega t + \frac{2\pi}{3}\right) \\ -\sin(\omega t) & -\sin\left(\omega t - \frac{2\pi}{3}\right) & -\sin\left(\omega t + \frac{2\pi}{3}\right) \\ \frac{1}{2} & \frac{1}{2} & \frac{1}{2} \end{bmatrix}$$

$$\times \begin{bmatrix} i_{aA} \\ i_{aB} \\ i_{aC} \end{bmatrix} = [B] \begin{bmatrix} i_{aA} \\ i_{aB} \\ i_{aC} \end{bmatrix} \tag{6.40}$$

where

$$[B] = \frac{2}{3} \begin{bmatrix} \cos(\omega t) & \cos\left(\omega t - \frac{2\pi}{3}\right) & \cos\left(\omega t + \frac{2\pi}{3}\right) \\ -\sin(\omega t) & -\sin\left(\omega t - \frac{2\pi}{3}\right) & -\sin\left(\omega t + \frac{2\pi}{3}\right) \\ \frac{1}{2} & \frac{1}{2} & \frac{1}{2} \end{bmatrix} \tag{6.41}$$

is the transformation matrix originally proposed by Blondel [57]. This is not a power invariant transformation matrix. The inverse transformation takes the form

$$\begin{bmatrix} i_{aA} \\ i_{aB} \\ i_{aC} \end{bmatrix} = \begin{bmatrix} \cos(\omega t) & -\sin(\omega t) & 1 \\ \cos\left(\omega t - \frac{2\pi}{3}\right) & -\sin\left(\omega t - \frac{2\pi}{3}\right) & 1 \\ \cos\left(\omega t + \frac{2\pi}{3}\right) & -\sin\left(\omega t + \frac{2\pi}{3}\right) & 1 \end{bmatrix} \begin{bmatrix} i_{ad} \\ i_{aq} \\ i_0 \end{bmatrix} = [B]^{-1} \begin{bmatrix} i_{ad} \\ i_{aq} \\ i_0 \end{bmatrix} \tag{6.42}$$

The inverse[2] of Blondel's transformation matrix

$$[B]^{-1} = \begin{bmatrix} \cos(\omega t) & -\sin(\omega t) & 1 \\ \cos\left(\omega t - \frac{2\pi}{3}\right) & -\sin\left(\omega t - \frac{2\pi}{3}\right) & 1 \\ \cos\left(\omega t + \frac{2\pi}{3}\right) & -\sin\left(\omega t + \frac{2\pi}{3}\right) & 1 \end{bmatrix} \tag{6.43}$$

Similar equations as (6.40) and (6.42) can be written for voltages and magnetic fluxes.

The transpose[3] of Blondel's transformation matrix (6.41)

$$[B]^T = \begin{bmatrix} \cos(\omega t) & -\sin(\omega t) & \frac{1}{2} \\ \cos\left(\omega t - \frac{2\pi}{3}\right) & -\sin\left(\omega t - \frac{2\pi}{3}\right) & \frac{1}{2} \\ \cos\left(\omega t + \frac{2\pi}{3}\right) & -\sin\left(\omega t + \frac{2\pi}{3}\right) & \frac{1}{2} \end{bmatrix} \tag{6.44}$$

The power input is proportional to the sum $v_{1A}i_{aA} + v_{1B}i_{aB} + v_{1C}i_{aC}$, i.e.,

$$p_{in} = v_{1A}i_{aA} + v_{1B}i_{aB} + v_{1C}i_{aC} = \begin{bmatrix} v_{1A} & v_{1B} & v_{1C} \end{bmatrix} \begin{bmatrix} i_{aA} \\ i_{aB} \\ i_{aC} \end{bmatrix} \tag{6.45}$$

Putting v_{1A}, v_{1B}, v_{1C} and i_{aA}, i_{aB}, i_{aC} from eqn (6.42), the power is

---

[2] The product $[A][A]^{-1} = [I]$, where $[A]^{-1}$ is the inverse matrix and $[I]$ is the square identity matrix. The identity matrix is a diagonal stretch of 1s going from the upper-left-hand corner to the lower right, with all other elements being 0.

[3] The transpose $[A]^T$ of matrix $[A]$ is formed by interchanging elements a_{ij} with elements a_{ji}.

$$p_{in} = \frac{3}{2}(v_{1d}i_{ad} + v_{1q}i_{aq} + 2v_0i_0) = \begin{bmatrix} v_{1d} & v_{1q} & 2v_0 \end{bmatrix} \begin{bmatrix} i_{ad} \\ i_{aq} \\ i_0 \end{bmatrix} \qquad (6.46)$$

According to Park [165], the proportionality coefficient must be 3/2 for any instant during normal operation at unity power factor.

6.4 Thrust and Speed Control of PM Motors

Thrust-speed envelopes of PM LSMs are classified into two categories: constant thrust (Fig. 6.4a) and constant power (Fig. 6.4b).

For *constant thrust* requirements, the thrust-speed envelope is of a rectangular shape. The maximum thrust F_{xmax} should be obtained at all linear speeds up to the speed v_b. Such envelope is required for linear servo drives and actuators.

For *constant power* requirements, the thrust-speed envelope is of a hyperbolic shape because $F_x v = P_{out} = const$. The constant power trajectory is maintained over a wide speed range from the base speed v_b to the maximum speed v_{max}. Such envelope is required for traction linear motors that are used in, e.g., linear-motor-driven vehicles. A constant power operation is implemented by weakening the excitation flux. Since PM motors do not have field excitation windings, the magnetic flux is weakened by applying appropriate control techniques. It can also be done by a hardware, i.e., using a hybrid excitation system that consists of PMs and additional excitation coils placed around PMs.

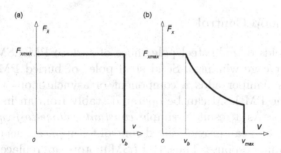

Fig. 6.4. Thrust-speed envelopes of LSMs: (a) constant thrust, (b) constant power.

Control algorithms are, in prinicple, similar for both sine-wave and square-wave PM motors. Fig. 6.5 shows how control loops can be developed to achieve thrust, speed and position control in motion control systems with PM LSMs. These cascaded control structures (Fig. 6.5) for LSMs have been based on control structures for rotary PM synchronous motors [98].

The following are some examples of control of PM LSMs and LBMs.

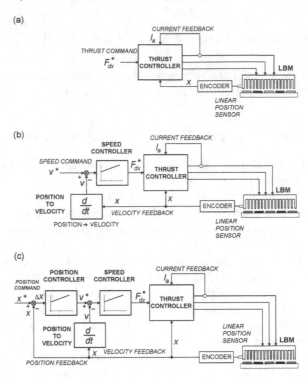

Fig. 6.5. Cascaded control loops for PM LSMs: (a) current-regulated thrust control, (b) velocity and thrust control, (c) position, velocity, and current control.

6.4.1 Open Loop Control

Aluminum shields or mild-steel pole shoes of surface PM LSMs are equivalent to damper cage windings. Solid steel poles of buried PMs behave also as dampers. A damper adds a component of asynchronous thrust production so that the PM LSM can be operated stably from an inverter without position sensors. As a result, a simple *constant voltage-to-frequency* control algorithm (Fig. 6.6) can provide speed control for applications that do not require fast dynamic response. Thus, PM LSM motors can replace LIMs in some variable-speed drives to improve the drive efficiency with minimal changes to the control electronics.

6.4.2 Closed Loop Control

To achieve *high performance motion control* with a PM LSM, a position sensor is typically required. The rotor position feedback needed to continuously perform the self-synchronization function for a sinusoidal PM motor is significantly more demanding than that for a square-wave motor. An absolute

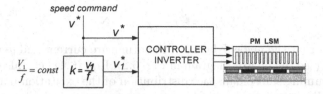

Fig. 6.6. Simplified block diagram of open loop *voltage-to-frequency* control of a PM LSM with damper.

encoder or resolver is typically required. The second condition for achieving high-performance motion control is high-quality phase current control.

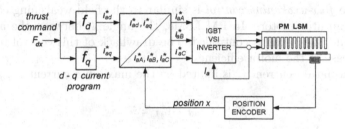

Fig. 6.7. Block diagram of high-performance thrust control scheme for PM LSM using vector control concept.

One of the possible approaches is the *vector control* shown in Fig. 6.7. The incoming thrust command F_{dx}^* is mapped into commands for i_{ad}^* and i_{aq}^* current components according to eqn (6.35).

The current commands in the moving reaction rail d-q reference frame (seen as d.c. quantities for a constant thrust command) are then transformed into the instantaneous sinusoidal current commands for the individual armature phases i_{aA}^*, i_{aB}^*, and i_{aC}^* using the reaction rail or armature position feedback and reverse Park's transformation equations [60, 96]. Current regulators for each of the three armature current phases then operate to excite the phase windings with the desired current amplitudes. The most common means of mapping the thrust command F_{dx}^* into values for i_{ad}^* and i_{aq}^* is to set a constraint of maximum *thrust-to-current* operation that is nearly equivalent to maximizing operating efficiency [97].

6.4.3 Zero Direct-Axis Current Control

To obtain the thrust proportional to the armature current $i_a = i_{aq}$ and free from the demagnetization of PMs, the PM LSM is driven by the d-axis current $i_{ad} = 0$ control, i.e.,

$$F_{dx} = \frac{3}{2}p\frac{\pi}{\tau}\psi_f i_{aq} \quad N \tag{6.47}$$

This means that the angle Ψ between the armature current and q-axis always remains $0°$ (see the phasor diagram in Fig. 6.2). Eqn (6.47) can also be simply derived assuming sinusoidal space distribution of the excitation magnetic flux density and sinusoidal time-varying armature currents including appropriate phase shifts.

6.4.4 Flux-Weakening Control

High coercivity rare-earth PMs are not affected by the armature reaction flux, and they cannot be permanently demagnetized by the armature flux. The d-axis reaction flux can be used for weakening the flux Φ_f produced by PMs. The *flux-weakening control* is similar to the field weakening control of d.c. commutator motors. In PM LSMs, the angle Ψ between the armature current and the q-axis is controlled. The drawback of this control technique is a decrease in the motor efficiency.

The armature current i_a is limited by the maximum current i_{amax}, i.e.,

$$i_a = \sqrt{i_{ad}^2 + i_{aq}^2} \leq i_{amax} \tag{6.48}$$

The maximum current i_{amax} is a continuous rated armature current for continuous duty cycle or a maximum available current of the inverter during short-time duty cycle [182].

The terminal voltage v_1 is limited by the maximum voltage v_{1max}, i.e.,

$$v_1 = \sqrt{v_{1d}^2 + v_{1q}^2} \leq v_{1max} \tag{6.49}$$

The maximum voltage v_{1max} is the maximum available output voltage of the inverter, which depends on the d.c. link voltage [182].

According to eqns (3.41) and (6.35), the thrust has two components: synchronous and reluctance thrust. Fig. 6.8 shows the thrust components of the LSM versus angle Ψ for the flux-weakening control. The synchronous thrust is proportional to $\cos\Psi$ with maximum at $\Psi = 0°$. The reluctance thrust takes its maximum value at $\Psi = 45°$. There is a critical angle Ψ_{cr} that corresponds to the maximum total thrust. The critical angle $\Psi > 0$ only if the reluctance thrust is produced. In an LSM with $X_{sd} \neq X_{sq}$ the maximum thrust for $\Psi_{cr} > 0$ is greater than the thrust for $i_{ad} = 0$ ($\Psi = 0^0$) control.

The thrust-speed characteristics of an LSM for $i_{ad} = 0$ control and flux-weakening control are shown in Fig. 6.9.

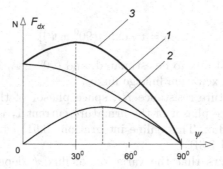

Fig. 6.8. Thrust plotted against the angle Ψ: 1 — synchronous thrust, 2 — reluctance thrust, 3 — total thrust.

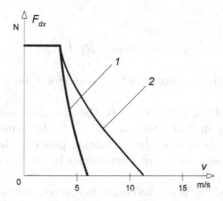

Fig. 6.9. Thrust–speed characteristics: 1 — $i_{ad} = 0$ control, 2 — flux-weakening control.

6.4.5 Direct Thrust Control

The basic idea of the *direct thrust control* (DTC) is to calculate the armature linkage flux directly from the electromagnetic induction law. The *space phasor* form of the armature voltage equation in the stationary reference frame is

$$\mathbf{v}_1 = R_1 \mathbf{i}_a + \frac{d\psi}{dt} \tag{6.50}$$

The armature linkage flux can be found as [115, 254]

$$\psi = \psi_0 + \int_{t_0}^{t_1} (\mathbf{v}_1 - R_1 \mathbf{i}_a) dt \tag{6.51}$$

Since the armature linkage flux is calculated as an integral, an initial value of the armature linkage flux ψ_0 is required. The initial value is

$$\psi_0 = \psi_f \exp[j(90^0 \pm \varPsi)]$$

where ψ_f is the PM flux to be estimated, and $(90° \pm \varPsi)$ is the angle between the armature and excitation linkage fluxes.

Only the armature resistance R_1, space phasor of the armature voltage vector $\mathbf{v}_1$ and space phasor of the armature current $\mathbf{i}_a$ is needed to find the armature linkage flux. This online integration (6.51) can be made with the aid of a high-speed DSP.

Eqn (6.51) shows that the variation of flux ψ depends on the voltage phasor $\mathbf{v}_1$. Thus, the flux can be controlled using the voltage phasor. In DTC, an optimum voltage phasor that makes the flux rotate and produce the desired thrust needs to be chosen.

Each voltage phasor is constant during each switching interval, and eqn (6.51) can be written as

$$\psi = \psi_0 + \mathbf{v}_1 t - R_1 \int_{t_0}^{t} \mathbf{i}_a dt \tag{6.52}$$

If R_1 is negligible, the armature linkage flux ψ will be following the voltage phasor $\mathbf{v}_1$.

In induction motors, when $\mathbf{v}_1 = 0$, the armature linkage flux ψ is in a stationary position. In PM LSMs at $\mathbf{v}_1 = 0$, the armature flux ψ is not stationary, because there is a relative motion between the armature and the excitation system. Zero voltage phasors cannot be used to control the rotation of ψ of PM LSMS.

The thrust of PM LSMs can be controlled by controlling the angle $(90^0 \pm \varPsi)$ between the armature and excitation linkage fluxes [115].

Depending on whether the actual thrust is smaller or larger than the reference thrust, the voltage phasors turn the linkage flux in the appropriate direction to increase or decrease the angle $(90^0 \pm \varPsi)$ and the thrust. The armature linkage flux always rotates in the direction determined by the output of the hysteresis controller of the thrust [115].

The linkage fluxes $\psi_d(k)$ and $\psi_q(k)$ at the kth sampling instant are

$$\psi_{dk} = \psi_{dk-1} + (v_{1dk-1} - R_1 i_{ad})t_s \tag{6.53}$$

$$\psi_{qk} = \psi_{qk-1} + (v_{1qk-1} - R_1 i_{aq})t_s \tag{6.54}$$

$$\psi_k = \sqrt{\psi_{dk}^2 + \psi_{qk}^2} \tag{6.55}$$

where the subscript $k - 1$ denotes previous samples, and t_s is the sampling period.

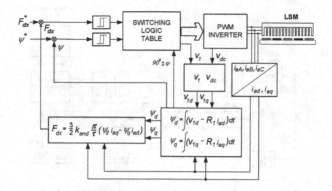

Fig. 6.10. Block diagram of direct thrust control of a PM LSM.

The electromagnetic thrust of an LSM is expressed by eqn (6.35). The end effect can be included by multiplying the thrust by a coefficient $k_{end} < 1$ that takes into account thrust reduction due to the end effect. Thus

$$F_{dx} = \frac{3}{2}pk_{end}\frac{\pi}{\tau}(\psi_d i_{aq} - \psi_q i_{ad}) \qquad (6.56)$$

The i_{ad} and i_{aq} currents are obtained from the measured three-phase currents i_{aA}, i_{aB}, i_{aC}, and the v_{1d} and v_{1q} voltages are calculated on the basis of the d.c. link voltage. The block diagram of the DTC control is shown in Fig. 6.10 [115].

6.4.6 Fuzzy Control

Fuzzy control of a PM LSM or LBM can be implemented according to the block diagram shown in Fig. 6.11. Two fuzzy controllers have been applied: for position and for speed control. A vector control is used in the power circuit with a current-controlled VSI.

6.5 Control of Hybrid Stepping Motors

6.5.1 Microstepping

When a stepping motor is driven in its *full-stepping* mode and two phases are energized simultaneously (Fig. 6.12a), the thrust available on each step is approximately the same. In the *half-stepping mode*, two phases and then only one are energized (Fig. 6.12b). If there is the same current in full-stepping and half-stepping modes, in each case a greater thrust is produced where two phases are energized simultaneously. This means that stronger thrust and weaker thrust are produced in each alternate step. The useful thrust is limited

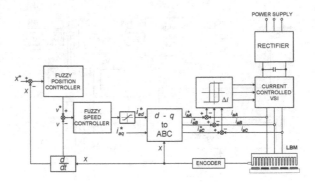

Fig. 6.11. Fuzzy control of a PM LSM or LBM.

by the weaker step; however, there will be a reduction of low-speed pulsations over the full-step mode.

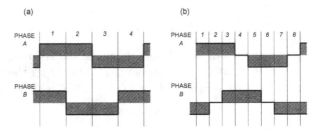

Fig. 6.12. Phase current waveforms: (a) full-stepping, two phases on, (b) half-stepping.

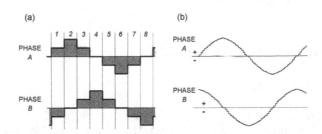

Fig. 6.13. Phase current waveforms to produce approximately equal thrust: (a) half-step current, profiled, (b) microstepping mode.

Approximately equal thrust in every step can be obtained by supplying a higher current when only one phase is energized (Fig. 6.13). The motor will not get overheated because it is designed to operate with two phases to be energized simultaneously. Since the winding losses are proportional to the current squared, approximately the same winding losses are dissipated if two phases are fed at the same time or if only one phase is fed with current increased by $\sqrt{2}$.

The same currents in both two phases produce an intermediate step equal to half of that for only one phase being energized. With unequal two-phase currents, the position of the platen will be shifted toward the stronger pole. This effect is utilized in the *microstepping* controller that subdivides the basic motor step by proportioning the current in the two-phase winding (Fig. 6.13b). Thus, the step size is reduced, and the operation at low speeds is very smooth. The motor in its microstepping mode operates similar to a two-phase synchronous motor and can even be driven directly from 50 or 60 Hz sinusoidal power supply, provided that a capacitor is connected in series with one phase.

The advantages of microstepping a linear stepping motor include [40]

- higher resolution for positioning accuracy,
- smoothness at low speeds,
- wide speed range,
- minimal force loss at resonant frequencies.

6.5.2 Electronic Controllers

To obtain the best performance at minimum settling time (Fig. 6.14), an *accelerometer feedback* is added, which provides electronic damping to the motor [40]. Accelerometer damping is recommended for HLSM applications that require

- very short settling time (Fig. 6.14),
- repetitive moves or periodic acceleration transients,
- maximum force utilization.

The block diagram of an HLSM controller is shown in Fig. 6.15. Power amplifiers of bipolar type are current controlled and use about 20 kHz fixed frequency and PWM. The resolution is from 50 to 125 microsteps per full step [40].

6.6 Precision Linear Positioning

Linear motors are now playing a key role in *advanced precision linear positioning* [22]. Linear *precision positioning systems* can be classified into open-loop

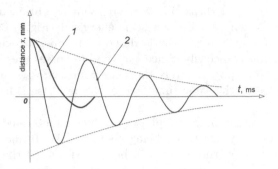

Fig. 6.14. HLSM settling time. 1 — with accelerometer, 2 — without accelerometer.

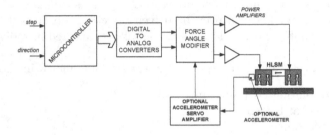

Fig. 6.15. Block diagram of an HLSM controller.

systems with HLSMs and closed-loop servo systems with LSMs, LBMs, or LIMs (Fig. 6.16).

A PM LBM (or LSM) driven *positioning stage* is shown in Fig. 6.17 [161]. A stationary base is made of aluminum, steel, ceramic, or granite plate. It provides a stable, precise, and flat platform to which all stationary positioning components are attached. The base of the stage is attached to the host system with the aid of mounting screws.

The *moving table* accommodates all moving positioning components. To achieve maximum acceleration, the mass of the moving table should be as small as possible, and usually aluminum is used as a lightweight material. A number of mounting holes on the moving table is necessary to fix the payload to the mounting table.

Linear bearing rails provide a precise guidance to the moving table. Minimum one bearing rail is required. Linear ball bearing or air bearings are attached to each rail.

The armature of a linear motor is fastened to the moving table and reaction rail (PM excitation system) is built in the base between the rails (Fig. 6.17).

A linear encoder is needed to obtain precise control of position of the table, velocity and acceleration. The readhead of the encoder is attached to the moving table.

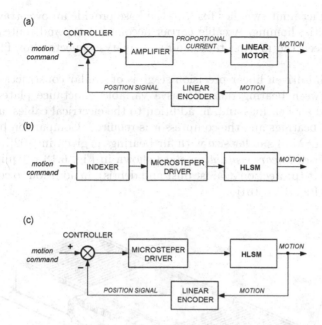

Fig. 6.16. Typical linear positioning systems: (a) closed-loop servo system with a.c. or d.c. linear motors, (b) open-loop positioning system with HLSMs, (c) closed-loop positioning system with HLSMs.

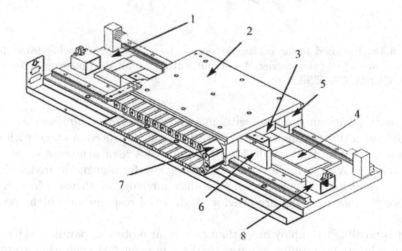

Fig. 6.17. Linear positioning stage driven by PM LBM: 1 — base, 2 — moving table, 3 — armature of LBM, 4 — PMs, 5 — linear bearing, 6 — encoder, 7 — cable carrier, 8 — limit switch. Courtesy of Normag, Santa Clarita, CA, USA.

Noncontact limit switches fixed to the base provide an over-travel protection and initial homing. A cable carrier accommodates and routes electrical cables between the moving table and stationary connector box fixed to the base.

An HLSM-driven linear precision stage is of similar construction. Instead of PMs between bearing rails, it has a variable reluctance platen. HLSMs usually need air bearings and, in addition to the electrical cables, an air hose between air bearings and the compressor is required. Comparison between an HLSM and LSM of similar size with air bearings is given in [102].

An *enclosed linear positioning stage* shown in Fig. 6.18 is equipped with bellows covers (protection against dust and debris), in addition to components sketched in Fig. 6.17 [161].

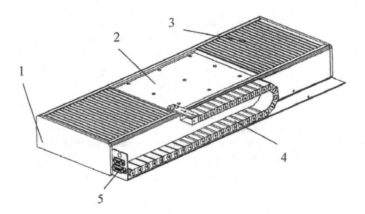

Fig. 6.18. Enclosed linear positioning stage. 1 — base, 2 — moving table, 3 — bellows cover, 4 — cable carrier, 4 — input/output terminals. Courtesy of Normag, Santa Clarita, CA, USA.

Linear positioning stages with moving *coreless armature windings* arranged vertically are shown in Fig. 6.19. Fig. 6.19a shows a stage with one moving armature [222], while Fig. 6.19b shows a *twin armature stage* [16]. A coreless moving armature does not have any ferromagnetic materials so that the positioning stage does not produce any cogging thrust. Moreover, a lightweight moving armature provides high-speed response and high acceleration.

It is possible to employ more than two linear motors in parallel. Fig. 6.20a shows a linear positioning stage with three moving PM excitation systems, while Fig. 6.20b shows a similar construction with four stationary PMs and three moving coreless armature windings [3]. Fig. 6.21 shows a *multilayer positioning stage* with five moving PM excitation systems [3].

The FEM modeling indicates that the magnetic flux density distribution along the air gap of multilayer LSMs is nonuniform [3]. Any nonuniformity

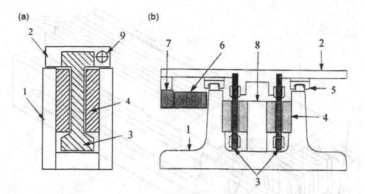

Fig. 6.19. Linear positioning stages with moving coreless armature windings: (a) single armature, (b) twin armature. 1 — base, 2 — moving table, 3 — armature of LBM, 4 — PMs, 5 — linear bearing, 6 — linear scale of encoder, 7 — readhead of encoder, 8 — yoke, 9 — cable.

in the magnetic field distribution causes different values of EMFs induced in coils that are distributed along the armature. Consequently, the current density distribution in the armature winding is also nonuniform, which can cause local overheating of the armature system.

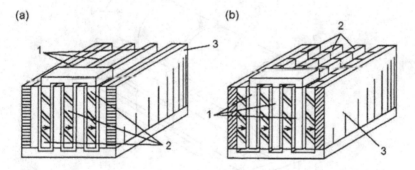

Fig. 6.20. Triple linear positioning stages with moving (a) PMs, (b) coreless armature windings. 1 — coreless armature winding, 2 — PMs, 3 — base. Courtesy of Technical University of Szczecin, Poland, and Institute of Electrodynamics of UNAS, Kiev, Ukraine.

Linear positioning stages are used in semiconductor technology, electronic assembly, quality assurance, laser cutting, optical scanning, water jet cutting, gantry systems (x, y, z stages), color printers, plotters, and Cartesian coordinate robotics [222].

Fig. 6.22 shows an application of a linear positioning stage, according to Fig. 6.19b, to a recorder [16]. The table driven by a twin armature LSM or

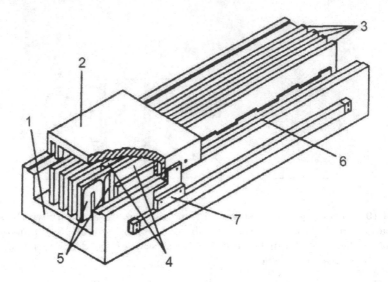

Fig. 6.21. Multilayer linear positioning stage. 1 — base, 2 — moving table, 3 — armature, 4 — PMs, 5 — armature coils, 6 — linear bearing, 7 — readhead of linear encoder. Courtesy of Technical University of Szczecin, Poland, and Institute of Electrodynamics of UNAS, Kiev, Ukraine.

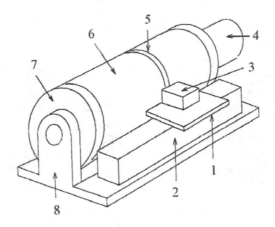

Fig. 6.22. Recorder with a linear positioning stage. 1 — moving table, 2 — base of positioning stage, 3 — recording head, 4 — rotary motor, 5 — recording track, 6 — recording film, 7 — rotary drum, 8 — pedestal. Courtesy of Hitachi Metals Ltd, Saitama, Japan.

LBM moves along the rotating drum. The optical recording head fixed to the moving table writes data on the track of the recording film. The mass of recording head is 2 kg, and the width of track is 1.4 mm. The recording head, while writing data, must keep constant position with high accuracy ± 1 μm. After writing, the head needs to move quickly to the next track and settle down within 20 ms. The maximum speed and maximum acceleration are 0.22 m/s and 44 m/s^2, respectively [16].

Examples

Example 6.1

Given are specifications of a 3-phase, 4-pole, single-sided LSM with surface configuration of PMs and the following specifications: $\tau = 40$ mm, $w_M = 36$ mm, $L_i = 80$ mm, $B_{mg} = 0.65$ T, $N_1 = 440$, $s_1 = 24$, $w_c = \tau$ (full-pitch coils, two-layer winding).

Find the EMF and electromagnetic thrust for sine-wave and square-wave control assuming that the rated armature current $I_a = I_a^{sq} = 8.0$ A and $f = 30$ Hz for both methods of control, and $\Psi = 0$ for the sine-wave motor.

Solution

The pole shoe width $b_p = w_M = 36$ mm, the number of slots per pole $Q_1 = 24/8 = 3$, the number of slots per pole per phase $q_1 = 24/(8 \times 3) = 1$, the winding factor calculated on the basis of eqns (2.39), (2.40), (2.41) $k_{d1} = \sin(\pi/(2 \times 3)/[1 \times \sin(\pi/(2 \times 3 \times 1)] = 1$, $k_{p1} = \sin(\pi \times 40/(2 \times 40) = 1$, $k_{w1} = 1 \times 1 = 1$, pole shoe width to pole pitch ratio according to eqn (3.11) $\alpha_i - 36/40 = 0.9$, and the synchronous speed $v_s = 2 \times 30 \times 0.04 = 2.4$ m/s. The number of coils per phase of a two-layer armature winding is $N_c = s_1/m_1 = 24/3 = 8$, and the number of turns per coil is $n_c = N_1/N_c = 440/8 = 55$.

Sine-wave motor

Form factor of the excitation field according to eqn (3.10)

$$k_f = \frac{4}{\pi} \sin\left(\frac{0.9\pi}{2}\right) = 1.258$$

Amplitude of the fundamental harmonic of magnetic flux density

$$B_{mg1} = k_f B_{mg} = 1.258 \times 0.65 = 0.817 \text{ T}$$

Fundamental harmonic of the magnetic flux according to eqn (3.26)

$$\Phi_{f1} = \frac{2}{\pi} 0.04 \times 0.08 \times 1.258 \times 0.65 = 1.665 \times 10^{-3} \text{ Wb}$$

It is assumed that $\Phi_f \approx \Phi_{f1} = 1.665 \times 10^{-3}$ Wb. Amplitude of the fundamental harmonic of EMF induced in a single conductor of the armature winding

$$E_{mf1} = B_{mg1}L_i v_s = 0.817 \times 0.08 \times 2.4 = 0.157 \text{ V}$$

Armature constant c_E according to eqn (6.2)

$$c_E = \frac{\pi}{0.04} \frac{1}{\sqrt{2}} 440 \times 1 = 24\ 435.9\ 1/\text{m}$$

Armature constant k_E according to eqn (6.3)

$$k_E = 24\ 435.9 \times 1.665 \times 10^{-3} = 40.691 \text{ Vs/m}$$

EMF per phase at synchronous speed $v_s = 2.4$ m/s

$$E_f = 40.691 \times 2.4 = 97.66 \text{ V}$$

The characteristic $E_f = f(v_s)$, neglecting the armature reaction and saturation, is a straight line. For example, at $v_s = 1$ m/s, $E_f = 40.69$ V, at $v_s = 2$ m/s, $E_f = 81.38$ V, at $v_s = 4.8$ m/s, $E_f = 195.3$ V, etc.
   Thrust constant c_F according to eqn (6.5)

$$c_F = 2c_E = 2 \times 24\ 435.9 = 48\ 871.7\ 1/\text{m}$$

Thrust constant k_F according to eqn (6.6)

$$k_F = 48\ 871.7 \times 1.665 \times 10^{-3} = 81.383 \text{ N/A}$$

Electromagnetic thrust at rated current $I_a = 8.0$ A according to eqn (6.7)

$$F_{dx} = \frac{3}{2} 81.383 \times 8.0 = 976.6 \text{ N}$$

Neglecting armature reaction and saturation, the characteristic $F_{dx} = f(I_a)$ is also a straight line. For $I_a = 2$ A, $F_{dx} = 244.15$ N, for $I_a = 6$ A, $F_{dx} = 732.44$ N, for $I_a = 1.2 \times 8.0$ A, $F_{dx} = 1172$ N, etc.

Square-wave motor

Magnetic flux calculated on the basis of eqn (6.10)

$$\Phi_f^{(sq)} = 0.9 \times 0.04 \times 0.08 \times 0.65 = 1.872 \times 10^{-3} \text{ Wb}$$

EMF constant c_{Edc} according to eqn (6.12)

$$c_{Edc} = \frac{4}{0.04} 440 \times 1 = 44\ 000\ 1/\text{m}$$

EMF constant k_{Edc} according to eqn (6.13)

$$k_{Edc} = 44\,000 \times 1.872 \times 10^{-3} = 82.368 \text{ Vs/m}$$

EMF at synchronous speed $v_s = 2.4$ m/s

$$E_f^{(sq)} = 44\,000 \times 2.4 = 197.68 \text{ V}$$

Please note that the above EMF is induced in two phases connected in series (two phases are conducting at the same time). For sine-wave motor, the EMF $E_f = 97.66$ is line-to-neutral EMF.

The thrust constants c_{Fdc} and k_{Fdc} are given by eqn (6.15), i.e.,

$$c_{Fdc} = c_{Edc} = 44\,000 \text{ 1/m}; \qquad k_{Fdc} = 44\,000 \times 1.872 \times 10^{-3} = 82.368 \text{ N/A}$$

Thrust at rated current $I_a = 8$ A according to eqn (6.14)

$$F_{dx}^{(sq)} = 82.368 \times 8.0 = 658.9 \text{ N}$$

Square-wave motor to sine-wave motor thrust ratio

$$\frac{F_{dx}^{(sq)}}{F_{dx}} = \frac{658.9}{976.6} = 0.675$$

Using eqn (6.17) for $I_a = I_a^{(sq)}$

$$\frac{F_{dx}^{(sq)}}{F_{dx}} \approx 0.6 \frac{1.872 \times 10^{-3}}{1.665 \times 10^{-3}} \approx 0.675$$

Example 6.2

For a 2nd order mass-spring-damper system described by eqn (1.26) find the transfer function.

Solution

Eqn (1.26) can be also written as

$$\ddot{x} + \frac{D_v}{m}\dot{x} + \frac{k_s}{m}x = \frac{k_s}{m}\frac{F_x(t)}{k}$$

Putting $D_v/m = 2\zeta\omega_n$, $k_s/m = \omega_n$, and $F_x(t)/k_s = y(t)$, the mass-spring-damper system equation can be written in terms of the damping factor ζ and undamped natural frequency ω_n, i.e.,

$$\ddot{x} + 2\zeta\omega_n\dot{x} + \omega_n^2 x = \omega_n^2 y(t)$$

Transfer functions are defined by the ratio between the Laplace transform of the input, i.e. the normalized force $y(t)$, and the output signal, i.e., the position of the mass $x(t)$. Taking the Laplace transform on both sides of the above equation

$$\mathcal{L}\{\ddot{x} + 2\zeta\omega_n\dot{x} + \omega_n^2 x\} = \mathcal{L}\{\omega_n y(t)\}$$

$$s^2 X(s) + 2\zeta\omega_n s X(s) + \omega_n^2 X(s) = \omega_n^2 Y(s)$$

$$\{s^2 + 2\zeta\omega_n s + \omega_n^2\} X(s) = \omega_n^2 Y(s)$$

the transfer function of the mass-spring-damper system is

$$T(s) = \frac{X(s)}{Y(s)} = \frac{\omega_n^2}{s^2 + 2\zeta\omega_n s + \omega_n^2}$$

The denominator of the transfer function is the same as the characteristic polynomial of eqn (1.26). The roots of the characteristic polynomial are called poles of the transfer function $T(s)$.

Example 6.3

A flat, single-sided, three-phase PM LSM has the following parameters: $R_1 = 1.6\ \Omega$, $X_{sd} = 5.0\ \Omega$, $X_{sq} = 4.9\ \Omega$ at $f = 50$ Hz. The axial length of the stack is 0.24 m, number of poles $2p = 8$, height of PM $h_M = 5$ mm, coercivity of PM $H_c = 850$ kA/m, EMF per phase $E_f = 80$ V. Find

(a) terminal voltage, input power, electromagnetic power, and electromagnetic thrust at $I_{ad} = -2.0$ A and $I_{aq} = 10$ A;
(b) terminal voltage, input power, electromagnetic power, and electromagnetic thrust as functions of I_{aq} for $I_a = 10$ A $= const$, $\phi_f = const$, and $1.0 \leq I_{aq} \leq 10.0$ A.

Assumption: LSM operates at steady-state conditions, i.e., $(d/dt)L_{sd}i_{ad} = (d/dt)L_{sq}i_{aq} = 0$.

Solution

(a) Terminal voltage, input power, electromagnetic power, and electromagnetic thrust at $I_{ad} = -2.0$ A and $I_{aq} = 10$ A

The angular frequency $\omega = 2\pi 50.0 = 314.16$ rad/s, pole pitch $\tau = 0.24/8 = 0.03$ m, linear synchronous speed $v_s = 2 \times 50.0 \times 0.03 = 3.0$ m/s, armature currents in $dq0$ reference frame $i_{ad} = \sqrt{2}(-2.0) = -2.828$ A, $i_{aq} = \sqrt{2}(10.0) = 14.142$ A, fictitious field excitation currents (equivalent MMF) of PM $I_f = 0.005 \times 850\ 000 = 4250$ A, and $i_f = \sqrt{2} \times 4250.0 = 6010.4$ A.

Synchronous inductances in the d- and q-axis

$$L_{sd} = \frac{5.0}{314.16} = 0.0159 \text{ H} \qquad\qquad L_{sq} = \frac{4.9}{314.16} = 0.0156 \text{ H}$$

Rotor linkage flux

$$\psi_f = \sqrt{2}\frac{80.0}{314.16} = 0.36 \text{ Wb}$$

Armature rms current

$$I_{ad} = \sqrt{(-2.0)^2 + 10^2} = 10.2 \text{ A}$$

Angle between armature current and q-axis

$$\Psi = \arccos\left(\frac{I_{aq}}{I_a}\right) = \arccos\left(\frac{10}{10.2}\right) = 11.21°$$

Voltages v_{1d} and v_{1q} in the d-q reference frame for steady-state conditions are calculated with the aid of eqn (6.31), in which $(d/dt)L_{sd}i_{ad} = (d/dt)L_{sq}i_{aq} = 0$, i.e.,

$$\begin{bmatrix} v_{1d} \\ v_{1q} \end{bmatrix} - \begin{bmatrix} R_1 & -X_{sq} \\ X_{sd} & R_1 \end{bmatrix}\begin{bmatrix} i_{ad} \\ i_{aq} \end{bmatrix} + \begin{bmatrix} 0 \\ \omega\psi_f \end{bmatrix}$$

$$= \begin{bmatrix} 1.6 & -4.9 \\ 5.0 & 1.6 \end{bmatrix}\begin{bmatrix} -2.828 \\ 14.142 \end{bmatrix} + \begin{bmatrix} 0 \\ 314.16 \times 0.36 \end{bmatrix} = \begin{bmatrix} -73.82 \\ 121.62 \end{bmatrix}$$

In phasor form with the d-q reference frame $V_{1d} = -73.82/\sqrt{2} = -52.2$ V, $V_{1q} = 121.62/\sqrt{2} = 86.0$ V, and the rms phase voltage $V_1 = \sqrt{(-52.2)^2 + 86.0^2} = 100.6$ V. Input power according to eqn (6.32)

$$p_{in} = \frac{3}{2}[(-73.82) \times (-2.828) + 86.0 \times 14.142] = 2893 \text{ W}$$

Electromagnetic power according to eqn (6.34)

$$p_{elm} = \frac{3}{2} \times 314.16[0.36 + (0.0159 - 0.0156)(-2.828)] \times 14.142 = 2394 \text{ W}$$

Electromagnetic thrust developed by the LSM is calculated on the basis of eqn (6.35), i.e.,

$$F_{dx} = \frac{2394}{3.0} = 798 \text{ N}$$

(b) Terminal voltage, input power, electromagnetic power, and electromagnetic thrust as functions of I_{aq} for $I_a = 10$ A $= const$ and $1.0 \leq I_{aq} \leq 10.0$ A

The d-axis current in phasor form as a function of I_{aq}

$$I_{ad}(I_{aq}) = \sqrt{I_a^2 - I_{aq}^2}$$

or

$$i_{ad}(I_{aq}) = \sqrt{2}I_{ad}(I_{aq})$$

The voltages in the d- and q-axis are calculated with the aid of eqn (6.31), i.e.,

$$v_{1d}(I_{aq}) = R_1 i_{ad}(I_{aq}) - X_{sq}\sqrt{2}I_{aq}$$

$$v_{1q}(I_{aq}) = X_{sd}i_{ad}(I_{aq}) + R_1\sqrt{2}I_{aq} + \omega\psi_f$$

Armature rms voltage

$$V_1(I_{aq}) = \frac{1}{\sqrt{2}}\sqrt{[v_{1d}(I_{aq})]^2 + [v_{1q}(I_{aq})]^2}$$

Table 6.1. Voltage, angle Ψ, input power, electromagnetic power, and thrust as functions of I_{aq} at $I_a = const$ and $\phi_f = const$. Example 6.2

I_{aq} A	I_{ad} A	V_1 V	Ψ degree	p_{in} W	p_{elm} W	F_{dx} N
1.0	9.95	131.81	84.26	723	243	81
2.0	9.80	132.32	78.46	966	486	162
3.0	9.54	132.50	72.54	1209	729	243
4.0	9.16	132.32	66.42	1451	971	324
5.0	8.66	131.73	60.0	1693	1213	404
6.0	8.0	130.66	53.13	1934	1454	485
7.0	7.14	128.95	45.57	2175	1695	565
8.0	6.0	126.32	36.87	2414	1934	645
9.0	4.36	121.98	25.84	2652	2172	724
10.0	0.0	107.78	0	2880	2400	800

Input power according to eqn (6.32)

$$p_{in}(I_{aq}) = \frac{3}{2}[v_{1d}(I_{aq})i_{ad}(I_{aq}) + v_{1q}(I_{aq})\sqrt{2}I_{aq}]$$

Electromagnetic power according to eqn (6.34)

$$p_{elm} = \frac{3}{2}\omega[\psi_f + (L_{sd} - L_{sq})i_{ad}(I_{aq})]\sqrt{2}I_{aq}$$

Angle between the current I_a and q axis

$$\Psi(I_{aq}) = \arccos\left(\frac{I_{aq}}{I_a}\right)$$

Electromagnetic thrust developed by the LSM is calculated on the basis of eqn (6.35), i.e.,

$$F_{dx}(I_{aq}) = \frac{p_{elm}(I_{aq})}{v_s}$$

The results of the calculations are listed in Table 6.1.

Example 6.4

For $f = 60$ Hz, $t = 0.006$ s, rms currents $I_{aA} = I_{aB} = I_{aC} = 10$ A, and rms voltages $V_{1A} = V_{1B} = V_{1C} = 230$ V, find the $dq0$ components of currents and voltages. Prove that the product of transformation matrices $[B][B]^{-1} = [I]$, where $[I]$ is the square identity matrix. Calculate the power in ABC and $dq0$ reference frames. The phase angle between the current and voltage is $30°$.

Solution

Angular frequency

$$\omega = 2\pi f = 2\pi 60 = 376.99 \text{ rad/s}$$

The angle Θ between A-phase and d-axis

$$\Theta = \omega t = 376.99 \times 0.006 = 2.262 \text{ rad} = 129.6°$$

Phase instantaneous currents

$$i_{aA} = \sqrt{2} \times 10\cos(129.6° - 30°) = -2.358 \text{ A}$$

$$i_{aB} = \sqrt{2} \times 10\cos(129.6° - 120° - 30°) = 13.255 \text{ A}$$

$$i_{aC} = \sqrt{2} \times 10\cos(129.6° + 120° - 30°) = -10.897 \text{ A}$$

Phase instantaneous voltages

$$v_{1A} = \sqrt{2} \times 230 \cos(129.6°) = -207.334 \text{ V}$$

$$v_{1B} = \sqrt{2} \times 230 \cos(129.6° - 120°) = 320.714 \text{ V}$$

$$v_{1C} = \sqrt{2} \times 230 \cos(129.6° + 120°) = -113.38 \text{ V}$$

Blondel transformation matrix given by eqn (6.41)

$$[B] = \frac{2}{3} \begin{bmatrix} \cos(\Theta) & \cos\left(\Theta - \frac{2\pi}{3}\right) & \cos\left(\Theta + \frac{2\pi}{3}\right) \\ -\sin(\Theta) & -\sin\left(\Theta - \frac{2\pi}{3}\right) & -\sin\left(\Theta + \frac{2\pi}{3}\right) \\ \frac{1}{2} & \frac{1}{2} & \frac{1}{2} \end{bmatrix} = \begin{bmatrix} -0.425 & 0.657 & -0.232 \\ -0.514 & -0.111 & 0.625 \\ 0.333 & 0.333 & 0.333 \end{bmatrix}$$

Inverse of matrix $[B]$ according to eqn (6.43)

$$[B]^{-1} = \begin{bmatrix} \cos(\Theta) & -\sin(\Theta) & 1 \\ \cos\left(\Theta - \frac{2\pi}{3}\right) & -\sin\left(\Theta - \frac{2\pi}{3}\right) & 1 \\ \cos\left(\Theta + \frac{2\pi}{3}\right) & -\sin\left(\Theta + \frac{2\pi}{3}\right) & 1 \end{bmatrix} = \begin{bmatrix} -0.637 & -0.771 & 1 \\ 0.986 & -0.167 & 1 \\ -0.349 & 0.937 & 1 \end{bmatrix}$$

Matrix $[B]$ multiplied by matrix $[B]^{-1}$ must give a square diagonal identity matrix, i.e.,

$$[B][B]^{-1} = \begin{bmatrix} -0.425 & 0.657 & -0.232 \\ -0.514 & -0.111 & 0.625 \\ 0.333 & 0.333 & 0.333 \end{bmatrix} \begin{bmatrix} -0.637 & -0.771 & 1 \\ 0.986 & -0.167 & 1 \\ -0.349 & 0.937 & 1 \end{bmatrix} = \begin{bmatrix} 1 & 0 & 0 \\ 0 & 1 & 0 \\ 0 & 0 & 1 \end{bmatrix}$$

Currents in $dq0$ reference frame

$$\begin{bmatrix} i_{ad} \\ i_{aq} \\ i_0 \end{bmatrix} = [B] \begin{bmatrix} i_{aA} \\ i_{aB} \\ i_{aC} \end{bmatrix} = \begin{bmatrix} -0.425 & 0.657 & -0.232 \\ -0.514 & -0.111 & 0.625 \\ 0.333 & 0.333 & 0.333 \end{bmatrix} \begin{bmatrix} -2.358 \\ 13.255 \\ -10.897 \end{bmatrix} = \begin{bmatrix} 12.247 \\ -7.071 \\ 0.0 \end{bmatrix}$$

Voltages in $dq0$ reference frame

$$\begin{bmatrix} v_{1d} \\ v_{1q} \\ v_0 \end{bmatrix} = [B] \begin{bmatrix} v_{1A} \\ v_{1B} \\ v_{1C} \end{bmatrix} = \begin{bmatrix} 0.425 & 0.657 & -0.232 \\ -0.514 & -0.111 & 0.625 \\ 0.333 & 0.333 & 0.333 \end{bmatrix} \begin{bmatrix} -207.334 \\ 320.714 \\ -113.38 \end{bmatrix} = \begin{bmatrix} 325.269 \\ 0.0 \\ 0.0 \end{bmatrix}$$

If ABC current and voltage systems are balanced, the currents and voltages in $dq0$ reference frame are independent of time.

Transformation from $dq0$ to ABC frame will confirm the correctness of calculations, i.e.,

$$\begin{bmatrix} i_{aA} \\ i_{aB} \\ i_{aC} \end{bmatrix} = [B]^{-1} \begin{bmatrix} i_{ad} \\ i_{aq} \\ i_0 \end{bmatrix} = \begin{bmatrix} -0.637 & -0.771 & 1 \\ 0.986 & -0.167 & 1 \\ -0.349 & 0.937 & 1 \end{bmatrix} \begin{bmatrix} 12.247 \\ -7.071 \\ 0.0 \end{bmatrix} = \begin{bmatrix} -2.358 \\ 13.255 \\ -10.897 \end{bmatrix}$$

$$\begin{bmatrix} v_{1A} \\ v_{1B} \\ v_{1C} \end{bmatrix} = [B]^{-1} \begin{bmatrix} v_{1d} \\ v_{1q} \\ v_0 \end{bmatrix} = \begin{bmatrix} -0.637 & -0.771 & 1 \\ 0.986 & -0.167 & 1 \\ -0.349 & 0.937 & 1 \end{bmatrix} \begin{bmatrix} 325.269 \\ 0 \\ 0 \end{bmatrix} = \begin{bmatrix} -207.334 \\ 320.714 \\ -113.38 \end{bmatrix}$$

The power on the basis of eqn (6.45)

$$p_{in} = (-207.334) \times (-2.358) + 320.714 \times 13.255 + (-113.38) \times (-10.897) = 5975.57 \text{ VA}$$

The power on the basis of eqn (6.46)

$$p_{in} = \frac{3}{2}[325.269 \times 12.247 + 0.0 \times (-7.071) + 2 \times 0.0 \times 0.0] = 5975.57 \text{ VA}$$

Both equations give the same results. For balanced current and voltage systems in ABC reference frame, the power is independent of time.

7

Sensors

7.1 Linear Optical Sensors

The encoder functioning as a feedback device is one of the basic components of motion control systems. There are four different sensor technologies used in linear servo applications, i.e., resistive, inductive, magnetic, and optical. The least complicated structure has a linear position transducer. Its principles of operation are based on the conversion of linear motion into rotation that is measured with the use of a precision potentiometer, tachometer, or digital encoder. However, the earliest linear encoders utilized in high-precision machines, e.g., in metal-cutting industry, were optical. Although other techniques are also now available, optical encoders are still predominant in industrial applications. In linear motor drives, where precision actuation and measurement are involved, most designers employ an incremental optical encoder as a well-accepted part of the electromechanical drive system. The typical optical encoder makes use of a graduated *scale* that is scanned by a movable optical *readhead*. The most important advantage of optical encoders is their easily achievable noncontact operation that eliminates friction and wear and permits reliable high-speed performance in workshop environments. Linear optical encoders are capable of achieving very *high resolution*, in some cases comparable to the *laser interferometry* technology. Their accuracy is a few orders of magnitude higher than that of similar resistive, magnetic, or inductive linear encoders. This is possible due to the superior precision of *interpolation* performed on much smaller-scale grating periods. The interpolation is a self-subdivision process of the signal representing the scale period.

7.1.1 Incremental Encoders

There are two basic methods of generating optical encoder signals. In the first method, the *transmitted light* is processed (Fig. 7.1a), while the second method employs *reflected light* (Fig. 7.1b).

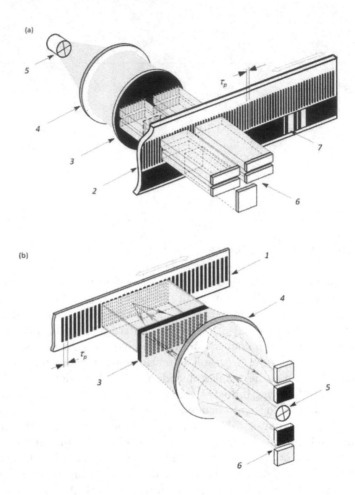

Fig. 7.1. Typical scanning methods: (a) transmitted light method, (b) reflected light method. 1 — scale reflective tape, 2 — scale transparent glass, 3 — scanning reticle, 4 — condenser lens, 5 — light source (LED), 6 — photodetectors, 7 — reference mark.

The simplest configuration of an optical encoder is described here. The light emitted by an LED either travels through, or reflects off, the scale. It is then directed through an identical index grating and onto photo detectors that generate electrical currents. When the scanning units move, the scale modulates the light, producing sinusoidal outputs from the sensors. There are five windows in the scale reticle. Four of them are phase-shifted 90° apart. The readhead's electronics combines the phase-shifted signals to produce two sinusoidal outputs. Twin signals s_1 and s_2, 90° out of phase, representing sine and cosine waveforms are generated. A fifth window on the scanning reticle has a random pattern graduation that creates a reference signal when aligned with identical pattern on the scale. The simplified optical incremental encoder consisting of light source, glass scale, scanning reticle, and two photodetectors shown in Fig. 7.2 illustrates the principles of forming s_1 and s_2 signals. The photo detector reads the maximum luminous intensity when transparent slots of the scale fully align with transparent slots of the scanning reticle. The light source and the photo detectors move along the glass scale grating. In consequence, the transparent slots of the scanning reticle change periodically their position relative to the stationary slots of the scale. Therefore, the light intensity detected by the photo sensors (photo elements) changes its value from maximum to zero according to a sinusoidal function (Fig. 7.2). Because the photo elements are displaced by the distance equal to one quarter of the scale grating period τ_p, when one photo element detects the maximum, the other one reads only half of the maximum. This displacement effectively shifts the two signals s_1 and s_2 by 90° in phase within the frequency domain. The 90° electrical separation or one quarter of the period between the two signals is referred to as the *quadrature*. Signals in the quadrature permit determination of the motion direction and speed at the same time allowing additional resolution through *edge counting*.

Linear motors employed in the x-y positioning stages and used in harsh factory or workshop environments require precise resolution with high reliability. The combination of a reflective, flexible scale tape placed along the track, and the readhead moving over the tape, offers many unique features.

The scale is made out of steel ribbon 5 to 10 mm wide and 0.2 mm thick, which has relatively low stiffness. Other materials, such as glass, mylar, or nonferrous metal tapes, can also be used. The scale is grated with alternating reflective strips (often made out of gold) and light-absorbing spaces. The grating period ranges from 100 to less than 20 μm, and after interpolation, resolution up to 0.1 μm is possible. The scale can be secured to the most commonly used materials (metals, composites, and ceramics) by means of a double-sided, elastic adhesive tape to accommodate the thermal expansion of the base. However, the mounting surface should be relatively smooth, clean, and parallel to the axis of motion with the scale ends rigidly fixed to the axis of the substrate. The location of stationary scale and moving readhead mounted on a positioning stage is shown in Fig. 7.3. The differential movement between the scale and the substrate should be close to zero, even in the presence of large temperature gradients. Usually, the scale tape is protected by varnish coating to facilitate easy cleaning. Scale tapes are generally supplied on a reel for 'cut–to–suit' convenience. For comparison, glass scales reach maximum length of 3 m in a single piece. Usually, the incremental tapes are installed together with the reference marks. Limit switches are separately installed next to the scale itself. These home position and/or zero point indicators for the end of travel are also sensed by the scanning readhead. The signals are synchronized with the incremental channels to guarantee repeatability.

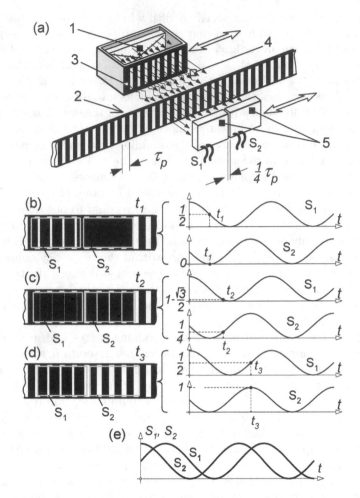

Fig. 7.2. Generation of photo detector signals: (a) photoelectric scanning setup; (b), (c), (d) examples of relative scale grating and scanning reticle positions; (e) corresponding signal variation. 1 — light source (LED), 2 — glass scale grating, 3 — scanning reticle, 4 — light rays, 5 — photodetectors.

The principle of the scanning process can be exemplified by the LIDA linear encoder, which was introduced to the market in 1977 by Heidenhain GmbH, Traunreut, Germany [84]. The operation of the LIDA encoder is explained in Fig. 7.1b. The redhead travels along the scale, while a light beam emitted by an LED source is directed onto the incremental scale grating through a condenser lens and scanning reticles. Then, it is reflected, and after passing back through the reticles, it is focused onto the photoelectric cells. Photo sensors detect changes in the light intensity caused by the interaction between the scale and reticle gratings. The four sinusoidal scanning signals,

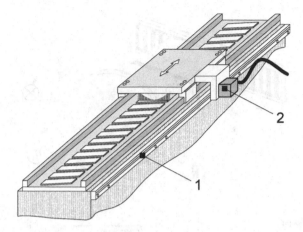

Fig. 7.3. Typical location of the linear encoder installed in the positioning stage driven by a linear motor. 1 — scale, 2 — readhead.

corresponding to the changes in the light intensity, are produced by the sensors. These waveforms with 90° phase shift enable formulation of symmetrical encoder zero output signals.

To improve the accuracy, especially in high-precision positioning applications within microelectronics industry, encoders frequently use interferential scanning principles. The diffracted (interferential) light method is required in the optical encoders with grating periods $\tau_p < 8$ μm. These devices employ a reflection-type diffraction grating fixed to the carrier (Fig. 7.4). An infrared LED emits light onto angular scale facets where it is scattered back into the readhead through the transparent grating. The periodic pattern on the scale and the periodic indexing of the grating produce sinusoidal interference fringes at the photodetector plane. The fringes move across the detector plane as the readhead moves along the scale. The arrangement of interlaced groups of photodetectors positioned in repeating patterns generates electric signals related to the fringe movement. The readhead electronics processes these signals and generates two sinusoidal waveforms of equal amplitude with the phase shift of 90°. In RGH encoders manufactured by Renishaw plc, the signal is averaged from over 80 facets in the detector plane. Therefore, the loss of a number of scale facets has only a marginal effect on the signal's amplitude and does not affect the counting process. Furthermore, because the signal is often subjected to disturbing effects (contamination or minor damage to the scale), the filtering and averaging process ensures its stability. In essence, the electronics within the readhead eliminates signals that do not match the scale period of 20 μm.

The electronics embedded in the readhead converts scanned incremental signals into analog or digital sinusoidal waveforms in the quadrature (Fig.

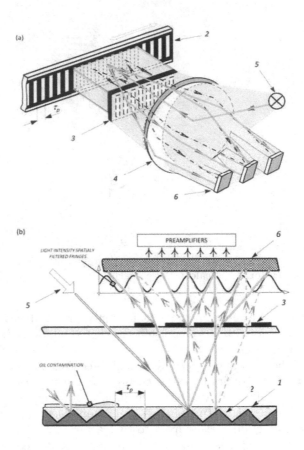

Fig. 7.4. Interferential measuring: (a) photoelectronic scanning, (b) optical filtering principle. 1 — scale, 2 — scale facets, 3 — phase gratings (readhead window), 4 — condenser lens, 5 — oblique illumination from LED, 6 — photodetectors.

7.5c). The signal period is equal to the scale pitch. The wave formats follow industry standard outputs: microcurrent (in μA) or voltage (1 V peak-to-peak). The readhead generates incremental square pulse trains in the quadrature, which conform to the standard EIA/RS422 differential line drive output (Fig. 7.5b). These fine-resolution digital waveforms are obtained by the subdivision of the analog signal passed from the readhead optics. In this context, the resolution is defined as the distance between consecutive edges of the digitized pulse trains. Commercially available readheads typically achieve the resolution of 5.0, 1.0, 0.5, or 0.1 μm. The readhead interpolation is ratiometric, i.e., it is independent of the signal amplitude.

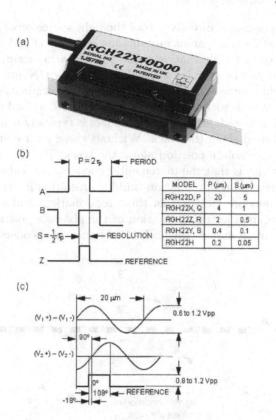

Fig. 7.5. Incremental readhead manufactured by Renishaw: (a) RGH22X readhead, (b) incremental 2 channels A and B in quadrature produced by digital readhead type, (c) incremental 2 channels V_1 and V_2 differential sinusoids in quadrature produced by analog readhead type. Courtesy of Renishaw plc, Gloucestershire, UK.

For digital output readheads, the recommended counterclock frequency for a given traversing speed is

$$f = \frac{v_{tr}}{s_r} k_{sf} \tag{7.1}$$

where v_{tr} is the traversing speed, s_r is the readhead resolution, f is the counterclock frequency of interpolation electronics, and k_{sf} is the safety factor, typically $k_{sf} = 4$. If v_{tr} is in m/s and s_r is in μm, the counterclock frequency f is in MHz.

In less demanding applications that do not require high resolution (e.g., material handling or distribution centers), the parts location is defined by simple optical encoders rather than sophisticated scanning interferometers for both simplicity and cost reduction. In that case, the positioning relies on the counting of the square pulse sequence. The pulses are produced by the optical sensor positioned directly across the light source separated by a linear array of alternating transparent and opaque windows (Fig. 7.6). As the light source and detector (or alternatively, the window array strip) move, the output signal from the detector is switched sequentially ON and OFF. Addition of a second light source and detector set affords determination of movement direction. For systems with long travel displacements, a third light source and sensor set equipped with incremental encoders is typically utilized to indicate markers distributed along the track. Without these road markers, it is difficult to define an absolute position along the track. The disadvantage of this approach, however, is that the incremental encoders are vulnerable to power interrupts, noise, and contamination buildup, resulting in erroneous position information. Therefore, the need for these road markers and associated external counters required for determination of the absolute position between the markers constitute a major drawback of incremental encoders.

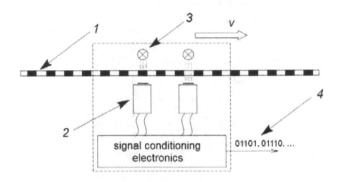

Fig. 7.6. Simplified incremental optical encoder arrangement. 1 — scale grating, 2 — detector ON/OFF, 3 — light source, 4 — counting bits.

Specifications of incremental self-adhesive scale RGS20-S produced by Renishaw plc, Gloucestershire, UK, is shown in Table 7.1. Dimensions of similar scale RGSZ20-S are shown in Fig. 7.7. Specifications of analog and digital readheads RGH series used in conjunction with RGS-S tape are presented in Table 7.2 and Table 7.3. The edge separation characteristics typical of digital readheads are shown in Fig. 7.8.

The optical readhead working with reflective scale of an incremental optical encoder is shown in Fig. 7.9.

The commonly used scale RGS20-S and one of the RGH22 series readheads (Fig. 7.10) comprise the RG2 system, i.e., non-contact, optical encoder designed for position feedback solutions (Renishaw). The readhead can be chosen with either sinusoidal or square wave output. The selected type depends on the application, electrical interfacing, and required resolution. Typically, the RG2 encoders are employed in' linear motor-driven machines such as tool presetters, measuring and layout equipment, and other high-speed systems

Table 7.1. Self-adhesive scale RGS20-S for RG2 encoder system manufactured by Renishaw plc, Gloucestershire, UK

Parameter	Specification
Scale type	Reflective gold-plated steel tape with lacquer coating and self-adhesive backing
Scale pitch	20 μm
Available lengths	Continuous length up to 50 m Longer than 50 m by special order
Measuring lengths	User selectable "cut–to–requirements" at the place of installation
Accuracy	Typical 15 μm/m without compensation
Linearity	$\pm$3 μm/m, $\pm$0.75 μm/60mm
Substrate materials	Metals, ceramics, and composites with expansion coefficient less than 22μm/m/$^\circ$C
Reference mark	Magnetic actuator adhesive- or screw-mounted. One or more at user-selected locations. Repeatability of position within temperature range $\pm$10°C from installation.

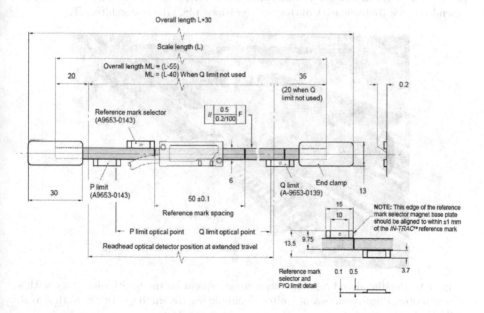

Fig. 7.7. Dimensions of RGSZ20-S scale and associated components. Courtesy of Renishaw plc, Gloucestershire, UK.

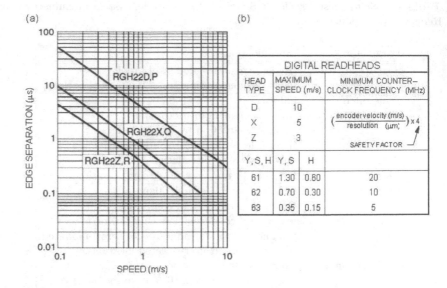

Fig. 7.8. RGH digital readheads characteristics: (a) edge separation, (b) recommended clock frequencies. Courtesy of Renishaw plc, Gloucestershire, UK.

Fig. 7.9. Readhead and reflective scale arrangement on the positioning stage with a linear motor. Photo courtesy of United Technologies Research Center, East Hartford, CT, USA.

Table 7.2. Analog readheads manufactured by Renishaw plc, Gloucestershire, UK

Parameter	Specifications	
Type	RGH22C 12 μA differential	RGH22B 1 V (peak–to–peak) differential
Signals	Incremental 2 channels I_1 and I_2 differential sinusoids in quadrature (90° phase shift). Signal period 20 μm	Incremental 2 channels V_1 and V_2 differential sinusoids in quadrature (90° phase shift). Signal period 20 μm
Output	7 to 16 μA	0.6 to 1.2 V (peak-to-peak)
Reference	Differential pulse 10 μA Duration 126°	Differential pulse Duration 126°
Power supply	5V ±5%, 120 mA (typical)	
Speed	1m/s at 50 kHz maximum	5m/s at 250 kHz maximum
Temperature	−20 to +70°C storage 0 to +55°C operating	
Humidity	10 to 90% RH noncondensing	
Sealing	IP54	
Operating acceleration	30 g	
Shock acceleration	100g (11 ms, one half of sinusoid)	
Vibration under operation	10 g at 55 to 2000 Hz	
Mass	Readhead: 45 g, Cable: 32 g/m	
Cable	Available lengths 0.5, 1.0, 1.5, 3.0, and 5.0 m Flexible life > 10^7 cycles at 50 mm bend radius for integral cable and > 10^6 cycles at 75 mm bend radius for extension cable 14 core, double shield, outer diameter 7.2 mm	

in which interpolation is provided by subsequent electronics. In the environments subjected to severe radio frequency interference (RFI), the RGH22B readhead with analog differential output voltage is preferred to the RGH22C model having the output current signal (Fig. 7.10).

The x-y motion stages applied in clean-room environments, e.g., semiconductor industry or ultra-precision machine tools such as grinders for ferrite components and diamond lathes for optics, are equipped with integrated two-coordinate encoders. The incremental x-y encoder contains a 2D phase-grating structure on a glass substrate (Fig. 7.11). Specifications of the two-coordinate PP 281 R encoder manufactured by Heidenhain, GmbH, Traunreut, Germany, are listed in Table 7.4. The measurement in a plane is possible through an in-

Table 7.3. Specifications of digital readhead manufactured by Renishaw plc, Gloucestershire, UK

Parameter	Specifications
Output signal	Square differential line driver to EIA RS422 Incremental channels A and B in quadrature (90° phase shift)
Signal period	20 μm for D type 4 μm for X type 2 μm for Z type 0.4 μm for Y type Resolution for all models 0.25×period
Alarm signal	Separate alarm channel or three state alarm Incremental channels force an open circuit for reliable operation when signal is too low
Power supply	5V ±5%, 120 mA (typical) 150 mA for Y type only
Operating acceleration	30g
Shock acceleration	100g (11 ms, 1/2 sine)
Vibration under operation	10g at 55 to 2000Hz (ICE 68-2-6)
Temperature	−20 to +70°C storage 0 to +55°C operating
Humidity	10 to 90% RH non-condensing
Mass	Readhead: 45 g, Cable: 32 g/m
Signal terminations	Standard RS422A line receiver circuitry RC filter is recommended Resistance 120 Ω Capacitor 4.7 nF for cable length < 25 m and 10 nF for cable length > 25 m
Cable	Available lengths 0.5, 1.0, 1.5, 3.0 and 5.0 m Flexible life > 10^7 cycles at 50 mm bend radius for integral cable and > 10^6 cycles at 75 mm bend radius for extension cable 14 core, double shield, outside diameter 7.2 mm

terferential scanning method. Two reference marks, one in each measurement direction, serve to define accurately the zero positions. The 8 μm grating period with fine interpolation and high uniformity of scanning is capable of 10 nm resolution.

Table 7.4. Specifications of PP 281 R two-coordinate incremental encoder manufactured by Heidenhain, GmbH, Traunreut, Germany.

Parameter	Specification
Grating period	$\tau_p = 8\ \mu\text{m}$
Coefficient of thermal expansion	8 pikomillimeter/K
Accuracy	$\pm 1\ \mu\text{m}$
Measuring range	68 mm × 68 mm; other ranges also available
Vibration	$< 80\ \text{m/s}^2$ at 55 to 2000 Hz
Shock vibration	$< 100\ \text{m/s}^2$ (11 ms)
Operating temperature	0 to +50°C
Mass of grid plate	75 g
Mass of APE and cable	120 g
Mass of scanning head	170 g
Power supply	5 V ±10%, 100 mA (without load)
Output signal	1 V (peak-to-peak)
Signal period	$4\ \mu\text{m}$

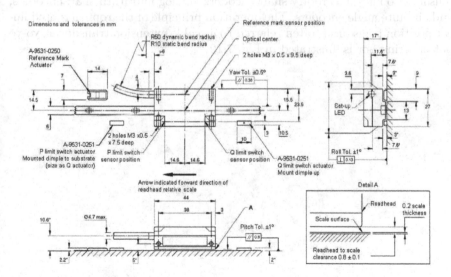

Fig. 7.10. Dimensions of the analog readhead RGH. Courtesy of *Renishaw plc*, Gloucestershire, UK.

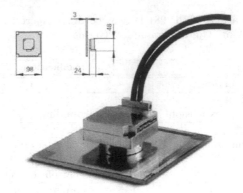

Fig. 7.11. Two-coordinate incremental encoder PP281R. Photo courtesy of Heidenhain, GmbH, Traunreut, Germany.

7.1.2 Absolute Encoders

The obvious method of measuring linear position, velocity, or both is conversion of linear movement into rotary motion. The rotation is measured by angle sensors also known as rotary shaft encoders, analog multiturn shaft encoders, and absolute angle encoders. The operation principle of the rope-actuated linear position transducer, often referred to as cable extension transducer, yo-yo pot or string pot is illustrated in Fig. 7.12.

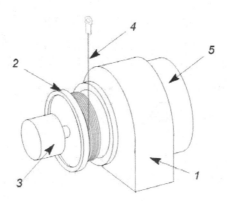

Fig. 7.12. Cable extension transducer. 1 — base, 2 — capstan, 3 — sensor, 4 — stainless steel wire rope, 5 — tension spring.

The transducer is fixed with the extensible wire rope attached to a movable object. As a result of object movement, the rope extends, rotating an internal transducer capstan. The sensing device then generates an electrical output signal proportional to the wire rope extension and/or to its veloc-

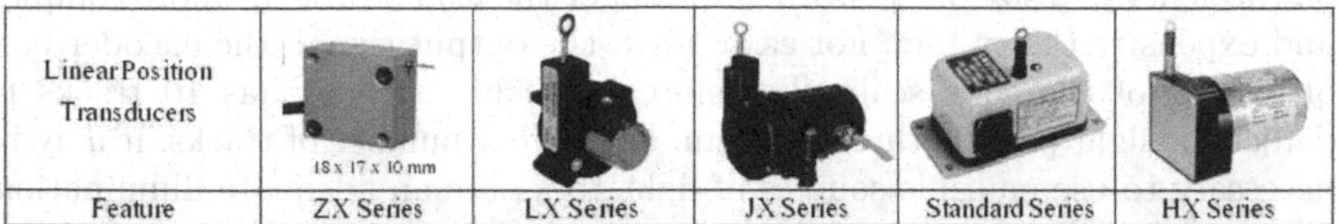

Linear Position Transducers					
	18 x 17 x 10 mm				
Feature	ZX Series	LX Series	JX Series	Standard Series	HX Series
Measured Parameter	Linear Position	Linear Position	Linear Position	Linear Position Liner Velocity	Linear Position Liner Velocity
Measurement Ranges	0 to 38 mm	0 to 50 mm to 0 to 1.25 m	0 to 50 mm to 0 to 2 m	0 to 50 mm to 0 to 2 m	0 to 50 mm to 0 to 50 m
Housing Construction	Aluminum	Thermoplastic	Thermoplastic	Aluminum	Aluminum Stainless Steel Polyurethane
Ingress Protection	NEMA 1. IP-40	NEMA 1. IP-40	NEMA 12. IP-52 NEMA 4X. IP-65	NEMA 1. IP-40 NEMA 1. IP-50	NEMA 4. IP-65 NEMA 6. IP-68
Linearity (analog output)	±1% Full Scale (FS)	±0.25% to ±1% FS	±0.25% to ±1% FS	±0.25% to ±1% FS	±0.25% to ±1% FS
Linearity (digital output)	—	±0.1% FS	±0.1% FS	±0.03% FS	±0.03% FS
Output Stages Options	A	A.E	A.E.F.G.H	A.B.C.D.E.F.G.H	A.B.C.D.E.F.G.H

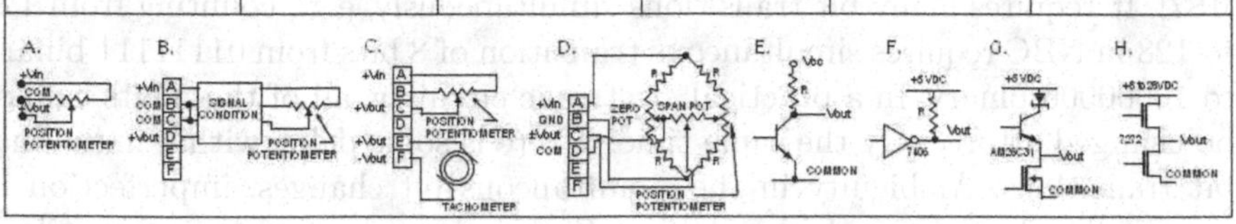

Fig. 7.13. Linear position cable extension transducers produced by UniMeasure Inc., Corvallis, OR, USA.

ity. The tension and retraction of the wire rope is achieved by an internal torsion spring mechanism. The sensing device may be a precision potentiometer, digital encoder, or a tachometer for velocity measurement. Liner position transducers proved to be a successful approach in a multitude of applications. With relatively non-critical alignment requirements, compact size, and ease of installation, these wire-rope-actuated transducers provide an extremely cost-effective method providing linear displacement feedback. Five different series of transducers manufactured by UniMeasure Inc., Corvallis, OR, USA are shown in Fig. 7.13.

Typically, *absolute encoders* are utilized in the devices inactive for long periods of time or moving at low speeds. They are also applied to systems where linear position must be maintained regardless of power interruptions, or where safe and failure-free operation is required. Primarily, machine tools and robotics applications make use of absolute position encoders. These devices supply a whole output word with unique binary code pattern representing each position. This code is derived from independent tracks on the linear scale detected by individual photodetectors. The output from these detectors would then be *high* or *low* depending on the code pattern read off the linear scale for the particular position. Absolute encoders are similar to incremental devices; however, they contain more sensors. The overall complexity depends on the generated size of the word. The longer the *logic* word, the more complex and expensive the system. For each *bit* in the output signal, the encoder uses one track of the code scale. Therefore, a 10 bit encoder has 10 tracks to detect the light passing through them. For higher number of tracks, it may be necessary to use multiple sources of light to assure an adequate illumination. The principle of operation of the linear absolute encoder is illustrated in Fig. 7.14. Although the information read from data tracks can be converted into position signals using many different codes, natural binary code (NBC), gray, gray excess, and binary coded decimal (BCD) codes are most common.

The NBC derives the numerical value from exponents with base 2. For example, the number 179 is expressed as $1 \times 2^7 + 0 \times 2^6 + 1 \times 2^5 + 1 \times 2^4 + 0 \times 2^3 + 0 \times 2^2 + 1 \times 2^1 + 1 \times 2^0$. In other words, the NBC value for 179 is 10110011.

The *binary code* is a *polystrophic code* characterized by multiple bit changes [187]. It requires many bit transitions simultaneously, e.g., counting from 127 to 128 in NBC requires simultaneous transition of 8 bits from 01111111 binary to 10000000 binary. In a practical electronic circuitry, all of these bits cannot be changed at precisely the same time. There is some delay within individual bit transitions. Ambiguity in the simultaneous bit changes, imperfection in the readhead mechanical installation, hysteresis and noise comprise only a few factors that affect the accuracy of the position detection. The potential error in the reading of the most significant bit can result in 180° feedback signal error.

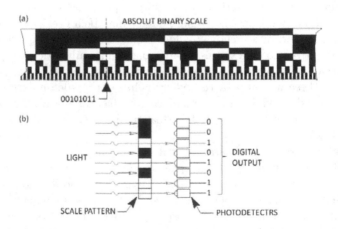

Fig. 7.14. Principle of operation of an absolute encoder: (a) absolute binary scale, (b) detection of bits.

More sophisticated scanning methods are used in modern absolute-position encoders. Two of them, the V-scan and the U-scan, allow for reliable simultaneous bit transitions. In the V-scan method, the sensors are positioned in the V-shape arrangement in two sensor banks (Fig. 7.15). Such a distribution makes room for error tolerances in the encoder system. The less significant bit is used to define in which direction the scale is moving, i.e., what kind of transition is performed (high–low or low–high).

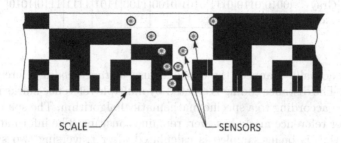

SCALE⎯   ⎯SENSORS

Fig. 7.15. Arrangement of sensors in V-scan method.

Another nonambiguous method is the gray code, particularly well suited to optical encoders. In this *monostropic code*, only two neighboring position values differ in exactly one binary digit, i.e. only one track changes at a time. This limits any decision during edge transition to plus or minus one count. Therefore, the maximum error when moving from one position to the next is 1/4 of the grating period of the finest track.

The gray excess code consists of a section from the middle of the gray code pattern. This permits a position value other than 2 and yet remains a unit-distance code (monostropic). An example of the gray excess code is: 4-bits of gray code provide 16 absolute position values, and to solve 10 positions, the first and last 3 values are omitted from the graduation pattern to produce the *10–excess–3* gray code. In the end, these codes (gray code, gray excess code, or any other appropriate code) are converted by the subsequent electronics (microprocessor) into the NBC. Differences between the binary and gray codes are shown in Table 7.5.

The application of absolute encoder scales to industrial linear motion systems characterized by extended single axis length (several tens of meters) typical of packaging, automation, and assembly lines is hindered by insufficient step resolution. For example, the scale with 12 tracks can generate 12-bit position information. This translates to 4096 unique encodings per scale length, providing approximately 250 μm resolution for 1 m of travel distance. In some cases, this is insufficient. An increase in the number of tracks can overcome this problem, but it results in higher complexity and costs of the encoder system. In practice, the total travel distance is subdivided into sections instead. Each section has the same absolute linear scale. To detect which

Table 7.5. Decimal, binary, gray, and gray excess codes

Decimal code	0	1	2	3	4	5	6	7	8	9
Binary code	0000	0001	0010	0011	0100	0101	0110	0111	1000	1001
Gray code	0000	0001	0011	0010	0110	0111	0101	0100	1100	1101
Gray excess	0010	0110	0111	0101	0100	1100	1101	1111	1110	1010

section is actually scanned, the encoders use *distance coded* reference mark (DCRM). The distance coding is created by multiple reference marks individually spaced according to a specific mathematical algorithm. The span between every other reference mark, however, remains constant. The information as to which section is being sampled is calculated after traversing two successive reference marks. Rather than scanning the entire scale length, this method can reduce the search interval to 100 mm or less. Moreover, the DCRM scale system enables quick recovery of position information after power interruption or system shutdown. The absolute position is established without a need to traverse the entire scale [179]. The maximum length of coded scale and minimum traversed distance depend on the basic increment κ_{sp}, representing the distance between odd reference marks.

The value of κ_{sp} must be divisible by 2 times the grating scale τ_p (i.e. $\kappa_{sp}/(2\tau_p)$ with no remainder). With application of magnetic encoders, the scale grating τ_p is equal to the pole pitch of the magnetic tape. The maximum codable length L_{max} allowing absolute position determination is calculated from

$$L_{max} = \kappa_{sp} \left(\frac{\kappa_{sp}}{2\tau_p} - 2 \right) \tag{7.2}$$

This is illustrated in Fig. 7.16 showing how the reference marks are distributed along the scale. The absolute position of the first traversed reference mark n is calculated according to the following formula [85]:

$$n = [|2\Delta n - k| - \text{sign}\,(2\Delta n - k) - 1]\,\frac{k}{2} + [\text{sign}\,(2\Delta n - k) \pm 1]\,\frac{\Delta n}{2} \tag{7.3}$$

where Δn is a number of signal periods between two successively traversed reference marks, k is a basic spacing expressed in number of signal periods, $k = \kappa_{sp}/\tau_p$. This formula yields n, a number of signal periods between the first mark on the scale (scale beginning) and the first traversed reference mark. The operator "sign" returns 0 if argument equals 0, 1 if argument is greater than 0, and -1 otherwise. The traversed direction is accounted for by a proper choice of sign in the second term of eqn (7.3). The "−" sign represents forward motion, while the "+" sign stands for backward motion.

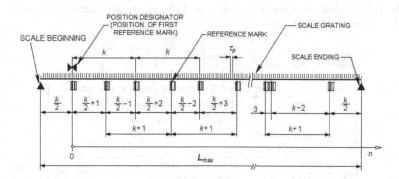

Fig. 7.16. Representation of an incremental scale with distance-coded reference mark.

In motion systems where safety and failure-free operation are not a priority, incremental rather than the absolute optical scales can be employed for distance-coded reference marks. Here, the absolute position is determined by counting the number of steps from each reference mark. In case of power interruption, however, the position information may be lost and can only be recovered by traversing at least two neighboring reference marks.

Linear encoders used in the metal cutting industry, i.e., LBM drives for machine tool tables, must meet the following requirements [84]:

- High counting accuracy at high speeds, e.g., from 0.1 to 0.25 m/s in milling of gray-cast iron and aluminum (this translates into wide frequency range of position loop-control and, therefore, fast feed-forward control).
- High acceleration capability, typically from 10 to 40 m/s^2 and even higher.
- High maximum rapid-travel speeds, typically from 1 to 1.5 m/s, sometimes even 2 m/s.

Manufacturers have developed two types of constructions that can meet these requirements: (a) exposed, and (b) sealed encoders. Exposed encoders are recommended in clean environments without a danger to contaminate the optics. However, in machines either completely encapsulated or using coolant and/or lubricant, sealed encoders are preferred. The advantage of the sealed system lies in the reduction of requirements for finishing the mounting surface. Furthermore, sealed linear encoders are characterized by simple mounting and higher protection rating. On the other hand, the advantages of exposed encoders include higher traversing speed, no friction, and better accuracy. Therefore, exposed encoders most often find applications in precision machines, measuring systems, and production equipment for the semiconductor industry. On the other hand, sealed linear encoders are widely utilized in metal-cutting machines.

Table 7.6. Sealed absolute linear encoder LC 181 manufactured by Heidenhain, GmbH, Traunreut, Germany

Parameter	Specifications
Measuring standard	DIADUR glass scale with 7 tracks with different grating periods
Data interface	Synchronous serial (EnDat)
Incremental signal	1 V (peak-to-peak) signal period 16 μm
Accuracy grades	$\pm$5 μm, $\pm$3 μm
Measuring steps	1 μm, 0.1 μm
Measuring length	240 to 3040 mm
Length of sealed scale	Measuring length +119 mm
Width of sealed scale	40 mm
Height of sealed scale	62.5 mm
Height of sealed scale and readhead	85 mm

The Heidenhain LC 181 sealed absolute position encoder data is shown in Table 7.6. It generates the absolute position value from seven incremental tracks. The grating periods of the tracks differ in a manner that makes it possible to evaluate the measuring signals of all seven tracks. This allows identification of any location on the scale within the measuring length of 3 m. In addition to the absolute position information, the LC 181 encoder provides sinusoidal incremental signal with its period of 16 μm at 1 V (peak-to-peak).

7.1.3 Data Matrix Code Identification and Positioning System

Optical laser and digital photo technologies are now widely applied in many areas of industrial production providing accurate information about the flow of goods and commodities. Depending on application, this might include determining the position of moving entities (e.g., work pieces, tool carriers), monitoring a car on a suspended rail system (e.g., in warehouses, distribution centers), or monitoring a conveyer (e.g., in warehouses, production facilities). The most advanced position encoding/identification system suitable for long travel-path applications is data matrix coding [169]. These systems allow for identification and synchronization of many objects transported over distances exceeding 300 m with the precision on the order of a fraction of a millimeter. Complex and extensive motion systems containing turns, junctions, and gradients, e.g., crane positioning, elevators, galvanization stations, or studio technology constitute the primary target for data matrix installations.

The data matrix technology is based on scanning of red light reflection from a coded label akin to the barcode identification systems. The main difference is that the barcode is an one-dimensional code. Meanwhile, the data matrix code is a two-dimensional representation of encoded information with capacity up to 1.5 kB condensed onto a very small footprint area. The data matrix codes can be printed directly on plastic or metal substrates. The bit code with built-in error correction is set in a chessboard fashion. A data matrix reader captures the code in a image, evaluates it internally, and sends the decoded information as a text command to the controller via Ethernet. This enables high passing speeds and allows code readout even when the information is partially lost. An example data matrix reader, ODT-MAC 400 produced by Pepperl+Fuchs GmbH, Manneheim, Germany, is capable of reading objects at speeds up to 20 m/s and at frequencies up to 60 scans/s. Detailed parameters of MAC 400 readers series are specified in Table 7.7. The exemplary application of data matrix system is illustrated in Fig. 7.17.

Table 7.7. Data matrix reader MAC400 manufactured by Papperl+Fuchs, GmbH, Mannheim, Germany.

Parameter	Specifications
Reading distance	60 mm ±3 mm
Reading field	30 × 20 mm
Resolution	752 × 480 pixel
Min. module size	>0.2 mm
Target velocity	Triggered ≤20 m/s
Scans per second	≤60
Video output	VGA-interface
Memory: RAM/Flash	32 MB/4 MB; Expandable using MMC card
Trigger sensors	4 × 24 V DC inputs and outputs
Interfaces	RS232 to the PC Ethernet TCP/IP to the control
Web server function	Access via Ethernet interface
Dimensions of metal housing	90×60×60 mm in straight housing 120×60×60 mm in right angle housing
Item number	Description
OTD-MAC400-ND-RD	Straight housing
OTD-MAC401-ND-RD	Right-angle housing
OTD-MAC401-LD-RD-MC	Reads all common 1D and 2D Codes, 45×30 mm reading field at 100 mm distance, right-angle housing

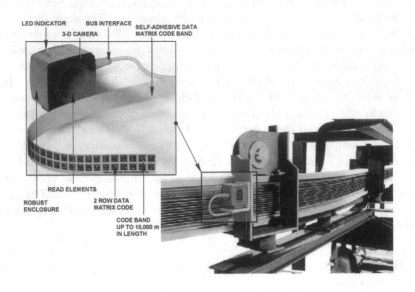

Fig. 7.17. Application of data matrix positioning system. Courtesy of Pepperl+Fuchs, GmbH, Mannheim, Germany.

7.2 Linear Magnetic Encoders

7.2.1 Construction

As compared with optical sensors, their magnetic counterparts are characterized by simplicity, reduced sensitivity to contamination, robustness, and low cost. Magnetic sensors can work in the presence of heavy liquid and chip buildup. Made out of metal, they can withstand more severe vibrations and are perceived to be more reliable. In addition, these devices have lower power requirements, good performance characteristics, and are well suited for large-volume manufacturing technology.

Magnetic encoders utilize *magnetoresistive* (MR) sensing elements and *magnetically salient targets*. The magnetically salient target is a long, alternatively magnetized ruler. The MR elements (sensors) change their resistance under the influence of magnetic flux density and can sense flux densities above 0.005 T [187]. The principle of operation of the magnetic linear encoder with the MR sensors is explained in Fig. 7.18. The MR sensor resistance changes approximately ±1.6% as the magnetic field excited by the passing salient target changes its polarity. Four sensors are electrically connected to a resistive bridge polarized by 5 V d.c. source. The bridge output voltage varies sinusoidally within the amplitude of 0.08 V (peak-to-peak) reflecting changes of sensor resistances. The two magnetic poles affect the sensors in the same way but with opposite polarity. Therefore, when the alternatively magnetized ruler moves one pole pitch τ_p, the output signal will complete one cycle. Sine and cosine signals are produced as the sensors traverse the scale. These analogue signals are interpolated internally to produce the resolution up to 4 μm.

Magnetic encoders employing MR sensors are capable of producing output signals with frequencies up to 200 kHz. High frequency response requires very high resolution, which is a function of the air gap size. The smaller the gap, the higher the resolution. This gap size should be approximately 80% of the pole pitch. For example, if a motion system requires the resolution of 0.05 mm (20 kHz frequency with 1 m/s linear speed), the encoder with 4× interpolation should contain the magnetic target with 0.2 mm pole pitch. The air gap in such a system is about 0.15 mm.

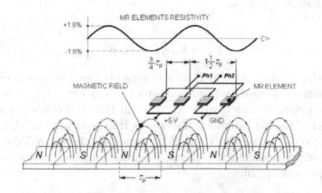

Fig. 7.18. Magnetic encoder with MR sensors.

Sensing elements and magnetically salient targets with stick-on reference marks are typically supplied as separate components. The scale can be supplied on a reel or cut to a specific length and protected by a nonmagnetic stainless cover strip. Occasionally, motion system hardware may be adopted by the end user to serve as a long salient target. In such a system, air gaps between sensors and the target are usually on the order of a few millimeters. The magnetoresistive (MR) linear magnetic encoders, produced by Merilna Tehnika, Ljubljana-Dobrunje, Slovenia, are presented in Fig. 7.19, [180].

In magnetic encoders with large air gaps the *Hall elements* are more suitable than MR sensors. These are true solid-state devices with good operating temperature limits, typically from −40 to 150 °C, long life expectation (20 billion operations), and which can work at zero speeds. The linear (analog) Hall element has a wide range of output signals (from 1.5 to 4.5 V) and a reasonable frequency response (100 kHz). Its output voltage is

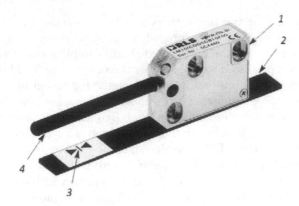

Fig. 7.19. RLS incremental magnetic encoder. 1 — readhead, 2 — magnetic scale, 3 — stick-on reference mark, 4 — cable. Photo courtesy of RLS Merilna Tehnika, Ljubljana-Dobrunje, Slovenia.

$$V_H = k_H \frac{1}{\delta} I_c B \sin \theta \qquad (7.4)$$

where I_c is the applied current, $B \sin \theta$ represents the component of the magnetic flux density vector perpendicular to the current path, θ is the angle between the magnetic flux density vector and the Hall element surface, δ is the thickness of the Hall element, and k_H is Hall constant $(\mathrm{m}^3/\mathrm{C})$.

Specifications of a typical Hall effect sensor are listed in Table 7.8. This sensor is used to scan moving electromagnetic objects, preferably toothed ferromagnetic racks [130].

Table 7.8. Specifications of the IHRM 12P15001 sensor employing magnetically biased Hall element manufactured by BEI Corporation, Industrial Encoder Division, Tustin, CA, USA

Parameter	Specifications
Voltage supply range	8 to 28 V d.c.
Supply current	20 mA
Max. switching current	100 mA
Max. switching frequency	20 kHz
Voltage drop	< 3 V d.c.
Air gap	2.5 mm
Temperature range	$-40°$ to $120°$C
Temperature coefficient	$-3\%/$K
Short-circuit protection	Yes
Reverse polarity protection	Yes
Housing	Stainless steel
External dimensions	M12x1 (thread) × 60 mm length

Encoder systems with Hall effect devices are arranged in a different way than those comprising MR elements. Hall sensors are typically placed between a moving, magnetically salient target, e.g., ferromagnetic ruler with teeth, and a bias PM that excites the magnetic field. In the case of low resolution of positioning systems, a long *flexible magnetic strip* distributed along the motion track serves as the magnetically salient target. This strip is made out of ferrite material or low-energy NdFeB PMs mixed with rubber, and is usually alternatively magnetized, i.e., N, S,...,N, S with pole pitch of a few millimeters. The alternatively magnetized flexible strip permits achieving repeatability up to ±5 μm (1.22 μm resolution) with 4096× multiplier (electronic circuit). The relative position is determined by counting the number of poles or target saliencies (steel teeth) moving through the sensor, while the speed is obtained from the frequency at which they pass. Meanwhile, the movement direction is obtained from the relative timing of two sensors in the quadrature with target saliency. This flexible-strip-based linear encoder is used in LEU, LEM and LZ series linear motors with inner air-cored armature winding manufactured by Anorad, now a branch of Rockwell Automation [12].

In linear motors utilized for propulsion with an array of magnetic poles N, S,...,N, S and short pole pitch, the installation of an additional magnetic strip for the encoder is not necessary. Encoder sensors are located near the surface of the guideway, and the field produced by PMs is used as the magnetic target.

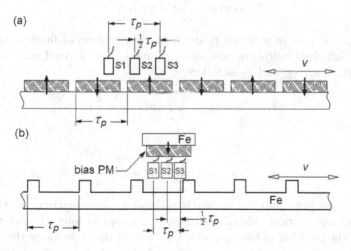

Fig. 7.20. Noise canceling magnetic sensors: (a) PM reaction rail, (b) ferromagnetic reaction rail with saliency. S1, S3 — antiphase sensors, S2 — noise-canceling sensor.

7.2.2 Noise Cancelation

One of the disadvantages of linear magnetic encoders is their sensitivity to external magnetic fields and temperature changes. Sometimes, the *magnetic noise* can exceed the sensor-generated signal up to one order of magnitude. Therefore, noise cancelation techniques aimed at suppression of unwanted disturbance signals are required.

One of the simple noise cancelation methods is based on an array of three magnetic sensors [P70]. Two of them are situated half of the magnetic period apart (in antiphase relationship, i.e., 180° out–of–phase), while the third sensor is placed between the two remaining. The three-sensor array is shown in Fig. 7.20.

The individual sensor output signal is a function of the magnetic flux density created by passing magnetic targets and the ambient noise of magnetic origin. For magnetic poles of alternative polarity (Fig. 7.20a), the three sensor output signals can be expressed as

$$s_1 = \sin \frac{\pi}{\tau_p} x + N \tag{7.5}$$

$$s_2 = \sin(\frac{\pi}{\tau_p} x + \frac{1}{2}\pi) + N \tag{7.6}$$

$$s_3 = \sin(\frac{\pi}{\tau_p} x + \pi) + N \tag{7.7}$$

where x is the pole or saliency position in the direction of motion, and N is the noise signal. Quadrature positions s_{1c} and s_{2c} are derived from the three sensor signals s_1, s_2, and s_3 as follows:

$$s_{1c} = s_1 - s_2 = \sqrt{2}\cos(\frac{\pi}{\tau_p} x + \frac{1}{4}\pi) \tag{7.8}$$

and

$$s_{2c} = s_3 - s_2 = \sqrt{2}\cos(\frac{\pi}{\tau_p} x + \frac{3}{4}\pi) \tag{7.9}$$

The sensor output quadrature signals s_{1c} and s_{2c} are digitized for the complete noise cancelation enhancement. In some applications, the noise N depends on the position of the magnetic pole within the strip along the reaction rail. This results in incomplete noise cancelation. However, digitization with zero-crossing detection occurring at points of geometrical symmetry will fully cancel the noise signals. Fig. 7.21 depicts the sensor output signal with resulting zero-crossing digitization.

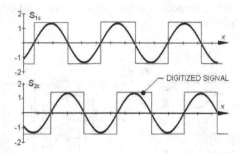

Fig. 7.21. Quadrature sensor output signals s_{1c} and s_{2c} digitized by sensing zero-crossings.

7.2.3 Signal Interpolation Process

The interpolation is a process of an encoder signal subdivision into phase-shifted copies. It can be applied to sinusoidal outputs in the quadrature only. For example, the transistor–transistor logic (TTL) signals cannot be interpolated. To enhance the resolution effectiveness, i.e., the overall accuracy, the interpolated signals are recombined in electronic circuitry.

The sinusoidal signals formed by the incremental encoders are processed by the digitizing electronic units. These are often incorporated into a numerical motion controller and enclosed in a separate housing. Three of the commonly used interpolation methods, i.e., (a) analog–digital interpolation using resistor networks, (b) digital interpolation with look-up and tracking counter, and (c) digital interpolation with arc-tangent calculator, have successfully been applied to the Heidenhain, GmbH encoders.

The first method makes use of the trigonometric identity $\sin(\alpha + \beta) = \sin\alpha\cos\beta + \cos\alpha\sin\beta$ to develop the phase-shifted copies of the original signals.

The encoder LS 774/LS 774C manufactured by Heidenhain, GmbH is based on the analog–digital conversion with 5-fold interpolation (the so called $5\times$ interpolator). The scanning signals s_{1c} and s_{2c} are amplified and interpolated in the resistor network that generates collateral phase-shifted signals by using vector algebra. The 5-fold interpolation process is shown in Fig. 7.22. Ten signals are produced with a phase shift ranging from $0°$ to $162°$ electrical. After conversion to the quadrature, these signals are combined into two square-wave trains by exclusive-OR (XOR) gates. The trains of impulses have frequency five times greater than that of the scanning input signals and are phase-shifted by quarter of the period. Each edge of the signals S_1 and S_2 can be used as a counting pulse within one period. The reference pulse S_0 is gated between the two successive edges of S_1 and S_2. The 20 μm grating period of the encoder, which combines 5-fold interpolation with 4-fold electronic evaluation, is capable of achieving 1 μm measuring step. A similar process can be used for 10- or 25-fold interpolation that results in 1/40 or 1/100 measuring step of the grating period.

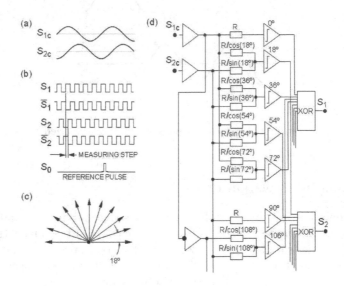

Fig. 7.22. Interpolation process with resistor network: (a) scanning signals, (b) measuring signals after 5-fold interpolation and digitization, (c) signal vectors diagram, (d) electronic circuit.

In interpolation processes utilizing higher subdivisions (50-fold and above), digital methods are required. Two scanning signals are first amplified, then quantified in the sample-and-hold circuitry, and finally digitized into regular intervals in the A/D converter. These digitized voltages define a single address (row and column) in a lookup table describing an instantaneous position (Fig. 7.23a). The actual position is compared with the value determined in the previous cycle that is stored in a tracking counter. The tracking counter produces incremental square-wave signals ($0°$ and $90°$) from the differences between previous and current positions. The lookup table interpolation method is used in the EXE 650 (50-fold interpolation) and EXE 660 (100-fold interpolation) encoders manufactured by Heidenhain, GmbH.

The most advanced interpolators use microprocessor technology. Fig. 7.23b illustrates the digital interpolation process employing an *arctangent* calculator. The microprocessor calculates the tangent S_1/S_2 from two digitized input voltages. The corresponding angle value (*arctangent*), which indicates the position within one signal period, is derived from the table stored in EPROM. The analog input signals s_{1c} and s_{2c} are simultaneously converted into quadrature waveforms, and signal periods are determined. The actual position is derived from the evaluated period and the calculated angle. Finally, to compensate for system errors, appropriate correction values are read from the table stored in the RAM. After the error correction, the digital signal is transmitted to the motion control unit.

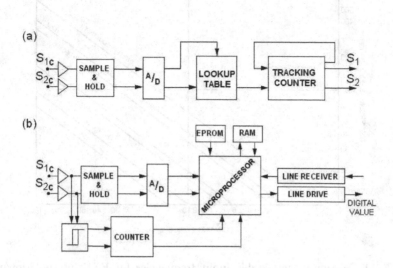

Fig. 7.23. Digital interpolation methods: (a) using lookup table and tracking counter, (b) using microprocessor to compute arctangent.

7.2.4 Transmission of Speed and Position Signals

The speed and position control systems are limited by the *pulse per meter* counts and the maximum linear *speed/frequency* response rate, which depends on

- mechanically permissible traversing speed,
- minimum possible edge separation of the square-wave output signals S_1 and S_2,
- maximum input frequency of the interpolating and the digitizing electronics.

The maximum traversing speed

$$v_{tr} = \tau_p f \tag{7.10}$$

depends on the maximum input frequency f of the interpolation and digitizing electronics and the scale grating period τ_p. If f is in kHz, and τ_p is in μm, the speed v_{tr} is in mm/s. An example of the relationship between maximum traversing speed and the grating period at various maximum permissible input frequencies is illustrated in Fig. 7.24.

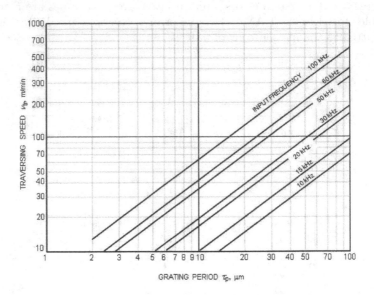

Fig. 7.24. Maximum permissible input frequencies for EXE interpolation unit. Courtesy of Heidenhain, GmbH, Traunreut, Germany.

The encoder signal is subdivided in subsequent electronics. The subdivision factor should remain in reasonable proportion to the accuracy of the encoder. For example, the subdivision factor of 1024 applied to the 10 μm or 40 μm signal period gives the resolution of approximately 10 nm and 40 nm, respectively.

Feed drives of machine tools can reach linear speeds over 2 m/s, while the handling equipment can reach over 5 m/s. It can be calculated that the velocity of 1 m/s and measuring step of 0.1 μm (after a 4-fold evaluation) result in the input frequency of 2.5 MHz. Owing to large distances separating the encoder and the processing electronics (up to 50 m), the interpolating and digitizing circuit is often connected as a separate unit between them. For signals with frequencies above 1 MHz, short cables need to be employed in order to preserve good transmission quality. Therefore, high-speed motion systems utilize encoders containing interpolation and digitizing circuits. If the high-frequency transmission signal is unavoidable, e.g., in the system with high traversing speed and small measuring steps, a linear encoder with sinusoidal output signals should be used. This sinusoidal signal should be 1 V (peak-to-peak) at the cutoff frequency of 200 kHz with amplitude of −3 dB. In this case, the cable length can reach 150 m.

Low traversing speed with high uniformity of motion requirement sets another limit for the measurement system. To maintain adequately uniform speed, a resolution of 0.1 μm and higher may be required.

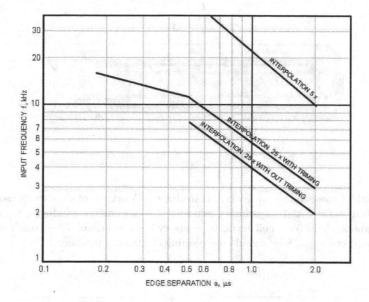

Fig. 7.25. Edge separation diagram applied to non-clock EXE interpolation unit. Courtesy of Heidenhain, GmbH, Traunreut, Germany.

In general, the control electronics limits the minimum edge separation for square-wave output signals. The relationship between input frequency f and the edge separation a for a given interpolation factor is shown in Fig. 7.25. The input frequency f can be found from eqn (7.10).

7.2.5 LVDT Linear Position Sensors

The *Linear Variable Differential Transducer* (LVDT) is a common electrome-chanical converter producing an electrical signal in response to rectilinear motion [137]. This linear position sensor is capable of measuring translation from a fraction of a micrometer to approximately half a meter.

The internal structure of the LVD transducer (Fig. 7.26) consists of primary winding centered between two identical secondary windings equidistant from the center. All three coils are wound on a single hollow tube made of thermally stable glass reinforced polymer. Typically, the coils assembly is wrapped in a high-permeability magnetic shielding and secured in the cylindrical stainless steel housing. This coil assembly constitutes the stationary element of the position sensor. The sensor's translating member is made up of the ferromagnetic cylinder core moving axially within a hollow bore of the coil assembly. During operation, the primary coil is excited by an a.c. current of appropriate frequency and magnitude. The transducer's electrical output

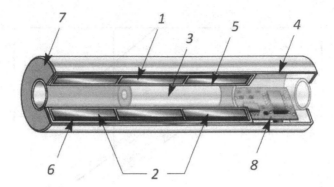

Fig. 7.26. Cross-section of LVDT transducer. Courtesy of Macro Sensors™, Pennsauken, NJ, USA. 1 — primary winding, 2 — secondary windings, 3 — core, 4 — magnetic shell, 5 — coil form, 6 — epoxy encapsulation, 7 — stainless steel housing and end caps, 8 — signal-conditioning electronics module.

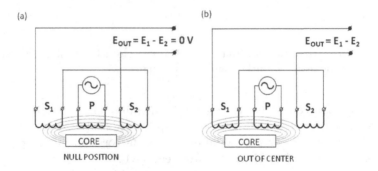

Fig. 7.27. Principle of generating output signals in the LVDT linear transducer.

signal is an alternating voltage between the two secondary coils. The magnitude of the output a.c. voltage depends on the axial position of the core within the coil assembly. This a.c. voltage is converted by suitable electronic circuitry to the high-level d.c. signal. The LVDT's functioning principles are illustrated in Fig. 7.27.

The primary coil P is energized by an a.c. current of constant amplitude. This generates magnetic flux through the ferromagnetic C-shaped core and mutually coupling the secondary windings S_1 and S_2. When the core is located exactly midway between S_1 and S_2 (Fig. 7.27a), the mutual inductances between the primary coil P and both the secondary coils S_1 and S_2 are the same. Consequently, the voltages E_1 and E_2 induced in the windings are equal to each other. At this null point (also known as the reference position), the differential voltage output ($E_1 - E_2$) equals zero. As the core moves closer

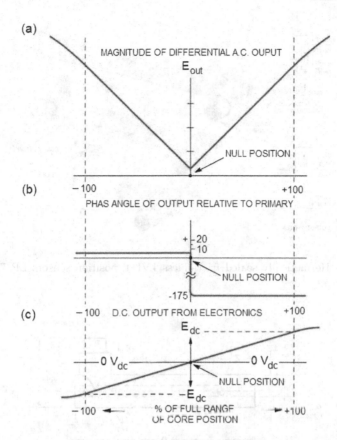

Fig. 7.28. Output characteristics of an LVDT position sensor: (a) magnitude of differential a.c. output as a function of the core position, (b) phase of the output a.c. signal relative to the primary coil excitation P, (c) electronic conditioning unit d.c. output signal.

to one of the secondary windings, it generates a stronger magnetic coupling between the two. Subsequently, the induced voltages E_1 and E_2 are not the same, and the sensor output voltage is nonzero. The differential signal magnitude $(E_1 - E_2)$ is proportional to the core distance from the reference position as illustrated in Fig. 7.28a. The value of E_{out} at the maximum core displacement is on the order of several rms V. It depends on the amplitude of the primary excitation voltage and the sensitivity of the LVDT structure. The E_{out} phase angle referenced to the primary excitation voltage remains constant until the core center passes the reference position, at which point the phase angle changes abruptly by 180° (Fig. 7.28b). This phase shift is used to define the motion direction.

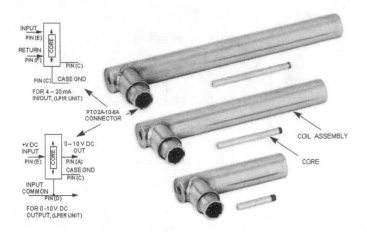

Fig. 7.29. Hermetically sealed frictionless LVDT position sensors LP 750 series, Macro Sensors™.

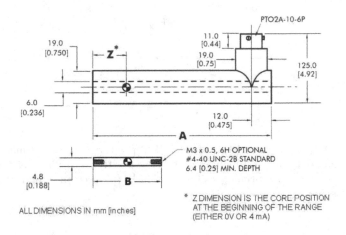

Fig. 7.30. Geometrical details of LP 750 series sensors.

Fig. 7.29, Fig. 7.30, and Table 7.9 show specifications of commercial LVDT series, manufactured by Macro Sensors™, Pennsauken, NJ, USA [129].

Table 7.9. Specifications of LP 750 series LVDTs manufactured by Macro Sensors™, Pennsauken, NJ, USA

Parameter	LP 750 -18750	LP 750 -3750	LP 750 -7000	LP 750 LP -9000	LP 750 -12000
Frequency response	50 Hz (nominal) (−3dB) (higher bandwidth available)				
Operating temperature	−40°C to +85°C				
Linearity error	≤ ±0.25% of FSO (≤ ±0.1% of FSO optional)				
Repeatability error	< 0.025% of FSO				
Hysteresis error	< 0.025% of FSO (FSO — full-scale output)				
Thermal coefficient scale factor	−0.027% (nominal)				
Vibration tolerance	10 g to 2 kHz				
Shock survival	100 g to 11 ms				
Dimensions A, B, Z, mm	140, 20 13.7	265, 42 24.6	465, 48 49.3	595, 114 63.2	790, 135 71.9
Loop-powered units (LPIR)					
Supply voltage	10 to 28 V d.c.				
Loop resistance	50 Ω				
Output current	4 to 20 mA				
Scale factor, mA per mm	0.168	0.084	0.046	0.036	0.027
Voltage output units (LPER)					
Input power	24 V d.c. (nominal) (3.5 − 26.5 V d.c.) 30 mA (nominal)				
Output voltage	0 to 10 V d.c.				
Scale factor, V per mm	0.105	0.052	0.029	0.022	0.017
Dimensions A, B, and Z are defined in Fig. 7.30					

Examples

Example 7.1

An application of linear encoder with scale grating $\tau_p = 5$ mm requires measurement of position along the 15 m track length. Calculate the minimum basic spacing k between odd reference marks warranting recovery of the unique absolute position with application of DCRM over the entire axis length. Moreover, determine the scanning head position after power interruption if the head was traversing in a backward direction and 59 signal periods were counted between two successive reference marks.

Solution

The millimeter length units are used throughout the following calculations. The basic increment κ_{sp} is found by rearranging eqn (7.2) and solving the resultant quadratic equation

$$\frac{1}{2\tau_p}\kappa_{sp}^2 - 2\kappa_{sp} - L_{max} = 0$$

Substituting known quantities

$$\frac{1}{2 \times 5}\kappa_{sp}^2 - 2\kappa_{sp} - 15\,000 = 0$$

There exists only one proper solution to this quadratic equation $\kappa_{sp1} = 397.43$ mm. However, the basic increment must be dividable by two times the scale grating without remainder, $(2\tau_p = 10)$. The correct value of κ_{sp} is found using function $ceil\,[\kappa_{sp1}/(2\tau_p)]$ that returns the last integer equal to or greater than the argument

$$\kappa_{sp} = 2\tau_p ceil\left(\frac{\kappa_{sp1}}{2\tau_p}\right) = 2 \times 5 \times ceil\left(\frac{397.43}{2 \times 5}\right) = 400 \text{ mm}$$

The actual codable length according to eqn (7.2) for $\kappa_{sp} = 400$ mm is

$$L_{max} = 400 \times \left(\frac{400}{2 \times 5} - 2\right) = 15\,200 \text{ mm}$$

The position of readhead after power recovering is calculated using eqn (7.3). First, the basic increment $\kappa_{sp} = 400$ mm should be converted into basic spacing k counted in the number of signal periods τ_p:

$$k = \frac{\kappa_{sp}}{\tau_p} = \frac{400}{5} = 80$$

With $+1$ for backward traversing direction, $\Delta n = 59$, and $k = 80$ stated in (7.3), the absolute position of the first traversed reference mark is

$$n = [|2 \times 59 - 80| - sign\,(2 \times 59 - 80) - 1] \times \frac{80}{2} + [sign\,(2 \times 59 - 80) + 1] \times \frac{59}{2}$$

$$= 1499$$

counted from the first reference mark on the scale, i.e. from the position designator. Finally, the absolute position of the first traversed reference mark counted from the scale beginning is (Fig. 7.16).

$$x = \left(n + \frac{k}{2}\right)\tau_p = \left(1499 + \frac{80}{2}\right) \times 5 = 7695 \text{ mm}$$

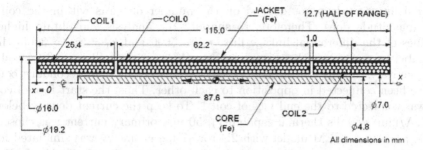

Fig. 7.31. Schematic of the analyzed LVDT. Example 7.2.

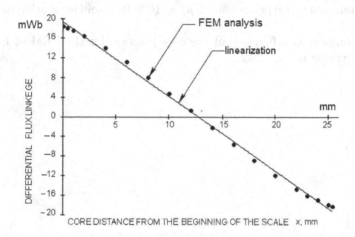

Fig. 7.32. Signal output linearity. Example 7.2.

Example 7.2

The schematic and dimensions of an exemplary LVDT used for conversion of linear position to voltage signal are specified in Fig. 7.31. The core is shown positioned at the null point. The full measurement range is 25.4 mm. Determine the linearity of the output signal as a function of displacement.

Solution

The examination is conducted using the Magnet Infolytica FEM package [54]. Even though the linear differential transformer uses alternating current supplied to the primary coil 0, the problem is modeled as a static because the solution does not depend on eddy currents in the core. The output voltage is a function of the coil 1 and 2 flux linkage, and can be obtained from the static field solution. The differential flux linkage between the sensing coils is calculated at successive positions of the core, starting at position $x = 0$ mm and ending with the core shifted by 25.4 mm to the right. While the output signal linearity does not depend on the number of turns within the coils, its magnitude does. Therefore, to simplify the analysis and obtain higher values of the differential linkage flux, $n_0 = 5000$ and $n_1 = n_2 = 3000$, the number of turns have been assumed for the primary and secondary coils, respectively. Within the framework of the FEM model, the secondary coils have been connected in opposition to each other. Thus, the start face of coil 2 was connected to the end face of coil 1. To keep the current density below 2.5 A/mm^2) (coil's thermal capability), 50 mA primary current was chosen. The parametric FEM model with 2D rotational geometry was simulated for the following 18 core positions: 0, 0.4, 1.0, 2.0, 4.0, ... , 20, 22, 23, 24, 25, 25.4 mm. All magnetic elements of the LVDT have been assumed to possess high and constant magnetic permeability, $\mu_r = 1000$ (to avoid saturation in effect).

The flux linkage as a function of the core position for the FEM of Example 7.2 is illustrated in Fig. 7.32.

8

High-Speed Maglev Transport

8.1 Electromagnetic and Electrodynamic Levitation

Magnetic levitation (maglev) can provide a super high-speed ground transport with a nonadhesive drive system that is independent of frictional forces between the guideway (track) and vehicle bogies. Maglev trains, a combination of contactless magnetic suspension and linear motor technology, realizes super-high-speed running, safety, reliability, low environmental impact, and minimum maintenance. Two maglev transportation technologies emerged in the early 1970s: *electromagnetic* (EML) levitation, which utilizes attractive forces of electromagnets with controlled air gap, and *electrodynamic* (EDL) levitation, which utilizes repulsive forces and superconductivity.

Nowadays, the target speed of ground transport of economic superpowers is minimum 400 km/h. Research done in Germany and Japan shows that vehicles suspended magnetically and propeled by linear motors are the optimum solution to modern transport problems. Magnetic levitation trains can run at speeds up to 550 km/h, consuming less energy than aircraft and road vehicles. Speed above 500 km/h can also be achieved by wheel-on-rail trains (TGV Atlantique set the world speed record of 574.8 km/h in 2007[1]), but this kind of propulsion is not adhesion free and emits a high level of acoustic noise.

In EML levitation systems (Fig. 8.1), the attraction force between the steel yoke (guidance) and electromagnet poles lifts the vehicle. The electromagnet is fixed to the undercarriage. The current of the electromagnet is automatically controlled in proportion to the air gap. Assuming that the magnetic permeability of steel tends to infinity, there is no fringing effect and no leakage fluxes, the inductance of the electromagnet winding as a function of the air gap is simply

---

[1] April 3, 2007, Eastern France, TGV with 18.65-MW total power of electric motors and three double-decker cars.

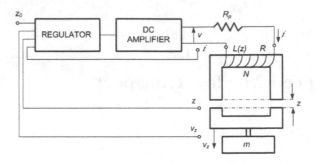

Fig. 8.1. Electromagnetic levitation system: z_o — required air gap, z — actual air gap, v_z — speed of the electromagnet in the z-direction, m — mass of yoke (part being suspended).

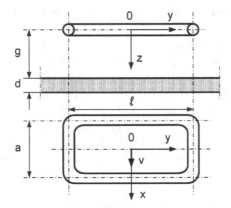

Fig. 8.2. Electrodynamic levitation system: a — coil width, l — coil length, d — aluminum plate thickness, g — air gap.

$$L(z) = L_g \frac{g}{z} \tag{8.1}$$

where the inductance at nominal air gap is

$$L_g = \frac{1}{2}\mu_0 \frac{N^2}{g} A \tag{8.2}$$

In eqns (8.1) and (8.2), μ_o is the magnetic permeability of free space, A is the area of the air gap under a single pole of the electromagnet, N is the number of turns of the coil, z is the axis perpendicular to the electromagnet pole shoes, and g is the nominal air gap. The *attraction force* can be found using eqn (1.15), Chapter 1. However, this equation does not include the magnetic voltage drop in the ferromagnetic core (magnetic saturation). With the magnetic flux path in the ferromagnetic core being included, eqn (1.15) for attraction force of an U-shaped electromagnet takes the form

$$F_z = \frac{1}{4} \frac{\mu_0 (Ni)^2}{[l_{Fe}/(2\mu_r) + g]^2} A \tag{8.3}$$

where l_{Fe} is the mean path of the magnetic flux in the ferromagnetic core including armature, and μ_r is the relative magnetic permeability of the ferromagnetic core (assuming the same magnetic flux in each portion of the core) for a given magnetic field intensity.

An EML levitation system needs a control system. When the air gap between the pole and the yoke increases, the current in the coil of the electromagnet must increase. When the air gap decreases, the current must decrease. In practice, to keep the required air gap $z = g = constant$ (about 10 mm), a control system with three feedback signals is used: displacement z, linear velocity v_z in the z direction, and current i.

In EDL levitation systems (Fig. 8.2), the repulsive forces between the SC electromagnet mounted on the undercarriage and aluminum plates or short circuited coils (guidance) fixed to the guideway are used. The air gap (100 to 300 mm) is much higher than that in EML levitation systems. Owing to the large air gap, the electrodynamically levitated trains can operate in severe climates with heavy snowfalls, ice formations, and white frost formations. The *repulsive force* between the d.c.-fed coil moving with velocity v and a nonferromagnetic conductive plate placed below the coil can be calculated using Hannakam's [82] formula, which has been modified by Guderjahn et al, [73] i.e.,

$$F_z = \frac{\mu_0 (Ni)^2}{\pi g} \left\{ \sqrt{\left(\frac{l}{2}\right)^2 + g^2} + \sqrt{\left(\frac{a}{2}\right)^2 + g^2} - 2g \right.$$

$$- \left[\sqrt{\left(\frac{l}{2}\right)^2 + \left(\frac{a}{2}\right)^2 + g^2} - \sqrt{\left(\frac{a}{2}\right)^2 + g^2} \right] \frac{g^2}{(a/2)^2 + g^2}$$

$$\left. - \left[\sqrt{\left(\frac{l}{2}\right)^2 + \left(\frac{a}{2}\right)^2 + g^2} - \sqrt{\left(\frac{l}{2}\right)^2 + g^2} \right] \frac{g^2}{(l/2)^2 + g^2} \right\} \frac{1}{1 + k^2} \tag{8.4}$$

If the thickness of conductive plate $d < \delta$

$$k = \frac{2}{\mu_0 v \sigma d} \tag{8.5}$$

and if the thickness of conductive plate $d > \delta$

$$k = \frac{2}{\mu_0 v \sigma \delta} \tag{8.6}$$

The parameter

$$\delta = \frac{1}{\sqrt{\pi f \mu_0 \sigma}} = \sqrt{\frac{\lambda}{\pi v \mu_0 \sigma}} \approx \sqrt{\frac{a}{\pi v \mu_0 \sigma}} \qquad (8.7)$$

is the equivalent depth of penetration of the electromagnetic field into the nonferromagnetic conductor with electric conductivity σ. The length of electromagnetic wave is $\lambda = v/f \approx a$ [73]. The coefficient $1/(1 + k^2)$ in eqn (8.4) can also be replaced by $exp\{-\xi/[1 + 2(g/l)^{3/2}]\}$, where $\xi = [4\pi/(\mu_o v \sigma g)]^{1/2}$. The coil moving with velocity v is subject to the drag (braking) force

$$F_x = k F_z \qquad (8.8)$$

More detailed discussions of eqn (8.4) are given in [29, 73].

8.2 Transrapid System (Germany)

8.2.1 Background

Research in transportation engineering carried out in Germany in the 1960s and early 1970s was focused on the energy consumption, costs, safety and impact on environment by trains, road cars, aircraft, and maglev vehicles [140]. Fig. 8.3a shows energy consumption per passenger per 1 km against speed of trains, cars, aircraft, and magnetic levitation trains [140]. The speed of magnetic levitation trains is less than that of airplanes, but the energy consumption is much lower. Maglev trains can enter city centers, and no time is wasted to travel from home to the airport and vice versa. Other advantages include low level of noise (Fig. 8.3b), high level of safety of riding (Fig. 8.3c), high comfort of riding, easy maintenance, low land absorption, adaptability to the landscape due to the high gradability of 10% and the small curvature radii of 2.25 km at 300 km/h, and no pollution to the natural environment.

8.2.2 Propulsion, Support, and Guidance

The *Transrapid* maglev system (Fig. 8.4) uses attractive forces produced by U-shaped electromagnets (EML levitation system) with current control and long-armature three-phase LSMs. The levitation electromagnets attached to the vehicle bogie are also the excitation electromagnets for LSMs. Another set of on-board electromagnets, i.e., guidance electromagnets, provides lateral stabilization. No SC coils are used. The support and guidance electromagnets and vehicle electric system are supplied by contactless linear generators.

Long-armature cores of LSMs are placed in two parallel rows at both sides of the guideway. Each armature has a laminated core with slots as in a typical a.c. linear motor. Laminations are stamped from an adhesive coated steel tape. For the manufacturing of the armature windings, not only electrical and

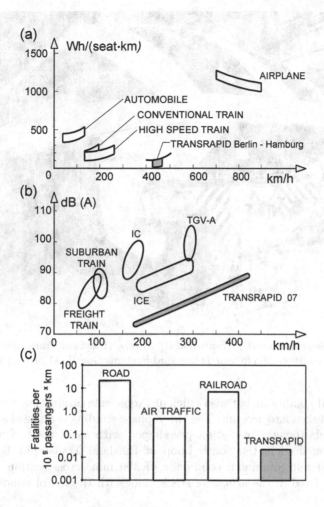

Fig. 8.3. Comparison of passenger transportation systems: (a) energy consumption per passenger per kilometer against speed; (b) maximum noise level at 25 m distance (IC — intercity train, ICE — intercity express; TGV-A — TGV Atlantique); (c) safety analysis — transportation system risk. Courtesy of Thyssen Transrapid System, GmbH, München, Germany.

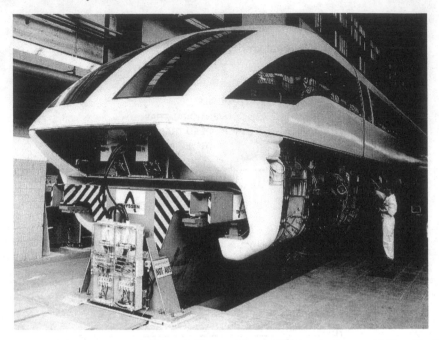

Fig. 8.4. (see color insert.) *Transrapid 07 Europa* (Emsland Transrapid Test Facility). Photo courtesy of Thyssen Transrapid System, GmbH, München, Germany.

geometrical conditions but also efficient large-scale production and assembly have been taken into account. The three-phase winding is made of a cable with synthetic elastomer insulation pre-shaped with small radii. Typical cable construction used in the South Loop of Emsland Transrapid Test Facility is a multistrand aluminum conductor of 300 mm^2 cross section. The cable winding is fixed in the armature stack slots with the aid of winding casings (snap locks).

The excitation system of LSMs and EML suspension of vehicles are integrated and consist of vehicle-mounted U-shaped electromagnets. Interaction of electromagnet poles and laminated cores of LSMs produce attractive forces that lift the vehicle. An electronic control system ensures a constant uniform air gap of about 10 mm (see also Fig. 8.1). Other sets of E-shaped electromagnets, so called guide electromagnets, face the side steel rails and provide lateral guidance (stabilization). Every section of the vehicle is equipped with 15 autonomous support and 13 autonomous guidance electromagnets. The cross section of the support, guidance, and propulsion system is shown in Fig. 8.5. There is a large clearance between the top of the guideway and bottom of the vehicle, so that the maglev train can also levitate over obstacles or a snow cover on the guideway.

The *attractive and lateral forces are controlled by currents of support and guide electromagnets*, respectively. The *thrust can be varied only by the mag-*

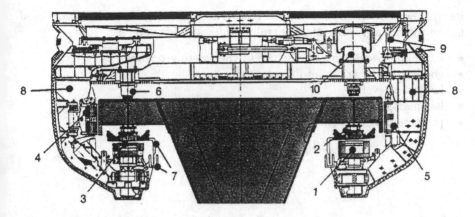

Fig. 8.5. Cross section of the support, guidance, and propulsion system: 1 — support electromagnet, 2 — LSM armature stack with windings, 3 — linear generator windings, 4 — guidance magnet, 5 — eddy-current brake electromagnet, 6 — support skids, 7 — *Inkrefa* sensor (vehicle location), 8 — levitation bogies, 9 — cabin suspension, 10 — pneumatic spring. Courtesy of Thyssen Transrapid System, GmbH, München, Germany.

nitude and phase angle of the armature current. By reversing the phase sequence, LSMs become synchronous generators, which then provide electrodynamic braking forces without any contact. The braking energy is fed back to the network. The long-armature LSM is characterized by the following features [221]:

- As a result of the combination of the suspension and drive systems, the mass of the vehicle determines the excitation of the LSM.
- The three-phase armature winding is fed from VVVF solid-state converters installed in substations that are distributed along the line.
- Control of the tractive effort keeps the air gap flux constant.
- Leakage reactance of the armature winding and of the feeder cable primarily determines the LSM characteristics since the individual sections of the armature windings are longer than the vehicle.

To reduce the energy consumption, the long-armature LSM of the guideway is divided into sections. Only that section in which the vehicle is running is switched on.

8.2.3 Guideway

The T-shaped elevated guideway has two rows of long stator (armature) LSMs with laminated cores on its bottom and lateral steel rails for guidance. The LSM armature core serves also as a suspension rail. Because the vehicle clasps its guideway, derailment is impossible. The construction of the *Transrapid* guideway and any other maglev train guideways demands

- minimum restrictions on the use of the existing terrain,
- good visual blending into the landscape,
- low noise level,
- low maintenance requirements and long life,
- protection against effects of the environment and vandalism.

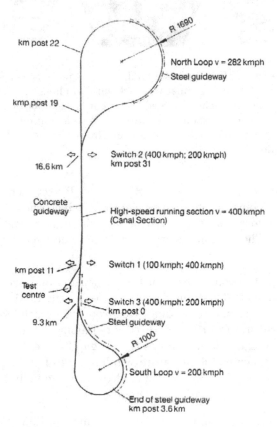

Fig. 8.6. The Emsland test line. Courtesy of Thyssen Transrapid System, GmbH, München, Germany.

The 31.5 km long *Transrapid* test line with two loops is located in the Emsland region (Fig. 8.6). The first 20.5 km section began operating in 1985. The Emsland *Transrapid* Test Facility (TVE) was completed in 1987. About 20 km of TVE was erected as an elevated concrete guideway, about 5 km as an elevated steel guideway, and the rest as a ground-level guideway.

8.2.4 Power Supply

The Emsland *Transrapid* Test Facility is supplied with power from the 110 kV public system (Fig. 8.7). The d.c. link circuit is supplied with 2.6 kV, 2×33 kA through a 110/20 kV transformer and two 20/1.2 kV rectifier transformers connected in parallel. The rectifier transformers each supply two connected-in-series fully controlled rectifier bridges to obtain a twelve-phase group. Smoothing reactors and protective d.c. high-speed circuit breakers are arranged at the input of the d.c. link circuit. The d.c. link voltage is converted by two PWM inverters into a three-phase VVVF changing from 0 to 2027 V and 0 to 215 Hz. The maximum LSM current is 1.2 kA. The energy consumption from the substation for the prototype vehicle is about 60 Wh/(seat×km) at constant speed of 400 km/h.

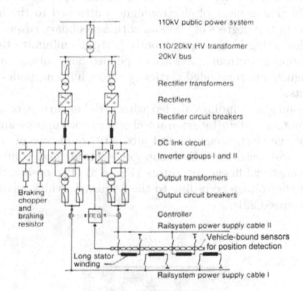

Fig. 8.7. Basic circuit of LSMs power supply of the Emsland *Transrapid* Test Facility. Courtesy of Thyssen Transrapid System, GmbH, München, Germany.

8.2.5 Vehicle

Specification data of the EML levitation vehicle *Transrapid 07* introduced by Thyssen Henschel in 1988 at the International Traffic Fair IVA'88 in Hamburg are presented in Table 8.1. A two-section train with passenger capacity 136 to 298 persons is formed only with two end cars. The measured aerodynamic drag for two-section vehicle at 400 km/h was originally 35.5 kN and then reduced to 33 kN.

Table 8.1. Technical data of *Transrapid 07*. Courtesy of Thyssen Transrapid System, GmbH, München, Germany

Length	26.99 m (end car), 24.77 m (intermediate car)
Width	3.7 m
Height	4.16 m
Mass	50.0 t (end car), 49.1 t (intermediate car)
Speed	300 to 500 km/h
Acceleration	Up to 1.5 m/s^2
Braking ability	Up to 1.5 m/s^2
Rated air gap	8 mm

The *support and guidance system* of the *Transrapid 07* is characterized by a chain-like arrangement of electromagnets attached to the hinge points and adjustable in two degrees of freedom with a secondary suspension system between the levitating bogie and car body [221]. To minimize the unsprung masses, the support electromagnets are suspended horizontally, and the guiding electromagnets are suspended vertically through linear guides and rubber spring elements.

The electromagnet windings are fed with variable current commanded by the air gap control system by separate choppers for support and guidance. The power for the electromagnets and auxiliary equipment of the vehicle is produced by *linear generators*. Each support electromagnet is fitted with two five-phase symmetrical linear generators (Fig. 8.8). The on-board boost converters adjust the voltage according to the frequency, which increases in proportion to the speed[221].

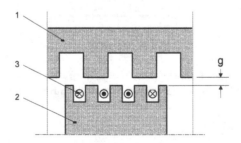

Fig. 8.8. Linear generator of *Transrapid* vehicles. 1 — armature (guideway), 2 — pole of a suspension electromagnet (vehicle), 3 — winding of the linear generator.

The speed record of 501 km/h was achieved by *Shanghai Transrapid* in November 2003. Prior to that, the speed record was 450 km/h set by *Transrapid 07* in June 1993.

Fig. 8.9. (see color insert.) *Transrapid 08*. Photo courtesy of *Thyssen Transrapid System, GmbH*, München, Germany.

Further optimization of the *Transrapid 07* was carried out to lower the manufacturing cost and to improve the ride comfort and safety. The new vehicle *Transrapid 08* shown in Fig. 8.9 [234] was a prototype of the fleet for the Berlin-Hamburg route (Table 8.2).

Table 8.2. Technical data of *Transrapid 08*. Courtesy of Thyssen Transrapid System, GmbH, München, Germany

Specifications	Transrapid 08	Berlin–Hamburg line 4-section train	5-section train
Length, m	79.7	103.5	128.3
Mass of empty train, t	149.5	198.2	247.8
Payload, t	39.0	54.8	70.2
Total mass, t	188.5	253.0	318.0
Number of seats	245	336	444

8.2.6 Control System of Electromagnets

The mechanical clearance between suspension electromagnets and guideway rails is kept constant by means of the *electromagnet current control* system (Fig. 8.10). The contactless *gap sensor* integrated in the pole face of the support electromagnet determines an electrical signal proportional to the distance between the electromagnet and steel rail. From the measured signals proportional to the electromagnet current, acceleration, and air gap, the rated electromagnet current is adjusted in the control loop and transmitted to the chopper as input signal.

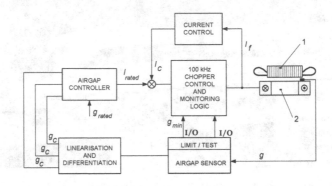

Fig. 8.10. Air gap control system of *Transrapid* vehicles. 1 — armature (guideway), 2 — suspension electromagnet (vehicle).

8.2.7 The Future of Transrapid System

In March 1994, the plan of construction of the 292 km maglev line from Berlin via Schwerin (main control center) to Hamburg (Table 8.3) was approved by the German Federal government. Five stations have been planned, i.e., Berlin Lehrter Bahnhof, Berlin Spandau, Schwerin, Hamburg Moorfleet, and Hamburg Hauptbahnhof. This decision was made due to plans to transfer the lower house of parliament (Deutsche Bundestag) and part of the German government to Berlin. The route Bonn–Berlin had also been investigated, but it was deferred due to unacceptable high costs [237].

For a five-section train, the energy consumption had been foreseen as follows [147]:

- 33–38 Wh/(seat×km) at constant speed 300 km/h;
- 57–65 Wh/(seat×km) at constant speed 430 km/h;
- 158–182 Wh/(seat×km) at acceleration from 0 to 430 km/h in 3.7 min (distance 16.5 km);
- 219–252 Wh/(seat×km) at acceleration from 0 to 300 km/h in 1.9 min (distance 4.8 km).

For comparison, IC trains consume 56 Wh/(seat×km), and TGV superexperss train consumes 108 Wh/(seat×km) at constant speed.

The financial concept was based on the separation of the guideway and operator companies. The federal government (*Deutsche Bahn AG*) was responsible for the financing and construction of the guideway (DM 6.1 billion

Table 8.3. Berlin–Hamburg *Transrapid* line.

Route length	292 km 131 km elevated guideway 161 km ground-level guideway
Span length of guideway girders	3 to 30 m
Maximum gradient	10%
Height of gradient	1.25 to 20.0 m
Maximum superelevation	12^0 (special applications 16^0)
Curve radii at superelevation 16^0 and lateral acceleration 1.5 m/s^2	705 m at 200 km/h 4415 m at 500 km/h
Number of stations	5
Main control center	Schwerin, 100 km from Hamburg
Main maintenance center	Perleberg, 268 km from Hamburg
Revenue speed	430 km/h
Maximum speed	500 km/h
Acceleration	$\leq 1.0(1.5)$ m/s^2
Deceleration (braking)	$\leq 1.0(1.5)$ m/s^2
Traveling time between Berlin Lehrter and Hamburg Hauptbahnhof	53 min
Train configuration	4 to 5 sections per trainset 336 to 444 seats per trainset
Trains fleet	20
Interval between trains	20 min
Number of passengers per year (2010)	11.4 to 15.2 million
Traffic density (passengers × km/year)	2.6 to 3.5 billion
Total investment (DM 1.7 = $ 1.0 at prices in 1996)	DM 9.8 billion
Operation cost per year	DM 250 million
Revenue per year	DM 700 to 950 million

in 1996). The operator companies, i.e., Adtranz (ABB Daimler Benz), Siemens and Thyssen Transrapid System, GmbH, were supposed to finance and construct the operating system (DM 3.7 billion in 1996). The total investment was estimated as DM 9.8 billion as calculated in 1996. It has been assumed that the Berlin–Hamburg maglev line will be completed by 2005. The concept of Maglev link between Berlin and Hamburg was canceled in 2001.

Shanghai was chosen as the site of the construction of the *Transrapid* project in June 2000. Shanghai's Maglev train (Fig. 8.11) opened for service in November 2004, and makes the 32 km trip between Pudong Airport (PVG) and downtown Shanghai (Longyang Road for transfer to Metro Line 2) in only 8 min. The approved speed is 430 km/h.

Fig. 8.11. (see color insert.) *Transrapid* in Shanghai, China.

The new 175 km *Transrapid* line from Shanghai to Hangzhou (Zhejiang province) was approved in 2006. With trains traveling at up to 430km/h, the projected journey time was 27 min.

Chinese Chengdu Aircraft Industry Group is developing its own maglev train. The design of new trains does not use German technology. This part of the National 863 Project's *Dolphin* was for developing high-speed maglev vehicles (500 km/h) in 2008, and it was put into production in Chengdu. In 2009, Tongji University (Jiadong Campus) in Shanghai built a 1.7 km test track as a part of the *Dolphin* project.

The *Transrapid* was considered by the UK government for a 500 km/h link between London and Glasgow, via Birmingham, Liverpool/Manchester, Leeds, Teesside, Newcastle and Edinburgh, but was rejected in 2007.

8.2.8 History of Transrapid Maglev System in Germany

- 1922 — First consideration of EML levitation train by H. Kemper.
- 1939-43 — Basic work on EML levitation train with jet engine at the Aerodynamic Test Establishment in Goettingen.
- 1969 — Construction of the first practical EML levitation model vehicle *Transrapid 01* (TR 01) by Krauss–Maffei, Münich. Support and guidance according to H. Kemper. Propulsion by a short-armature linear motor.
- 1971 — First passenger-carrying EML levitation prototype vehicle built by Messerschmitt–Boelkow–Blohm (MBB) tested on a 660 m long track at Ottobrun. Propulsion by a short-primary LIM. Maximum speed 90 km/h.

Transrapid 02 operated by Krauss-Maffei on 0.93 km track with EML levitation support and maximum speed 164 km/h.

- 1972 — *Transrapid 03* operated by Krauss-Maffei on 0.93 km track with EML levitation support and short-primary LIM at maximum speed 140 km/h. Start of the development of EDL levitation system with SC coils by AEG-Telefunken, Brown Boverie & Cie AG (BBC) and Siemens. Construction of a 0.9 km circular track and EET 01 test vehicle at Enlargen.
- 1973 — *Transrapid 04* operated by Krauss Maffei on 2.4 km track with EML levitation support.
- 1974 — Merger of Krauss-Maffei and MBB, forming Transrapid EMS. Construction of *Komet* vehicle with EML support, rocket engines, and maximum speed 401.3 km/h by MBB.
- 1975 — First practical vehicle HMB 1 with long armature LSM and EML levitation support introduced by *Thyssen Henschel*, Kassel.
- 1976 – First passenger-carrying vehicle HMB 2 with long-armature LSM and EML levitation support introduced by Thyssen Henschel, Kassel.
- 1977 — Federal Ministry of Research and Technology decides to develop EML levitation systems and abandon EDL systems.
- 1978 — Foundation of the Magnetbahn Transrapid consortium by AEG-Telefunken, Brown Boveri & Cie AG, Dyckerhoff & Widmann, Krauss-Maffei, MBB, Siemens AG and Thyssen Industrie AG Henschel.
- 1979 — Emsland Transrapid Test Facility (TVE) construction work starts. International Traffic Fair (IVA'79) in Hamburg with first-in-the-world operation of *Transrapid 05* vehicle (EML and LSM) authorized to carry passengers at speed 75 km/h.
- 1980 — Construction of *Transrapid 06*. Emsland Transrapid Test Facility starts.
- 1981 — Foundation of Gesellschaft für Magnetbahnsysteme Transrapid International with Krauss-Maffei, Messerschmitt Boelkow Blohm, and Thyssen Industrie AG Henschel as partners.
- 1983 — First operation of *Transrapid 06* (EML and LSM).
- 1984 — Opening of the first 21.5 km section of Emsland Transrapid Test Facility (North Loop). *Transrapid 06* achieves the speed of 302 km/h.
- 1987 — Completion of the Emsland Transrapid Test Facility (South Loop). *Transrapid 06* achieves the speed of 406 km/h.
- 1988 — *Transrapid 06* achieves the speed of 412.6 km/h. *Transrapid 07* at the International Traffic Fair (IVA'88) in Hamburg.
- 1992 — Maglev link between Berlin and Hamburg in unified Germany indentified.
- 1993 — *Transrapid 07* achieves the speed record of 450 km/h.
- 1994 — Maglev link between Berlin and Hamburg approved by parliamentary bodies.
- 1995 — Public demonstration of *Transrapid 07* in Emsland starts.
- 1999 — First tests of *Transrapid 08*.
- 2000 — Shanghai was chosen as the site of the construction of Transrapid.

- 2001 — Maglev link between Berlin and Hamburg cancelled.
- 2003 – Speed record of 501 km/h was achieved by Shanghai Transrapid in November 2003.
- 2004 — Shanghai Transrapid line opened for service between Pudong Airport and downtown Shanghai.
- 2006 — *Transrapid* train collided with a maintenance vehicle at 170 km/h on Emsland elevated test track in Lathen (22 September). The accident was caused by human error. There were 23 fatalities and 10 severe injuries.
- 2006 — Extension of the 175-km Transrapid line from Shanghai to Hangzhou is approved by the Chinese State Council. Over 7 million passengers traveled with the Transrapid in Shanghai.
- 2007 — Delivery of the 1st section of the *Transrapid 09* to the Transrapid test facility in Emsland, Germany

8.3 Yamanashi Maglev Test Line in Japan

8.3.1 Background

The population of greater Tokyo, including Chiba, Kanagawa and Saitama prefectures, is now about 40 million inhabitants or almost one third of the total population of Japan (127.6 million). Most of the governmental, administrative, business, financial, and cultural institutions are located in Tokyo. To correct the imbalance created by the overcentralization of people, power, and resources, it is essential that the political, economic, and social functions served by the greater Tokyo metropolitan area be partially relocated and distributed through the nation.

At the present time, Tokyo and Osaka are connected by the *Tokaido Shinkansen* superexpress trains with the maximum speed of 300 km/h, carrying about 368,000 passengers in 283 trains a day, which is nearly the limit of this line [159]. There is a strong demand on another environmental friendly transportation system with higher speed. The *Chuo Shinkansen*, a new transportation artery between Tokyo and Osaka using superconducting (SC) technology, is expected to be implemented in the second decade of the 21st century. New *Chuo Shinkansen* trains achieving speed over 500 km/h are necessary to unload the limited capacity of the *Tokaido Shinkansen* line, preserve the natural environment, and limit the risks from natural disasters.

8.3.2 Location of Yamanashi Maglev Test Line

The Yamanashi Maglev Test Line is a part of the future *Chuo Shinkansen* line between Tokyo and Osaka. It is a joint project of the Central Japan Railway Company (JR Central), Railway Technical Research Insitute (RTRI) and Japan Railway Construction Public Corporation, which was approved by the Ministry of Transport in 1990. The 42.8 km test line will be constructed between Sakaigawa village, Higashi–Yatsushiro district, and Akiyama village, Minami–Tsuru district in Yamanashi Prefecture, west from Tokyo. At present, an 18.4 km Katsunuma-budokyo–Ohtsuki section has been completed (Figs 8.12 and 8.13). The maximum planned speed is 550 km/h (operation speed 500 km/h), minimum curve radius is 8 km, maximum gradient is 4%, and distance between the centers of adjacent parallel guideways is 5.8 m. A 12.8 km section is a double-track line where the dynamics of two trains passing each other at a relative speed of about 1000 km/h will be studied. Other specifications are given in Table 8.4. The cost of the 18.4 km test line, power conversion substation, control center, train, and train depot is about 230 billion yen (1996).

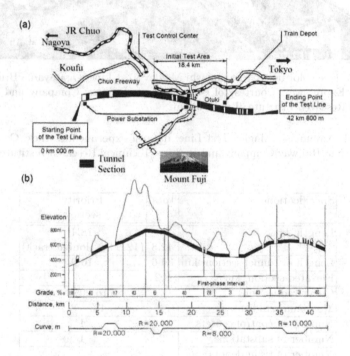

Fig. 8.12. Yamanashi Maglev Test Line: (a) outline, (b) profile. Courtesy of Central Japan Railway Company and Railway Technical Research Institute, Tokyo, Japan.

8.3.3 Principle of Operation

The experimental maglev train MLX01 for the Yamanashi Maglev Test Line is suspended on the principle of EDL levitation where the repulsive forces are produced between stationary short-circuited coils and moving SC electromagnets. The track is U-shaped and embraces the bottom of the vehicle.

Fig. 8.13. (see color insert.) Yamanashi Maglev Test Line: Ogatayama Bridge over the Chuo Expressway. Courtesy of Central Japan Railway Company and Railway Technical Research Institute, Tokyo, Japan.

Table 8.4. Yamanashi Maglev Test Line: data of experimental track. Courtesy of Central Japan Railway Company and Railway Technical Research Institute, Tokyo, Japan

Specifications	Total	Priority section
Length, km	42.8	18.4 (12.8 km double track)
Length of tunnel section, km	34.6	16.0
Elevated section, km	8.2	2.4
Curve radius, km		8 to 20
Maximum gradient		4%
Number of control centers	1	1
Number of substations	2	1
Number of train depots	1	1

At each side of the track, nonpowered short-circuited coils serving both as levitation and lateral guideway coils are mounted in a vertical position. The vehicle is equipped with SC electromagnets. In addition, a three-phase propulsion vertical winding fed with three-phase current is installed at each side of the track, which, together with the train electromagnets, forms an air-cored LSM. The three-phase stationary winding produces a traveling magnetic field and corresponds to the armature winding of a conventional synchronous motor. The vehicle's SC electromagnets correspond to the excitation system of a synchronous machine. When the train, propelled by the LSM, passes short-circuited coils at high speed, currents induced in these coils together with the magnetic field excited by SC electromagnets produce strong repulsive and lateral stabilization forces on the vehicle. The same SC electromagnets are used both for levitation, lateral guidance, and propulsion.

8.3.4 Guideway

The arrangement of ground and vehicle windings is shown in Fig. 8.14. Both propulsion coils and levitation–guidance coils are attached to the concrete side walls of the guideway (Fig. 2.25).

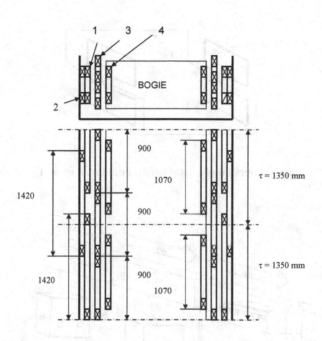

Fig. 8.14. Arrangement of propulsion, levitation-guidance and excitation coils: 1 — propulsion, front side, 2 — propulsion, reverse side, 3 — 8-shaped levitation and guidance coils, 4 — excitation coil (on-board SC electromagnet) [239].

All ground coils are made of aluminum conductors insulated with polyester (epoxy) resin. *Propulsion coils* have dimensions approximately 1.42×0.6 m. The 8-shaped *levitation* and *guidance coils* have dimensions approximately 0.9×0.9 m and are attached to the surface of the three-phase two-layer propulsion winding. The levitation and guidance coils consist of two sections: for levitation and for lateral stabilization (guidance) of the vehicle (Fig. 8.15). The guidance sections facing each other at two opposite sides are electrically

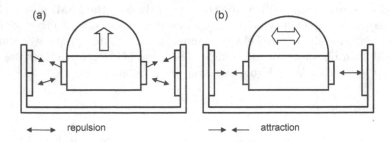

Fig. 8.15. Operation of 8-shaped coils: (a) levitation, (b) lateral stabilization (guidance) of the vehicle [239].

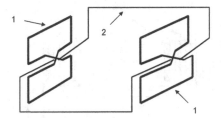

Fig. 8.16. Null-flux connection of levitation and guidance coils. 1 — coils, 2 — null-flux cable [239].

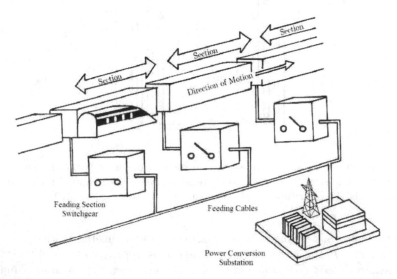

Fig. 8.17. Feeding system of propulsion winding sections. Courtesy of Central Japan Railway Company and Railway Technical Research Institute, Tokyo, Japan.

connected under the track, constituting a *null-flux connection* (Fig. 8.16). If the train deviates from the center of the guideway, the deviation is reversed by the attractive forces of the superconducting electromagnet on the distant side of the guideway and repulsive forces on the opposite (near) side.

In order to achieve passenger comfort when traveling at very high speeds, the ground coils must be installed more precisely than *Shinkansen* rails. The accurate attachment, easy construction, and simple maintenance require three types of guideways: (a) panel type, (b) side-wall beam type, and (c) direct attachment type (Chapter 2).

To save energy, a group of propulsion coils are connected in series and create a winding section. Only those sections carrying the train are powered through the feeding *section switchgears* (Fig. 8.17).

Table 8.5. Specification data of MLX01 Maglev Trains. Courtesy of Central Japan Railway Company and Railway Technical Research Institute, Tokyo, Japan

Specifications	MLX01 (first train)	MLX01 (second train)
Maximum speed	550 km/h	
Number of cars	3	4
Pole pitch of electromagnets	1.35 m	
Vehicle configuration	Articulated bogie system with superconducting electromagnets	
Car body structure	Semi-monocoque structure using aluminum alloy	
Levitation height	0.1 m at 500 km/h	
Width of car	2.9 m car body, 3.22 bogie	
Height of car	3.28 m while levitating 3.22 m on-gear running	
Length of end cars	28.0 m	
Length of intermediate car	21.6 m	24.3 m and 21.6 m
Cross section area of the car	8.9 m^2	
Maximum mass of fully loaded car	32 t end car 20 t intermediate car	33 t end car 22 t intermediate cars
Length of the train	77.6 m	101.9 m

8.3.5 Vehicle

The MLX01 test vehicle (Table 8.5), consisting of two end cars and one intermediate car, is shown in Fig. 8.18 and Fig. 8.19 [239]. The car body has

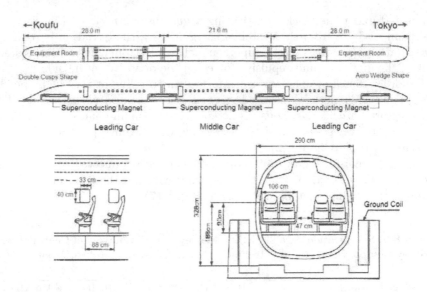

Fig. 8.18. Outline of the MLX01 Maglev Train. Courtesy of Central Japan Railway Company and Railway Technical Research Institute, Tokyo, Japan.

Fig. 8.19. (see color insert.) Double-cusp-shaped head car (facing Koufu) of the MLX01 Maglev Train at Expo 2005, Aichi Prefecture, Japan. Courtesy of Central Japan Railway Company and Railway Technical Research Institute, Tokyo, Japan.

been designed to obtain a mass reduction and provide a comfortable interior. Both the mass and cross section of the car are smaller than those of existing *Shinkansen* trains to reduce the air drag and improve dynamic performance. The structure of the body using the aircraft and rolling stock technologies has a light weight and enough strength to endure the repeat of large pressure fluctuations when passing through tunnels. Two types of nose shapes, i.e., *double cusp* and *aerowedge*, were developed, which considerably reduce air drag and aerodynamic noise.

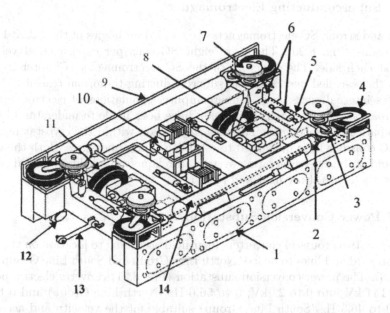

Fig. 8.20. Bogie of Yamanashi Maglev Test Line Vehicle: 1 — SC magnet, 2 — helium refrigerator, 3 — guiding stopper wheel, 4 — guiding gear, 5 — oil reservoir tank, 6 — dampers, 7 — air spring, 8 — hydraulic pressure unit, 9 — side cover, 10 — helium compressor, 11 — landing gear, 12 — emergency landing wheel, 13 — longitudinal anchor (to car body), 14 — liquid helium and nitrogen tanks. Courtesy of Central Japan Railway Company and Railway Technical Research Institute, Tokyo, Japan.

The bogie, on which the SC magnets are mounted, serves to transmit the propulsion and levitation forces to the vehicles (Fig. 8.20). A refrigeration system for freezing the helium is also mounted on the bogie. To improve traveling comfort, pneumatic springs for car body suspension and vibration control devices are incorporated in some bogies.

The bogie is fitted with landing and guide gear wheels that are necessary when traveling at low speeds. Hydraulic apparatus are used for raising and lowering these wheels.

Landing gears have been developed taking durability and mass reduction into consideration. Disk brakes and rubber tires are now capable of use at speeds over 500 km/h. To follow the track center at low speeds, the train is equipped with side guide gear wheels of smaller diameter than landing gear wheels.

The speed record of 581 km/h on the Yamanashi Maglev Test Line was achieved in 2003 (manned vehicle).

8.3.6 Superconducting Electromagnet

Light and strong SC electromagnets are carried on bogies of the MLX01 maglev trains (Fig. 8.20). There are eight SC coils per experimental vehicle, four at each side. The structure of the SC electromagnet (Chapter 2) prevents the so-called *quench* effect (superconducting-to-normal transition) and reduces internal excess heat. Supercomputer simulations of electromagnetic disturbances and mechanical vibration have been made to understand better these parasitic phenomena. The heat generation within the cryostat housing the SC coils has been quantified to establish countermeasures. Both the cryostat and on-board refrigeration system to reliquefy the helium gas vaporized within the cryostat are light and robust.

8.3.7 Power Conversion Substation

There are two groups of converters identified according to location on the substation yard and lines to be fed: North Line Group and South Line Group (Table 8.6). The power conversion substation (Fig. 8.21) converts electric power from 154 kV into 0 to 22 kV, 0 to 56.6 Hz (North Line Group) and 0 to 11 kV, 0 to 46.3 Hz (South Line Group) suitable for the velocity and acceleration or deceleration of the maglev train. At the frequency $f = 56.6$ Hz and pole pitch $\tau = 1.35$ m (Tables 8.5 and 8.6), the maximum train velocity is $v_{max} = 2f\tau = 2 \times 56.6 \times 1.35 = 152.82$ m/s ≈ 550 km/h. At $f = 46.3$ Hz, the maximum velocity is 450 km/h.

The main step-down transformer is rated at 60 MVA, 154/66 kV. The North Line Group consists of a 69 MW thyristor converter and 38 MVA GTO inverter with 500 Hz PWM. The South Line Group consists of 33 MW GTO converter with 350 Hz PWM and 20 MVA GTO inverter with 300 Hz PWM. High-power 4.5 kV, 4 kA GTO devices have been used [136].

Thus, the Yamanashi Maglev Test Line uses one of the largest inverters in the world. The inverter output current must be high (about 1 kA) because the insulation system of ground coils can withstand only limited voltage level. To reduce the input harmonic current at low speed operation and improve the power factor, converters are of multibridge structure.

Each group of three-phase inverters feeds three sections of the North Line and three sections of the South Line. The North Line with maximum speed of 550 km/h is fed with 22 kV. The South Line with maximum speed of 450

Table 8.6. Specifications of power converters. Courtesy of Central Japan Railway Company and Railway Technical Research Institute, Tokyo, Japan

Parameter	North Line North Group	South Line South Group
Converter		
• Rated power	69 MW	33 MW
• Input voltage	66 kV	66 kV
• Input frequency	50 Hz	50 Hz
• Output voltage	±3450 V d.c.	±2625 V d.c.
Inverter		
• Number of phases	3	3
• Rated power	38 MVA	20 MVA
• Output voltage	0 to 12.7 kV	0 to 6.35 kV
• Output voltage behind output transformer	0 to 22 kV	0 to 11 kV
• Output current	0 to 960 A	0 to 1015 A
• Output frequency	0 to 56.6 Hz	0 to 46.3 Hz
d.c. chopper		
• Rated power	19 MW	4.09 MW
• Input voltage	±3.55 kV	±2.625 kV
• Chopping frequency	300 Hz	350 Hz
• Control	PWM	PWM

km/h is fed with 11 kV. The feeding section switchgears turn on and off each time the train passes by. Vacuum switches have been developed for endurance through over one million test operations.

8.3.8 Brakes

The braking system of the prototype maglev train consists of ground-based brakes and on-board brakes.

The *ground-based brakes* incorporate regenerative braking and rheostatic braking. During regenerative braking, the current is reversed and returned to the power system. In rheostatic braking, the LSM operates as a generator and the kinetic energy of the train is converted into electric energy dissipated in the braking resistor.

The *on-board brakes* constitute a backup. To achieve a stable braking force from very high speed to standstill, two kinds of on-board brakes are installed. In high-speed range, *aerodynamic brakes* are effective, while in middle to low-speed range, the train is brought to a halt by built-in-wheel *disk brakes* (Fig. 8.22).

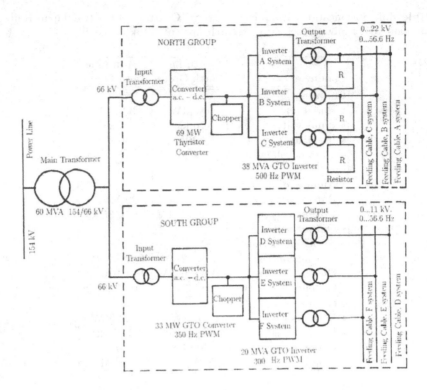

Fig. 8.21. Block diagram of power conversion substation [239].

8.3.9 Boarding System

An *indoor type passenger platform* has been designed (Fig. 8.23). The boarding system aligns with the passenger door in order to guide the passengers during boarding and provides a *magnetic shielding*. A narrow section of the guideway wall (where guide tires have a contact) facing the train door can fold down to allow passengers to board.

In order to make the interior of the passenger cabin more spacious while providing extra smoothness on the exterior of the cars and make them more airtight, special *upward-sliding doors* have been designed. Doors have been equipped both with infrared and contact sensors that sense passengers getting on and off the train, thus eliminating the danger of doors accidentally closing.

8.3.10 Control System

The experimental maglev train is operated automatically by ground-based control equipment. The whole control system (Fig. 8.24) consists of the three following systems: (a) traffic control, (b) safety control, and (3) drive control system. The train operation control center is located in the station building.

(a)   (b)

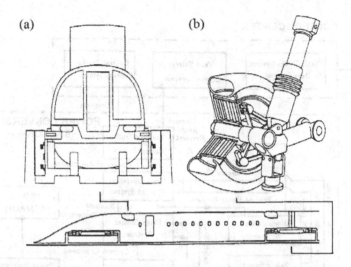

Fig. 8.22. On-board brakes: (a) aerodynamic brake, (b) disk brake. Courtesy of Central Japan Railway Company and Railway Technical Research Institute, Tokyo, Japan.

Fig. 8.23. Passenger boarding system. Courtesy of Central Japan Railway Company and Railway Technical Research Institute, Tokyo, Japan.

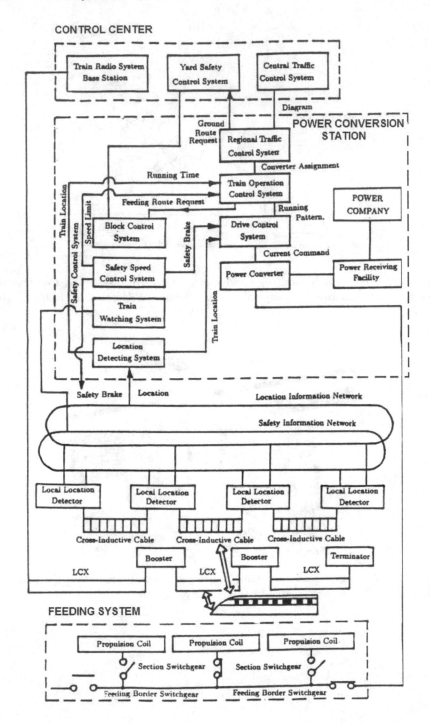

Fig. 8.24. Operation system of the experimental Yamanashi Maglev Test Line. Courtesy of Central Japan Railway Company and Railway Technical Research Institute, Tokyo, Japan.

Traffic control system

The *traffic control* system comprises the central traffic control system, the regional traffic control system, and the train operation control system. The central traffic control system generates a *schedule of all trains*, and the regional traffic control system regionally details it. The train operation system controls each train according to the direction of the regional traffic control system. The traffic control system also monitors all trains on the whole line.

Safety control system

The *safety control* system monitors the location and velocity of trains at all times and generates an acceptable speed limit for each train. Only within this limit are the traffic control system and the drive control system allowed to work. If the train runs out of this acceptable limit, the safety control system outputs the command of the safety brake and stops the train safely.

Since the maglev train is controlled on the ground, the safety control system uses cross-inductive cable installed along the whole line to detect the position of the train with high precision within several centimeters.

Drive control system

The *drive control* system corresponds to the driver in the conventional railway system. This system takes the running (rolling, gradient, acceleration, curve, and air) resistance and riding comfort into account and enforces the running pattern generated by the train operation system with the power electronics converters. It also switches on and off the section switchgears according to the position of the train.

8.3.11 Communication System

One leaky coaxial cable (LCX) per track has been installed for the train radio communication system. The train radio system uses millimetric waves from 30 to 300 GHz.

8.3.12 Experiments

The following experiments are carried out on Yamanashi Maglev Test Line:

- Basic running tests, i.e., levitation running tests, wheel running tests, speed increasing tests, maximum speed (approximately 550 km/h) verification tests
- General functional tests, i.e., high-speed passing tests, tunnel entering tests, substation crossover tests, multiple train control tests, emergency tests, etc.

- Reliability verification tests, i.e., high-speed continuous running tests, transportation capability verification tests, *etc.*;
- Other verification tests such as passenger physiology confirmation tests, station facilities verification tests, environmental impact verification tests, economy verification tests, maintenance standards verification tests, etc.

Experiments began in April 1997. Intensive experiments with two trains were carried out in 1998. High-speed running tests have been performed since 1999.

The main goal of the tests is (a) confirmation of possibilities of safe, comfortable, and stable run at 500 km/h, (b) confirmation of reliability and durability of the vehicle, wayside facilities, and equipment as well as superconducting magnets, (c) confirmation of structural standards, including the minimum radius of curvature and the steepest gradient, (d) confirmation of center-to-center track distance for safety of trains passing each other, (e) confirmation of vehicle performance in relation to tunnel cross section and pressure fluctuations in tunnels, (f) confirmation of performance of the turnout facilities, (g) confirmation of environmental impact, (h) establishment of multiple-train operation control systems, (i) confirmation of operation and safety systems and track maintenance criteria, (j) establishment of inter-substation control systems, and (k) pursuit of economic issues, construction and operation costs.

Central Japan Railway Co. (JR Tokai) recently got the green light to effectively proceed with development of its planned 450 km Chuo Shinkansen Line (Tokyo–Nagoya–Osaka) using magnetic levitation technology that will allow trains to run at 500 km/h (top speed 600 km/h). Construction costs (2009) are evaluated as 8 trillion yen ($94 billion). After 46 years of service, the existing Tokaido Shinkansen Line require a major overhaul and needs an alternative. The transport minister is expected in 2011 to allow JR Tokai to begin construction. Actual construction of the line is expected to start in 2014.

8.3.13 History of Superconducting Maglev Transportation Technology in Japan

- 1962 — Research in linear motor propulsion and noncontact suspension started.
- 1972 — Experimental SC maglev test vehicle ML-100 succeeded in 10 cm levitation on the yard of the Railway Technical Research Institute (RTRI) in Kokubunji, Tokyo.
- 1977 — Test run of ML-500 vehicle on inversed T-shaped guideway started.
- 1979 — Unmanned test vehicle ML-500 achieves world speed record of 517 km/h at the 7-km Miyazaki Maglev Test Track, Kyushu island.
- 1980 — Test run of MLU001 vehicle on U-shaped guideway started.
- 1987 — The speed of 400.8 km/h achieved by 2-car manned unit. RTRI reorganized. Test run of MLU001 started.
- 1990 — The Yamanashi Maglev Test Line construction plan approved by the Ministry of Transport.

- 1991 — Test run on sidewall levitation system started (Miyazaki Maglev Test Track). MLU002 was destroyed by a fire during a test run.
- 1994 — The speed of 431 km/h attained on Miyazaki Maglev Test Track by unmanned MLU002N vehicle.
- 1995 — The speed of 411 km/h attained on Miyazaki Maglev Test Track by manned MLU002N vehicle.
- 1996 — The 18.4 km section of Yamanashi Maglev Test Line completed. First train MLX01 (3 cars) delivered.
- 1997 — Running tests with MLX01 train commenced. Speed records on Yamanashi Maglev Test Line: 531 km/h on December 12 (manned train), 550.0 km/h on December 24 (unmanned train). Second train MLX01 delivered.
- 1998 — Maglev trains MLX01 passed each other at a relative speed of 966 km/h on Yamanashi Maglev Test Line.
- 1999 — Speed record of 552 km/h (manned train) on Yamanashi Maglev Test Line.
- 2003 — Manned speed of 581 km/h recorded. Total test run distance of over 300,000 km and over 50,000 test ride passengers recorded.
- 2005 — Two Maglev vehicles MLX01 with 3 sections achieved passing speed of 1026 km/h (575 + 451) on Yamanashi Maglev Test Line. MLX01 with HTS electromagnet achieved speed of 553 km/h.
- 2006 — JR Central's Board of Directors approved renewal and extension plan for 42.8 km Yamanashi Maglev Test Line (Yen 355 billion).
- 2010 — Green light to proceed with development of 450 km Maglev Chuo Shinkansen Line (Tokyo–Nagoya–Osaka).

8.4 American Urban Maglev

The cost and complexity of presently developed high-speed maglev systems as *Transrapid* and Yamanashi Maglev Test lines have slowed their deployment. A PM magnetic levitation system may offer an economic alternative to existing maglev systems. The so-called *Inductrack*, employing PM Halbach arrays, is an example of a practical cost-effective low-speed maglev transportation system, e.g., urban maglev systems, people movers, and point-to-point shipment of high-value freight [171, P111, P118].

A 120 m test track for testing an urban maglev vehicle was built by *General Atomics* in 2004 in San Diego (Sorrento Valley), CA, USA. The test vehicle consists of a single 5-m long chassis unit (Fig. 8.25). An EDL system with a flat PM LSM is used for levitation and propulsion. NdFeB PMs arranged into a Halbach array are mounted on the vehicle. Coreless levitation coils are installed in the guideway between the upper and lower Halbach arrays. When the vehicle moves, currents induced in shorted levitation coils interact with PMs to produce suspension forces.

Fig. 8.25. (see color insert.) Prototype of urban maglev vehicle built by General Atomics, San Diego, CA, USA. Photo taken by the first author.

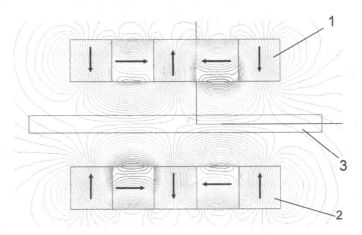

Fig. 8.26. PM configuration in General Atomics' maglev vehicle. 1 — upper Halbach array, 2 — lower Halbach array, 3 — copper coils.

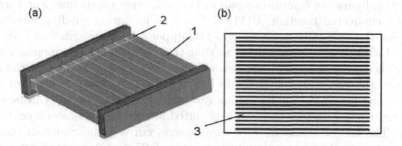

Fig. 8.27. Construction of active guideway of General Atomics' maglev vehicle: (a) "ladder guideway", (b) laminated guideway. 1 — Litz wire cable (rung of "ladder"), 2 — Cable ends soldered into shorting bus bars, 3 — copper or anodized aluminum sheets [171].

Fig. 8.28. (see color insert.) Principle of operation of General Atomics' maglev vehicle. 1 — upper Halbach array levitation magnets, 2 — lower Halbach array levitation magnets, 3 — Litz wire guideway, 4 — LSM armature winding, 5 — propulsion magnets. Photo taken by the first author.

PM cofiguration comprises a pair of Halbach array magnetically and structurally connected together [P111, P118]. The levitation winding is located between the pair of Halbach arrays. The upper and lower arrays of PMs are phased with respect to each other so that their vertical field components add, while their horizontal field components cancel (or nearly cancel) in the gap between the two arrays as shown in Fig. 8.26. The guideway is made of Litz wire shorted coils. A "ladder guideway" with close-packed rungs can be constructed using Litz wire cables encapsulated in thin-wall stainless steel tubes [171]. The use of "braided" Litz wire assures current uniformity in the cables and minimizes eddy-current losses (Fig. 8.27a). A laminated copper or aluminum guideway has also been considered (Fig. 8.27b). The laminated guideway is a high-efficiency alternative to the Litz wire "ladder guideway."

Halbach arrays of PMs have been configured to provide a nominal air gap of 25 mm. Upper levitation magnets and currents induced in short-circuited coils installed in the guideway produce attraction forces, while lower levitation magnets and currents induced in coils produce repulsive forces. As long as the vehicle keeps moving, these forces keep it airborne. When the vehicle slows down or comes to a stop, it settles back down onto its wheels, which are permanently deployed.

The thrust is provided by propulsion PMs (on the vehicle) that interact with the armature winding of a long laminated core LSM embedded in the guideway. Owing to large air gap, an LSM is fundamentally better suited to the needs of an EDS suspension system than an LIM. The LSM three-phase armature winding is simply made of copper cables. Levitation and propulsion components are shown in Fig. 8.28.

The lift F_z to drag F_x ratio as a function of the guideway circuit parameters is [170, 171]

$$\frac{F_z}{F_x} = \frac{2\pi f L}{R} = \frac{2\pi v}{l_a} \frac{L}{R} \tag{8.9}$$

where $f = v/l_a$, v is the linear velocity, l_a is the spatial period (wavelength) of Halbach array (from one pole to the next the same polarity pole), R and L are the guideway circuit resistance and inductance, respectively. The lift-to-drag ratio increases linearly with the velocity v and with the l/R ratio of the guideway circuit. Putting the output power P_{out} according to eqn(1.2) to eqn (8.9), the so-called *levitation efficiency* is

$$\eta_l = \frac{F_z}{P_{out}} = \frac{2\pi}{l_a} \frac{L}{R} \text{ N/W} \tag{8.10}$$

According to [171], typical value of $\eta_l = 1$ to 5, depending on the guideway design and arrangement of Halbach arrays.

The mass of empty vehicle is 9500 kg. The length of the vehicle is 12 m (two chassis units), the width is 2.6 m and the height is 3 m. The vehicle is supported on wheels when stationary, but levitates as it reaches the lift-off speed of about 2.5 m/s.

Testing with chassis weight up to 10,000 kg, to a speed of 10 m/s, air gaps up to 30 mm and acceleration up to 2.8 m/s^2 have been achieved [74]. Upon successful completion of tests, the test track will continue being used for system optimization, while a demonstration system is constructed at California University of Pennsylvania (CalU) in California, PA, located about 60 miles southwest of Pittsburgh. When completed, this system will be 7.4 km in length with four stations and 3 vehicles, connecting the upper and lower campus via a 7% grade. The system will serve the main campus, the city, and student housing/sports facilities on the upper campus [74].

8.5 Swissmetro

Switzerland is a small country (41,293 km^2) with a mountainous landscape and moderate density of population, i.e., 172 inhabitants per km^2 (compare with 234.5 inhabitants per km^2 in Germany and 333.3 inhabitants per km^2 in Japan). The highest population density spread over a distance of 300 km is in the Swiss central plateau with major cities Geneva, Lausanne, Bern, Luzern, Zürich, St Gallen (West to East), and Basel and Bellinzona (North to South). The distance between each of the neighboring major cities is 40 to 100 km. There is a high saturation of main transport routes in Switzerland. To solve the transportation problems and protect the natural environment, a high-speed *underground transportation network* seems to be the only solution (Fig. 8.29).

8.5.1 Assumptions

The *Swissmetro* network project assumes speeds around 400 km/h, about 12 min travel time between stations, 3 min stops at stations, 4 to 8 trains per hour, and trains with minimum passenger capacity of 200 persons. The travel time from Zürich to Geneva will take less than 1 h as compared with 3 h by surface trains.

The *Swissmetro* project is based on four complementary modern technologies:

- An underground infrastructure with two parallel tunnels of 5 m in diameter in each direction and stations linked to the existing public means of transportation (railways, roads, airports)
- A partial vacuum with the air pressure reduced to 8 to 10 kPa (0.08 to 0.1 atm) in tunnels to minimize the air resistance and thus reduce the energy consumption
- A propulsion system by using linear motors with guidaway mounted armatures
- Support and guidance system by using the EML levitation technology

8.5.2 Pilot Project

The line Geneva–Lausanne, about 60 km long, has been selected as the *pilot project*. The cross section of the tunnel and the vehicle is shown in Fig. 8.30. Specifications of the pilot vehicle are given in Table 8.7. To control the air gap of levitation and guidance electromagnets, air gap estimators based on determination of inductance will be used rather than position sensors.

Swissmetro vehicles require electric power for propulsion, guidance, and auxiliary equipment such as lights, airconditioning, communication, and safety equipment, which is estimated as 500 kW (power supplied to the vehicle, peak demand). The power will be transferred to the vehicle with the aid of high-frequency linear transformers. The long primary (about 1 km) of the linear transformer is fixed to the guidaway, and the secondary winding is fixed to the vehicle. At the frequency 2 kHz and air gap 20 mm, the predicted efficiency of the linear transformer is about 80%.

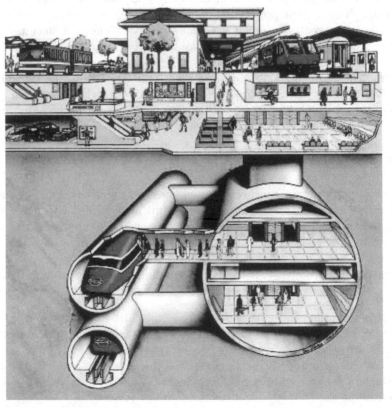

Fig. 8.29. (see color insert.) Cross section through the station of *Swissmetro*. Courtesy of Swissmetro, Geneva, Switzerland.

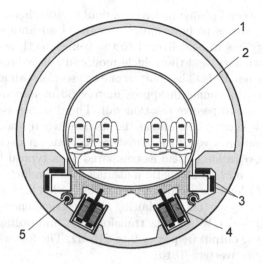

Fig. 8.30. Cross section of the tunnel and vehicle of *Swissmetro* pilot project: 1 — inner concrete tunnel ring (5 m diameter), 2 — passenger car body, 3 — EML levitation support and guidance system, 4 — LSM, 5 — linear transformer. Courtesy of Swissmetro, Geneva, Switzerland.

Table 8.7. Specification data of *Swissmetro* pilot vehicle. Courtesy of Swissmetro, Geneva, Switzerland

Maximum speed	500 km/h
Mechanical power demand	6 MW
Power supplied to the vehicle (levitation, guidance and auxiliary equipment)	500 kW
Diameter of vehicle	3.2 m
Length of vehicle	80 m
Number of passengers per vehicle	200
Total mass of vehicle	50 t
Initial acceleration from standstill to 290 km/h	1.3 m/s^2
Air gap for propulsion, levitation, guidance, and linear transformer	20 mm

Two configurations of iron-cored LSMs have been investigated [181]:

- guideway-mounted long armature with three phase winding and on-board excitation system;
- guideway-mounted short three-phase armature integrated with excitation system and vehicle-mounted passive reaction rail.

Both configurations use two LSMs with pole pitch $\tau \approx 0.323$ m located in parallel at both sides of the guideway. LSMs are rated at 4.5 kV/1.4 kA or 6.4

kV/0.9 kA. The second configuration, i.e., double-sided homopolar LSM (Fig. 1.14, Chapter 1) seems to have more economical advantages. No electrical propulsion energy has to be delivered to the vehicle as three-phase armature windings are stationary, and the vehicle-mounted passive reaction rail needs neither windings nor PMs. The short armature sections about 10 m long are distributed along the tunnel in approximately 50 m intervals, equal to the length of the on-board passive reaction rail. The distance between armature sections should be shorter near and at the stations where trains require higher power for acceleration. Even if the armature sections are oversized due to the presence of the excitation system as compared with typical LSMs, there is a substantial saving on ferromagnetic materials and conductors. Both electromagnetic and PM field excitation of homopolar LSMs have been considered.

The power supply system distributing the electrical energy along the tunnel consists of 125/6.1 kV, 50 Hz transformers, and voltage source PWM converters with the output frequency 0 to 215 Hz. The following requirements are to be met by converters [181]:

• maximum apparent power of 8.5 MVA at power factor $\cos \phi = 0.85$ (active power 7.225 MW);
• power factor on the line side $\cos \phi = 1.0$;
• operation in propulsion and regenerative braking mode with any desired power factor on the LSM side;
• from 290 km/h up to maximum speed 500 km/h, the acceleration at constant electric power of 7 MW with efficiency exceeding 85%.

The maximum d.c. link voltage between rectifier and inverter has been assumed to be 10 kV.

The predicted energy consumption at constant speed of 400 km/h of the *Swissmetro* system, taking into account the maintenance of vacuum, is 43 Wh/(seat×km). At the same speed, the *Transrapid 07* consumes 50.4 Wh/seat×km.

History of *Swissmetro*

• 1974 — Swiss engineer R. Nieth develops the concept of *Swissmetro* to provide high-speed transport between principal urban and rural areas.
• 1985 — Interest in the project is shown at a political level.
• 1989 — The Federal Department of Transport, Communication, and Energy (EVED) grants the Swiss Federal Institute of Technology (EPFL), Lausanne, a subsidy to conduct the preliminary study.
• 1992 — Founding of Swissmetro, S.A. in Geneva as a joint stock corporation.
• 1993 — The results of the preliminary study confirm the desirability, feasibility, and viability of the project. Inauguration of the first *Swissmetro* exhibition at Swiss Transport Museum in Lucerne.

- 1994 — Launch of the main study supported financially by major Swiss companies.
- 1998 — Main study of the system and components.
- 1999 — Swiss Government supports the project.
- 2000 — Study on the environmental impact of *Swissmetro*.
- 2009 — The *Swissmetro SA* in Geneva is going into liquidation because of lack of support. Around SFr 11 million ($10.77 million) was invested in the project, with federal authorities paying half. The rights were transferred to Federal Institute of Technology Lausanne (EPFL).

8.6 Marine Express

A reliable and weather-independent, short-distance inter-island or island-mainland link is still a difficult transportation and civil engineering task. Ferry connections are subject to the weather, and world statistics show numerous ferry sinkings with many casualties. Undersea tunnels are extremely expensive and demand expensive maintenance, e.g. to pump in fresh air and pump out the excessive water coming in all the time. For example, the world's longest 53.85 km Seikan Tunnel between Honshu and Hokkaido islands (Japan), took over 40 years to be completed since its idea was conceived and geological survey began.

To combat noise, save land, and protect the natural environment, it is wise, where possible, to locate future airports on the sea, minimum 10 km away from the mainland. This can be done by building artificial islands. Such an airport requires reliable and convenient 24 hour passenger transportation between the mainland and artificial island. A similar problem arises in the case of future undersea cities accommodating hundreds thousands of inhabitants.

Amphibious trains traveling both under the water and on the dry land would be an ideal solution [244]. Such an underwater train, also called *marine express*, is proposed to be levitated on an elevated guidaway and propelled by linear motors [244]. The elevated guidaway can be laid on the sea bottom or on land. To overcome the buoyancy of the train under the water the bogie needs to be electromagnetically attracted to the guidaway. On land, the train needs to be levitated by repulsive forces. In both cases, a constant air gap is to be maintained by controlling the input electrical variables. An LSM with SC excitation system has been identified as the best candidate due to its high thrust, large air gap, low excitation power, and controllability of vehicle buoyancy under water [244].

Water density is much higher than that of air, and consequently, water resistance force is much higher than air resistance force — see eqn (1.55). It has been found that, in order to keep the same thrust, a 100 m long marine express traveling on land with the speed of 420 km/h will have to reduce its speed to 61 km/h under water [244]. On the other hand, the power required to run on land is 9.3 MW at 420 km/h in comparison with only 1.4 MW at 62 km/h under water (the power is approximately proportional to v^3) [245].

A research team lead by K.Yoshida and T.Ota at Kyushu University, Fukuoka, Japan, has successfully tested small-scale prototypes of underwater train cars to prove the feasibility of the idea. An SC LSM has been replaced by a PM LSM [245]. When the buoyancy of the water is larger than the weight of the vehicle, an attractive-mode levitation and propulsion is realized by controlling the armature current and load angle.

A *pitching-damping control* method has been proposed to obtain stable levitation and propulsion. It is impossible to control the pitching angle completely together with the control of levitation and propulsion by controlling only the thrust F_x and normal force F_z [245]. On the other hand, by controlling the thrust F_x of an LSM, the pitching motion can be easily and effectively damped. The *decoupled control* of levitation and propulsion with pitching motion control is effective, and it has enabled a small-scale prototype vehicle to run successfully under water in a repulsive mode [245].

Examples

Example 8.1

Find the attraction force of a U-shaped electromagnet with the following dimensions: width of pole $a = 12$mm, length of core $l = 200$mm, length of the path of the magnetic flux in the ferromagnetic core including armature $l_{Fe} = 400$mm, nominal air gap $g = 20$mm. The total number of turns is $N = 200$, and the d.c. current in the coil is $I = 100$ A. Find the attraction force: (a) neglecting the magnetic voltage drop in the ferromagnetic core, and (b) including the magnetic voltage drop in the ferromagnetic core. The relative magnetic permeability of the ferromagnetic core is $\mu_r \approx 200$.

Solution

It can be assumed that the cross-section area of the ferromagnetic core is equal to the cross section of the air gap A, i.e.,

$$A = al = 0.012 \times 0.2 = 0.0024 \text{ m}^2$$

The attraction force, neglecting the magnetic flux path in the ferromagnetic core, is calculated using eqn (1.15), i.e.,

$$F_z = \frac{1}{4} \frac{0.4\pi \times 10^{-6}(200 \times 100)^2}{0.02^2} \times 0.0024 = 754.0 \text{ N}$$

Including the magnetic voltage drop in the ferromagnetic core, the attraction force according to eqn (8.3) is

$$F_z = \frac{1}{4}\frac{0.4\pi \times 10^{-6}(200 \times 100)^2}{[0.4/(2 \times 200) + 0.02]^2} \times 0.0024 = 683.9 \text{ N}$$

The corresponding magnetic flux density in the air gap

$$B_g = \frac{1}{2}\frac{\mu_0(NI)}{l_{Fe}/(2\mu_r) + g} = \frac{1}{2}\frac{0.4\pi \times 10^{-6}(200 \times 100)}{0.4/(2 \times 200) + 0.02} = 0.598 \text{ T}$$

Table 8.8. Calculation results of attraction forces and corresponding magnetic flux densities for constant current $I = 100$ A and air gap range $8 \le g \le 30$ mm. Example 8.1

Air gap	F_z according to eqn (8.3) including magnetic saturation	F_z according to eqn (1.15) neglecting magnetic saturation	Air gap magnetic flux density
mm	N	N	T
8	3723	4712	1.396
10	2493	3016	1.142
15	1178	1340	0.785
20	684	754	0.598
25	446	483	0.483
30	314	335	0.405

Table 8.8 shows attraction forces and corresponding magnetic flux densities for constant current $I = 100$ A and air gap range $8 \le g \le 30$ mm.

Example 8.2

A coil as shown in Fig. 8.2 with dimensions $a = 1.0$ m, $l = 0.6$ m, number of turns $N = 200$ carrying d.c. current $I = 1000$ A is moving with linear speed $v = 20$ m/s over a conductive aluminum plate with thickness $d = 25.4$ mm and electric conductivity $\sigma = 30 \times 10^6$ S/m. The distance between the coil and conductive plate is $g = 100$ mm. Find the repulsive force between the coil and aluminum plate.

Solution

Assuming $\lambda \approx a$, the frequency of the current in the conductive plate is $f = v/\lambda = 20/1.0 = 20$ Hz, and the angular frequency is $\omega = 2\pi f = 2\pi \times 20 = 125.7$ rad/s. The equivalent depth of penetration according to eqn (8.7)

$$\delta = \sqrt{\frac{2}{125.7 \times 0.4\pi \times 10^{-6} \times 30 \times 10^6}} = 0.021 \text{ m}$$

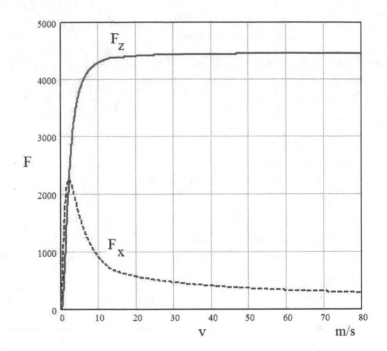

Fig. 8.31. Repulsive force F_z and drag (braking) force F_x plotted against the linear velocity v of the coil carrying constant d.c. current $I = 1000$ A over aluminum plate ($d = 25.4$ mm, $g = 100$ mm). Example 8.2.

is less than the aluminum plate thickness, i.e., $\delta < d$, so that

$$k = \frac{2}{0.4\pi \times 10^{-6} \times 20.0 \times 30 \times 10^6 \times 0.021} = 0.129$$

as given by eqn (8.6). Putting

$$A = \left(\frac{a}{2}\right)^2 + g^2 = \left(\frac{1.0}{2}\right)^2 + 0.1^2 = 0.26 \text{ m}^2$$

and

$$B = \left(\frac{l}{2}\right)^2 + g^2 = \left(\frac{0.6}{2}\right)^2 + 0.1^2 = 0.10 \text{ m}^2$$

the repulsive force, as calculated with the aid of eqn (8.4), is

$$F_z = \frac{0.4\pi \times 10^{-6}(60 \times 1000)^2}{\pi \times 0.1}$$

$$\times \left\{ \sqrt{0.26} + \sqrt{0.10} - 2 \times 0.1 - \left[\sqrt{0.26 + \left(\frac{0.6}{2}\right)^2} - \sqrt{0.26} \right] \right.$$

$$\left. \times \frac{0.10^2}{0.26} - \left[\sqrt{0.10 + \left(\frac{1.0}{2}\right)^2} \right] \frac{0.10^2}{0.10} \right\} \times \frac{1}{1 + 0.104} = 4402.8 \text{ N}$$

Similarly, on the basis of eqn (8.8) the drag force is

$$F_x = 0.129 \times 4402.8 = 568.4 \text{ N}$$

The repulsive and drag forces are plotted against the linear velocity from $v = 0$ to $v = 80$ m/s in Fig. 8.31.

Example 8.3

The MLX01 three-car maglev trainset with an SC LSM runs on a practically horizontal guidaway with the specific gradient resistance $k_g = 0$. The mass of the loaded train is $m = 84$ t, the front car is $w = 2.9$ m wide, and $h = 3.28$ m high with a wedge-shaped front nose. The following experimental test results on the Yamanashi Maglev Test Line are available [114]: (1) LSM propulsion force 200 kN at $v = 350$ km/h, and (2) acceleration $a = 0.2g$. Assuming air density $\rho = 1.21$ kg/m^3 at 20^{0}C and 1 atm, find the required output power and specific "rolling" resistance for the MLX01 maglev train.

Solution

The weight of the trainset

$$G = mg = 84,000 \times 9.81 = 824.04 \text{ kN}$$

The required output power of the LSM at constant speed $v = 350$ km/h is

$$P_{out} = F_x v = 200 \times 10^3 \frac{350}{3.6} = 19,444,444 \text{ W} \approx 19.45 \text{ MW}$$

For a wedge-shaped nose, the coefficient of proportionality in the air resistance force equation (1.55) is $C \approx 0.2$. The air resistance force at $v = 350$ km/h

$$F_{air} = 0.5 C \rho v^2 wh = 0.5 \times 0.2 \times 1.21 \times \left(\frac{350}{3.6}\right)^2 \times 2.9 \times 3.28 = 10,879 \text{ N}$$

The air resistance force obtained from eqn (1.55) seems to be too small because the air resistance due to the longitudinal surfaces of the train has not been taken into account. Also, the air resistance due to tunnels has been neglected.

The air resistance for MLX01 maglev train will probably be similar to that estimated for *Transrapid* maglev train, i.e., about 35 kN.

The specific acceleration resistance

$$k_a = \frac{a}{g} = \frac{0.2g}{g} = 0.2$$

On the basis of the traction effort equation (1.51)

$$k_r = \frac{1}{G}(F_x - F_{air}) - k_a$$

$$= \frac{1}{824.04 \times 10^3}(200 \times 10^3 - 35 \times 10^3) - 0.2 = 0.000233$$

This is rather a very rough estimation of k_r. This resistance corresponds to the "rolling" resistance of wheel-on-rail trains and is due to braking forces of electrodynamic nature, i.e., interaction of the magnetic field of on-board SC electromagnets and induced currents in levitation and guidance coils.

Building and Factory Transportation Systems

9.1 Elevator Hoisting Machines

Hoisting technology started when Archimedes constructed the first elevator that was based on pulleys and winches around 236 BC. However, early primitive elevators did not guarantee any safety for passengers. The situation changed with E. Otis' invention of a reliable *safety gear* in 1853 [P1]. In the first elevators, the *drum* was used to collect the *rope*. The major disadvantage was the necessity to lift the load together with the supporting structure. The next type, referred to as a *rope traction elevator*, has been constructed in such a way as to obtain the load balanced by the *counterweight*. This latter construction is widely used today.

Elevators can be classified into three major categories based on their size:

- High-rise elevators used in the tallest buildings in major cities and manufactured at a volume of about 2,000 units annually. These elevators add image and prestige to the company manufacturing them.
- Mid-rise elevators installed in office buildings, hotels, and other similar structures (annual market size of approximately 20,000 units). The appearance, comfort, and ride quality become most important for these installations.
- Low-rise elevators mostly installed in residential buildings (total annual sales of about 200,000 units worldwide).

There is an increasing demand to reduce both the space needed for *hoistways* in buildings and the size of elevator electrical supplies. These requirements have a strong influence on the selection of the hoisting machine, which can be remarkably improved by utilizing a linear motor [78]. Because the linear motor produces straightforward movement without mechanical transformations, thus improving the efficiency due to a smaller number of components, the usage of linear motors appears highly attractive. This technology matured in 1991 when *Nippon Otis* introduced the first commercial application of the *linear motor elevator* into the Japanese market [202]. Since then, the intensive research in this field was aimed at surpassing conventional technology in performance and cost. The type of electric motor, i.e., induction, switched reluctance, or synchronous, is important in the linear motor elevator technology. However, the elevator structure is even more important because requirements for the hoisting system vary with the arrangements of the elevator components. Sometimes the hoisting machinery must lift all traveling masses, whereas in other cases the counterweight, does part of the work. Moreover, in some elevators, the weight of the motor can be utilized as part of the mass balance.

9.1.1 Linear-Motor-Driven Elevator Cars

The earliest patent concerning a linear motor in elevators was granted to K. Kudermann in 1970 [P211]. The principle of that patent is illustrated in Fig. 9.1. The proposed system consists of a counterweight and two *linear motor armatures on both sides of the car*. The disadvantage of this setup is that the motor must also lift its own weight, thus increasing the demand for power supply during acceleration.

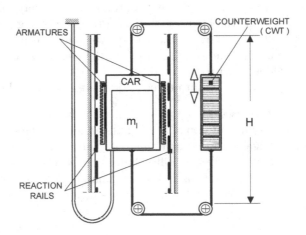

Fig. 9.1. Armature of a linear motor installed in the elevator car.

The best linear motor drive is a combination of the double-sided motor with the guide rail, which provides an advantage in balancing the attractive forces. The linear motor must be distributed on both sides of the car to obtain a symmetrical hoisting. The analysis of a few types of motors indicates that the LSM with ferrite PMs can offer the smallest cost-related mass, as shown in Table 9.1 [78]. The relative power used in Table 9.1 is defined as

$$p = \frac{P_{inm} + \Delta P}{P_{hoist}} \tag{9.1}$$

where $P_{hoist} = m_l g v$ is the hoisting power, v is the rated speed at rated load of the elevator, P_{inm} is the power supplied by the hoisting machinery, and ΔP is the power dissipated in the hoisting system. The relative power loss is the ratio of losses to the hoisting power

$$\Delta p = \frac{\Delta P}{P_{hoist}} \tag{9.2}$$

The P_{hoist} is not a theoretical minimum for the power of the hoisting machine because the counterweight does a part of the hoisting work, and the rest must by supplied by the hoisting machinery.

Table 9.1. Comparison of different linear motors applied in 1000 kg car traveling at 2 m/s for mid-rise elevator. Cost of related motor masses reflects material cost in 2010

Parameter	Motor		
	LRM	LSM, ferrite PMs	LSM, NdFeB PMs
Relative power, p	1.51	0.84	0.81
Relative power loss, Δp	0.57	0.15	0.14
Relative traveling mass	7.46	5.47	5.25
Relative moving mass of motor	1.29	0.36	0.55
Relative stationary mass of motor	0.66	0.90	0.30
Cost-related motor mass	5.82	3.65	4.25
Relative drive size	9.68	1.63	1.52

The relative masses are moving masses or stationary masses related to the mass of the rated load of the elevator m_l.

The relative drive size is defined as the ratio of the maximum electrical current demand I_{max} to the current corresponding to the hoisting power I_{hoist}.

The relative cost represents the weighted prices of the active material used for the elevator, excluding all the supportive structures and scaled to the mass of the whole hoisting system. The cost weighting coefficients have been assumed as follows: 2 for steel, 5 for aluminum, 8 for copper, 4 for ferrite PMs, and 40 for NdFeB PMs.

Coefficients listed in Table 9.1 lead to the conclusion that the PM LSM can offer the smallest cost-related mass, thus being the most promising alternative. On the other hand, it still remains uncertain whether the benefit of eliminating the rotary machine can be favorably compared to the increased cost of the linear motor.

9.1.2 Elevator with Linear Motor in the Pit

The alternative for a hydraulic elevator can be a linear motor with moving reaction rail as shown in Fig. 9.2.

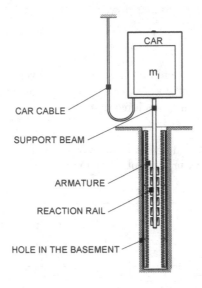

Fig. 9.2. Linear-motor-driven elevator with moving reaction rail.

The idea is to replace the piston of hydraulic machinery with the reaction rail, and the cylinder with the armature of the linear motor. Power can be supplied to the stationary winding. However, either a pit in the basement or the 1:2 roping ratio is required to obtain the required doubled force. Furthermore, the car travel distance is limited to maximum 20 m. Comparison of different linear motors installed in the pit for low-rise elevators is given in Table 9.2.

The structure has the same features as the hydraulic set-up but with a few additional advantages, i.e.,

- no oil and oil tank,
- all machinery is located in the shaft, so that a separate room is not needed,
- the regenerative braking can be applied for the car going downward, so that the lack of counterweight can be partially compensated.

The performance of this elevator with PM LSM compares well with the conventional hydraulic system, but the amount of magnetic material required makes this structure commercially unattainable.

Table 9.2. Comparison of different linear motors installed in the pit for low-rise ropeless elevator (1600 kg car, 4 m/s). Cost of related motor masses reflects material cost in 2010 [78]

Parameter	Motor		
	LRM	LSM, ferrite PMs	LSM, NdFeB PMs
Relative power, p	17.75	7.96	7.01
Relative power loss, Δp	8.05	1.90	1.60
Relative traveling mass	3.58	3.70	3.40
Relative moving mass of motor	0.33	0.45	0.25
Relative stationary mass of motor	4.28	1.66	1.51
Cost-related motor mass	15.64	7.00	6.74
Relative drive size	111.6	10.37	12.40

9.1.3 Linear Motor in Counterweight

The *counterweight* is the most natural place for the linear motor in a traction type elevator. In this way, the mass of the motor can be utilized as a part of the balance. However, the energy must be supplied to the motor through a cable, the length of which on the balance side varies with the counterweight position. Proper measures should then be taken to compensate for this variation of the balance weight. This counterbalance motor placement has been well known for some time and even commercially explored [P14, P50, 202]. Although, similar to the motor installation in the car, when the motor is installed in the counterbalance, the difference in the mass of the supply cable must be taken into account (Fig. 9.3a).

The mass of the counterweight for optimum balancing should be

$$m_c = m + \frac{1}{2}m_l + \frac{H}{4}(m_{ec} - m_e) \tag{9.3}$$

and the mass of balancing ropes (per unit length)

$$m_{brp} = m_{rope} - \frac{m_e}{4} \tag{9.4}$$

where m_c is the mass of the counterweight, m is the mass of the car, m_l is the mass of the rated load, m_{ec} is the mass per unit length (kg/m) of the traveling electric cable on the counterweight side, m_e is the mass per unit (kg/m) of the traveling electric cable on the car side, m_{rope} is the mass of rope per unit length (kg/m), and H is the total hoisting height. The possible linear motor placement in the counterweight is shown in Fig. 9.3b. The system can be designed with the armature in the counterweight and reaction rail on the wall, as well as with the reaction rail in the moving counterweight and the armature on the wall.

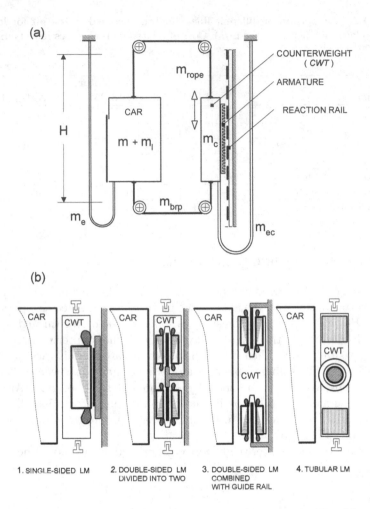

Fig. 9.3. Elevator system with the linear motor in the counterweight: (a) construction, (b) possible location of the linear motor.

Table 9.3 shows a comparison of different linear motors mounted in the counterweight of a low-rise elevator. PM LSMs require less power than conventional hoisting motors. Further analysis shows that the performance of a reluctance motor improves at higher speeds.

9.1.4 Conventional versus Linear-Motor-Driven Elevator

Table 9.4 illustrates the major parameters of a mid-rise elevator system with different motor types [78]. Both the conventional traction motor and linear

Table 9.3. Comparison of different linear motors mounted in the counterweight of a low-rise elevator (630 kg car, 1 m/s). Cost of related motor masses reflects material cost in 2010 [78]

Parameter	Motor		
	LRM	LSM, ferrite PMs	LSM, NdFeB PMs
Relative power, p	2.48	1.07	1.03
Relative power loss, Δp	1.13	0.26	0.24
Relative traveling mass	5.48	5.48	5.48
Relative moving mass of motor	1.13	0.36	0.27
Relative stationary mass of motor	0.33	0.45	0.25
Cost-related motor mass	4.62	2.46	2.40
Relative drive size	25.59	2.75	2.5

motor placed either in the car or in the counterweight offer approximately the same sizing of a hoisting machinery and the drive.

Table 9.4. Rotary motors versus linear motors for mid-rise elevators (1000 kg car, 2 m/s). Cost of related motor masses reflects material cost in 2010

Parameter	Direct hoisting (rotary motor)	Geared hoisting (rotary motor)	Linear reluctance motor in the car	PM (NdFeB) linear motor in counterweight
Relative supply power, p	1.03	1.12	1.98	1.98
Relative power loss, Δp	0.28	0.31	0.92	0.86
Relative traveling mass	4.82	4.81	6.59	5.53
Relative moving mass of motor	0.18	0.02	0.59	0.51
Relative stationary mass of motor	0.23	0.05	1.26	0.32
Cost-related motor mass	1.72	0.21	4.60	3.65
Relative drive size	2.21	1.97	7.44	6.84

9.2 Ropeless Elevators

9.2.1 Vertical Transport in Ultrahigh Buildings

Land in the world's biggest cities, e.g., New York or Tokyo, is extremely expensive, which drives the expansion of rentable spaces into higher and higher buildings and underground areas. However, the larger and taller the buildings, the more elevators are required to keep acceptable waiting time for dispatching. The increasing interest in *ultrahigh buildings*, also called *hyper buildings* or *vertical cities* that exceed 600 m in their height inspires elevator companies to intensify their research effort in alternative technologies for vertical transportation. This becomes apparent in Japan where building contractors, elevator companies, and other institutions are spending considerable effort in developing transportation concepts for these hyper buildings. Vertical transportation systems in ultrahigh buildings must address many technical issues, amongst others

- transport configuration with traffic flow within and between buildings;
- diverse building capacity incorporating residential, commercial, and service functions;
- use of alternative building transportation systems (roped elevators, ropeless elevators, escalators, people movers);
- the highest levels of reliability, safety, and passenger rescue;
- comfort of travel (air pressure changes, vibration, vertical and horizontal motions, travel time);
- elevator propulsion, guidance, brakes, power consumption, control, and communication.

Ultrahigh buildings pose new problems in the construction of high-speed elevator systems, i.e., vertical oscillations, horizontal swing, car noise, and cable length limitations. Because the ropes (steel cables supporting the car) are very long and usually have a low dumping coefficient, even small disturbances, e.g., traction machine torque ripple, can cause car oscillations. This vibration can further be amplified when the disturbance frequency coincides with the car's natural frequency. In some instances, this bounce may destabilize the system that controls the elevator speed. The horizontal swing of the elevator car can be caused by curvatures of the guiding rails or by imperfect rail segment junctions. Car noise is due to the roller-guides tracking the rails and by wind noise (air passing through the traveling car).

The car rise limit is imposed by the cable weight and strength and can be considered as a function of five variables: (a) safety factor, (b) rope strength, (c) rope mass per unit length, (d) number of ropes, and (e) mass of the car. Under the most favorable conditions, a cable-based elevator can achieve a rise of approximately 1200 m, based on 10 commonly used steel cables with 320 kN strength, mass per unit length of 2.14 kg/m, and safety factor of 10 [95].

Ropeless elevators with multiple cars in one shaft may be perceived as practical solution to ultratall buildings above 1000 m. The primary concern is that the roping technology may not be extensible to hoistways of that height, both from a rope strength standpoint, and safety margin considerations. Another problem is that roped elevator systems, understood to be based on sky lobbies, would consume too much space to make such a building financially viable. An analysis conducted by Mitsubishi Corporation [95] found that almost

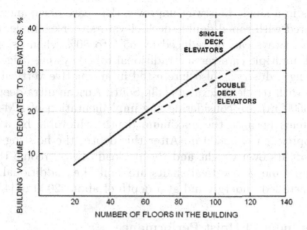

Fig. 9.4. Elevator system space occupancy ratio in a tall building.

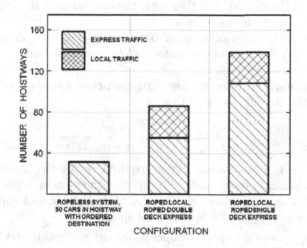

Fig. 9.5. Required number of hoistways in roped and ropeless elevator system capable dispatching 2000 passengers per minute. Study for 250-floor building, 1000 m tall, with 100,000 population.

30% of the total space in a 100-floor skyscraper must be devoted to elevators, including their hoistways, halls, and machine rooms (Fig. 9.4). Recently, the peak rents for Tokyo skyscrapers were estimated about US\$ $1000/m^2/$year. Clearly, the elevator space occupancy ratio has a significant financial impact on building utilization.

The one-shaft, multicar, ropeless elevator system is considered to be the most promising answer to these problems. It eliminates the suspension cables, and with them, the rope-strength and vertical-oscillation problems. Use of multiple cars in a single hoistway improves the space occupancy factor even when compared with roped double-deck elevators. For example, the required number of hoistways can be reduced by 65% to 80% when using a ropeless elevator with multiple cars versus traditional roped system (Fig. 9.5).

The mining industry is also interested in alternative vertical transportation systems with ropeless elevators [43]. South African ultra-deep gold mines (exceeding 3500 m) are considering the implementation of PM-LSM-driven hoisting systems because the maximum depth achievable by a roped hoist system is approximately 2800 m. After this depth, the hoisting rope can no longer support its own weight and the payload. To overcome this problem, at the present time, subvertical shafts are sunk, i.e., additional roped hoist system are installed underground at a depth of about 2000 m [43].

9.2.2 Assessment of Hoist Performance

The *hoist efficiency* without the rope is independent of the height (in a mine independent of the depth), and therefore, the operation of the lifting machinery is not limited by the rope mass.

The successful implementation of the linear motor hoist depends on two main factors: (a) the ratio of the *motor weight–to–thrust* that it can produce, and (b) the motor size (cost).

One of the main criterion for assessing the performance of the hoist is the efficiency of the overall system [44]:

$$\eta = \frac{P_{outm}}{P_{inm}} = \frac{m_l g v}{(m_l + m + m_{rope}) g v} \tag{9.5}$$

where P_{outm} is the mechanical power required to lift the payload, P_{inm} is the total mechanical power required to operate the system, m_l is the mass of the payload, m is the mass of the car, m_{rope} is the mass of ropes, g is the acceleration of gravity, and v is the linear speed.

Assuming a constant speed of operation and neglecting the friction, the efficiency of the hoist system without a counterweight can be expressed as follows [44]:

- for a conventional roped hoist

$$\eta = \frac{m_l}{m_l + m + m_{rope}} \tag{9.6}$$

- for a ropeless hoist

$$\eta = \frac{m_l}{m_l + m} \tag{9.7}$$

Assuming that the mass of the car and payload are the same in both cases, the system efficiency becomes entirely dependent on the mass of ropes.

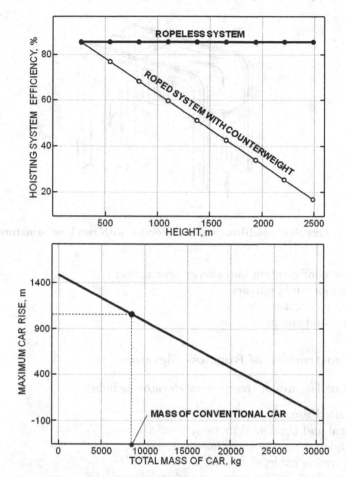

Fig. 9.6. Limits of rope lift system imposed by cable mass: (a) comparison of hoisting system efficiencies, (b) rope ability to support its own weight.

The taller (deeper) the hoist shaft, the longer and therefore the heavier the rope since $m_{rope} = 0.25\pi\rho n_{rope}d_{rope}^2 l_{rope}$ where ρ is the specific mass density of steel and n_{rope}, d_{rope}, and l_{rope} are the number of ropes, diameter, and length of the rope, respectively. Heavier ropes have larger diameter to withstand the increased tensile stresses. As a result, the increased cross-sectional area of ropes contributes further to their mass, affecting the overall system efficiency (Fig. 9.6a) [43].

The efficiency of roped hoist without counterweight tends to zero as the height (or depth) approaches the operating limit (about 2800 m for steel ropes).

Technologies crucial to the successful development of ropeless elevators can be divided into four major categories:

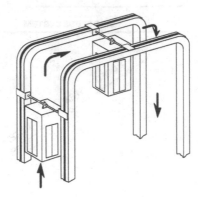

Fig. 9.7. One shaft, multicar, ropeless elevator with two long armature LSMs.

- system configuration and energy management
- propulsion and guidance
- safety and brakes
- control and communication

9.2.3 Construction of Ropeless Elevators

The unique features of the ropeless elevator include

- unlimited rise,
- vertical and horizontal motion,
- multiple cars in the same hoistway,
- no traveling cable,
- high traffic-handling capacity with minimum core space.

The ropeless elevator with multiple cars in one shaft can be built by hanging each cab from a bar fixed to the movers of a pair of LSMs (Fig. 9.7). Long armature windings are stationary, and PMs are integrated with movers. Each car hangs from a shaft fixed to the PM movers. The mover is a steel rail with PMs that is installed between two parallel stationary armature systems. Armature units are segmented into blocks (Fig. 9.8). The guidance system maintains a small and constant air gap between the armature cores and PMs. When the armature windings are excited, the interaction between the excitation flux and the flux of PMs produces linear thrust moving the car up or down. The long stationary armature is divided into sections, the minimum number of which is equal to the number of cars. Only one car can be permitted in each section at a time so that each car may be controlled independently of the others. The more sections there are, the more precisely the system can be controlled; however, the construction cost is higher since each section requires its own high-power converters.

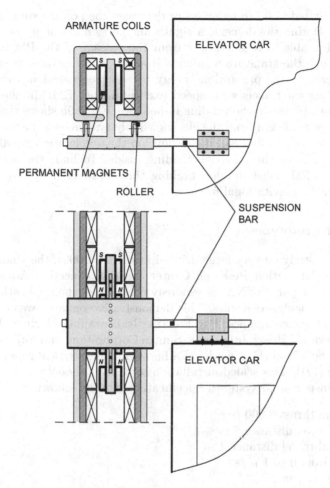

Fig. 9.8. Ropeless elevator with double-sided PM LSM.

9.2.4 Operation

The LSM-powered ropeless elevator consumes significantly more energy than an equivalent roped elevator to move a given number of passengers to a given distance. The reason for this is that a typical counterweighted elevator with ropes needs to drive only the offset load (typically only 12% to 14% of the total moving mass), while the ropeless elevator must lift the entire combined mass of the car, the frame, passengers, and the PM movers. Assuming motors and converters of similar efficiency, the ropeless elevator requires power supply 7 to 8 times that of its roped counterpart.

An ascending ropeless elevator requires almost constant power for a given load as it rises in the hoistway. When moving at constant speed, the power is consumed to overcome the force of gravity. Peak acceleration or deceleration, when limited to $0.11g$ increases or decreases the power required by only about 10%. During the descent, a significant portion of energy is recovered as the LSM is able to regenerate it from the passage of the PM excitation system through the armature windings. However, the recovered energy is always smaller than the propulsion energy because regeneration occurs when the descending car travels with speed exciting higher EMF in the winding than the drop voltage on the winding impedance. Fig. 9.9b shows that the car absorbs the power taken from the electric supply system as it starts traveling downward. A freely falling car would develop the acceleration equal to $1.0g$, that is too high for the passengers riding inside. To limit the acceleration to $0.11g$ the LSMs must develop braking thrust (directed upward) slightly smaller than the gravitational force.

9.2.5 First Prototypes

Completing a study on very large drives in the early 1990s, the Underground Development Utilization Research Center of the Engineering Advancement Association of Japan (ENNA) is seriously considering the application of linear motors to ropeless elevators. This demonstrative research was conducted by a group of five companies: Fuji Electric, Ishikawajima–Harima Heavy Industries, Kawasaki Heavy Industries, Simizu Corporation, and Fujitec. ENNA has already built a working model of a horizontal and vertical ropeless transport system [121]. This scaled installation serves as a model for future hyper building transportation systems. Specifications are the following:

- maximum thrust 3,000 N
- vertical travel distance 8 m
- horizontal travel distance 1 m
- velocity from 0 to 1 m/s
- acceleration 1 m/s^2
- moving mass 270 kg (including payload)
- car dimensions: 1.0 W ×0.9 D ×1.8 H m
- power and control: IGBT inverter
- position encoder: optical pulse generator

The overall view of the test system is shown in Fig. 9.10. The car is suspended as a pendulum from the traveling carriage. To prevent the car from swaying at the time of acceleration or when passing through a curved rail, a friction plate has been set between the carriage structure and the car. The car is guided by two rails: (a) a rigid H-shaped beam, which supports and guides the car (travel rail), and (b) a motor guidance maintaining a constant linear motor air gap. The system is capable of moving the car both along the straight and curved path. Such a transition takes approximately 10 s.

To meet the most important propulsion system requirements, i.e., lightweight and large motive force, an LSM with PM mover installed on the car and the

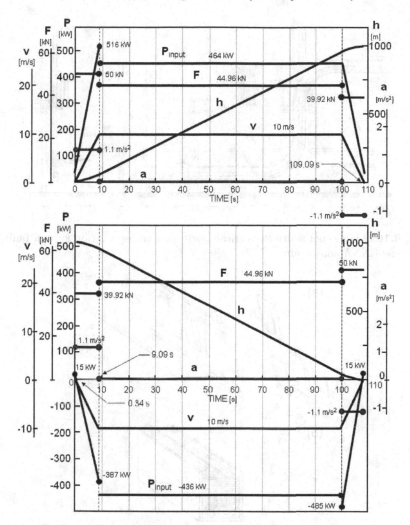

Fig. 9.9. Computed motion profile for a car of ropeless elevator moving 16 passengers with 10 m/s of contract speed in the 1000 m rise hoistway: (a) for car going up, (b) for car going down.

armature on the track has been adopted. The mover is placed between two sets of stationary armature systems. High-energy N32 NdFeB PMs (energy product 255 kJ/m^3) have been used. PMs are fixed to both sides of 10 mm thick steel rails and skewed by one slot pitch of armature core to reduce the thrust ripple (Fig. 9.11a). To minimize the attractive forces between PMs and armature cores, a double-sided armature configuration has been chosen (Fig. 9.11). This motor topology is referred to as a U-type LSM with short PM reaction rail and two long armature systems.

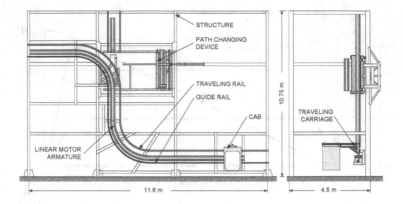

Fig. 9.10. Linear-motor-driven vertical–horizontal transportation system built by Japanese ENNA consortium.

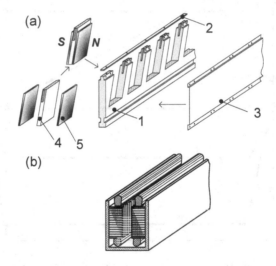

Fig. 9.11. Construction of PM LSM for ENNA ropeless elevator system: (a) PM mover, (b) double-sided armature. 1 — PM-mounting aluminum plate, 2 — locking bar, 3 — protective cover made of non-ferromagnetic stainless steel, 4 — steel plate, 5 — PM.

The armature of LSM is divided into 10 straight and curved sections, each about 2 m long. The segments are powered from sinusoidal IGBT inverters with low noise emission levels arranged in a tandem. The power from inverters is distributed by means of switches connecting individual armature sections in a sequence (Fig. 9.12). When the mover travels between the segments, both inverters operate; otherwise, only one is used. The presented system is able to fully control the movement of a single carriage. For higher number of cars, a more complicated power distribution scheme is needed. The speed and position are detected by a slotted plate on the car and photodetectors mounted on the track.

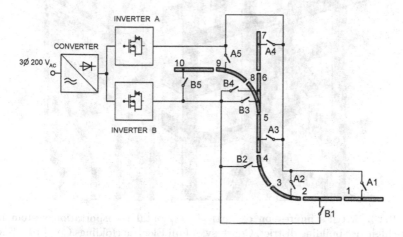

Fig. 9.12. Power distribution system.

The safety of the system is guaranteed by a braking device activated by a hydraulic system that is mounted at each end of the trace ends to hold the traveling carriage in position. Oil buffers are mounted to minimize a sudden impact in the case of an improperly decelerating carriage. An electrical dynamic braking is applied in the case of power failure to prevent a free fall of the car. The armature windings are automatically connected (shunted) to the external resistors. The power dissipated in the braking resistor provides smooth, controllable braking force, allowing the car to descend at stable crawl speed even when all safety monitoring equipment fail.

The ENNA project, completed in 1995, shows that, in the near future, the vertical transportation system would evolve into arteries in which cars could move three-dimensionally. Travel will not be limited to enclosed spaces, like the conventional elevator systems in use today, but multiple passenger cars will be moving freely between floors, buildings, in underground or in any subterranean space.

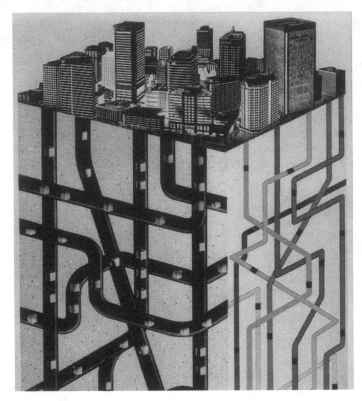

Fig. 9.13. Artist's impression of vertical–horizontal transportation system in a future high-rise building district. Courtesy of Fuji Electric Holdings Co., Ltd., Tokyo, Japan.

Fig. 9.13 shows an artist's impression of a future high-rise building district with vertical–horizontal ropeless elevators propelled by linear motors.

9.2.6 Brakes

One of the most important elevator subcomponents is a *braking system*. The brake must:

- maintain the car in a fixed vertical and horizontal (side-to-side and front-to-back) position while passengers are boarding and exiting;
- operate to stop the vertical motion of the car under complete power failure;
- have the capability of producing at least 1.0 m/s^2 deceleration for periods above 10 s when the car is traveling downward;
- supply auxiliary guidance to the car in the case of a complete power failure. This requirement is important for elevators with active magnetic guidance (AMG) [156].

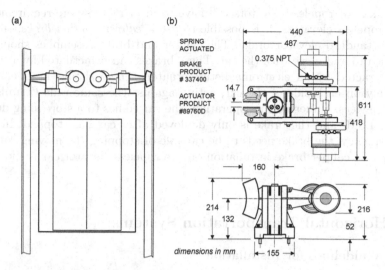

Fig. 9.14. Ropeless elevator brakes: (a) installation, (b) commercial brake caliper type from Nexen, Vadnais Heights, MN, USA (formerly Horton Industrial Products).

The primary function of a brake is to hold the car in a fixed position when passengers are entering and exiting. While this function could be accomplished using an electric motor, it is more efficient to use a friction brake to hold the car in position. In a roped elevator, this brake generally acts on a drum attached to the sheave. Therefore, a releveling process is required as the car load changes and the cable length between the brake and the car stretches or contracts. For the brake mounted directly on the car (ropeless design), the need for releveling is eliminated as there is no cable located between the brake and car stretch. The second function of the brake is to provide stopping the car in the case of a power failure. The braking force in a conventional elevator is generally applied by a spring pressure and released using an electrical actuator (solenoid). In the case of an electrical outage, the braking force is automatically produced by the spring pressure. Usually, elevator brakes are either fully applied or fully disengaged. Normally they would not be modulated or applied gently to generate, for example 50%, of possible braking force.

The brake in a standard roped elevator is used under steady-state conditions. It is applied after the car has completely stopped, and released only after the motor has been energized to support the car. This is done to prevent the car from lurching either up or down at the brake release. Conventional elevator brakes may also operate in a dynamic mode if power is lost to the hoistway or safety limits are exceeded. In this mode, the brake is required to stop the car safely while it is moving up to 120% of the contract speed, i.e., speed negotiated between the manufacturer and customer.

Brakes for ropeless elevators will have to meet more severe requirements. With ropeless elevators, it is possible to use a *caliper friction brake*, similar to the standard elevator brake. The caliper and brake assemblies should be attached to and move with the car. These brakes can be actuated by a spring and retracted by an electromagnetic actuator. There should be one brake assembly on each side of the car, acting against the same rails (T-rails) as safety devices. Under normal operation, the car comes to a stop being driven by the LSM, and the brake is only deployed after car has stopped. In case of emergency, the brake needs to be capable of stopping the moving car. An example of caliper brake installation on the ropeless elevator car is shown in Fig. 9.14.

9.3 Horizontal Transportation Systems

9.3.1 Guidelines for Installation

Linear motors can simplify the transfer bulk and loose materials, small containers, pallets, bottled liquids, parts, hand tools, documents, etc., in building and factory transportation systems. It is easy to adopt the linear motor system to the allowable space as linear motors are compact machines, do not have rotating parts, do not need maintenance and allow design of a transfer system that is in harmony with surrounding equipment. Linear motors are silent, and the noise emitted does not exceed 65 dB. Magnetic fields emitted by armature windings and PM excitation system are not dangerous to human organisms. Well-designed transportation systems emit very low level of RFI. High-intensity magnetic fields exist only in the air gap between the armature and reaction rail.

It is possible to design both on-floor and overhead transportation systems. Moreover, the linear motor transportation line can be sloped to create a link between two or more floors. Parallel transportation lines or line for double-deck carriages can also be designed. The rule is that transportation lines must never cross. If, say, a loop like "8" shaped line is needed, the crossing must be designed on two different levels. Possible designs are shown in Fig. 9.15 [106].

Stations for loading and unloading materials must be carefully planned, and additional space must be reserved. Linear motors cannot be installed near inlets and exhausts of air-conditioning systems because airborne dirty particles could deteriorate the sensors' performance and sometimes contaminate the air gap between the armature and reaction rail. In the case of PM excitation, the space around the reaction rail must be free of any ferromagnetic particles.

9.3.2 Construction

There are two basic constructions of building or factory horizontal transportation systems with LSMs:

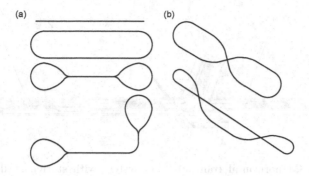

Fig. 9.15. Arrangement of linear motor transportation lines: (a) on one level, (b) on two levels.

- Long stationary reaction rail with PMs and moving short armature (Fig. 9.16)
- Stationary armature or armatures and moving reaction rail with PMs (Fig. 9.17) [184, 185]

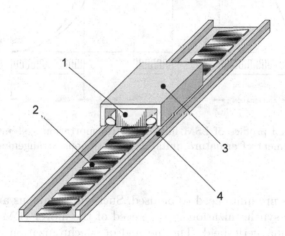

Fig. 9.16. Horizontal factory transportation system with moving short armature LSM and long PM reaction rail. 1 — armature, 2 — PMs, 3 — carrier, 4 — guiderail.

The stationary long reaction rail is expensive, and the moving armature winding needs a contactless energy transfer [201]. It is more economical to use stationary short armature units distributed along the track. The mover is accelerated by the short armature unit and is then driven by its own inertia. Since the speed decreases, the next unit must be installed at a certain distance to reaccelerate the mover (Fig. 9.18). The longer the travel distance,

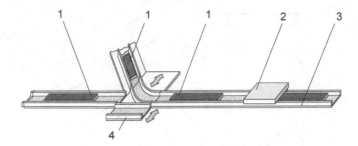

Fig. 9.17. LSM horizontal transportation system with stationary discontinuous armature and short moving PM reaction rail. 1 — armature unit, 2 — carrier with PM excitation system, 3 — guiderail, 4 — switch.

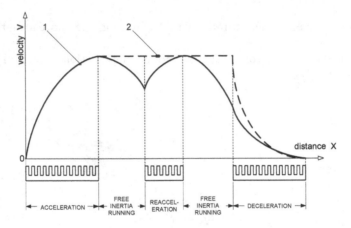

Fig. 9.18. Speed profiles of LSM horizontal transportation systems. 1 — discontinuous arrangement of armature units, 2 — continuous arrangement of armature units.

the more armature units need to be used. Such discontinuous arrangement of LSMs requires synchronization of the speed of the mover (PM reaction rail) and traveling magnetic field. The method of synchronization can use either feedback signals from position sensors or open-loop control of synchronization [184, 185]. The first methods are very reliable; however, they are costly. Open-loop methods are more economical (lower cost of sensors).

9.3.3 Applications

The horizontal transportation system with stationary short armature units can be used for *transfer of containers* in storage areas.

The system developed by Preussag Noell, GmbH, Würzberg, in cooperation with the Technical University of Braunschweig, Germany, consists of track-mounted, single-sided, iron-cored armature units and car-mounted movable PM excitation systems [93]. The length of the PM excitation system covers at least two armature units, including the distance between them. A group of armature units connected in series is fed from an IGBT VVVF inverter. The position of cars is detected with the aid of special position sensors. Cars of length up to 15 m run on standard gauge railway track. The air gap between the armature core and PMs is 13 to 15 mm. With the pole pitch from 80 to 100 mm and PM height from 9 to 15 mm, thrust density of 30 to 40 kN/m² can be obtained [93].

Another LSM system for container transportation has been developed at Kyushu University, Fukuoka, Japan [246]. The air-cored armature coils are distributed along the guideway beneath the container carrier, and the PM excitation system is integrated with the carrier. The air-cored LSM produces both the propulsion force (thrust) and normal repulsive force (no armature ferromagnetic core). The normal repulsive force (electrodynamic levitation) helps to reduce the mass of the payload by about 85% [246]. By applying the *decoupled control method*, the thrust and normal force can be controlled independently [246]. It is possible not only to reduce the mass of payloads but also to reduce friction, vibration, and noise caused by heavy goods.

Horizontal linear motor transportation systems can alleviate and simplify duties of medical personnel in hospitals. Clinical charts, x-ray films, chemicals, specimens and documents can be transported by linear motor conveyance systems in special containers. Shinko Electric Co. Ltd., Tokyo, Japan has developed *linear motor conveyance technology* for hospital staff [120]. This system, called *LimLinear*, can travel both vertically (5 m/s) and horizontally (3 m/s), connects several floors, and is controlled by computer.

Examples

Example 9.1

Consider a ropeless elevator introduced in *Example 1.6* (Chapter 1). The mass of the fully loaded car is $m = 4583$ kg (with 16 passengers), acceleration/deceleration $a = 1.1$ m/s², speed $v = 10.0$ m/s, and LSM efficiency $\eta = 0.97$. It is assumed that the hoistway efficiency is 100%. Analyze the motion, thrust, power, and energy profiles in the time t domain for a car ascending and descending the hoistway 1000 m high.

Solution

Car going upward—Acceleration

The time to reach the full speed $v = 10.0$ m/s, i.e., acceleration time

$$t_a = t_1 = \frac{v}{a} = \frac{10.0}{1.1} \approx 9.09 \text{ s}$$

The travel distance at the end of the acceleration phase

$$h_a = h_1 = \frac{1}{2}at_1^2 = \frac{1}{2} \times 1.1 \times 9.09^2 = 45.45 \text{ m}$$

The thrust required to accelerate the mass $m = 4583$ kg upward with acceleration a

$$F_{xa} = m(g + a) = 4583 \times (9.81 + 1.1) \approx 50,000 \text{ N}$$

The peak input power at the end of the acceleration phase $(t = t_1)$

$$P_{ina} = \frac{1}{\eta}F_{xa}at_1 = \frac{1}{0.97} \times 50,000 \times 1.1 \times 9.09 \approx 516.0 \text{ kW}$$

Energy delivered to the LSM during the up-acceleration phase

$$E_a = \frac{1}{\eta}F_{xa}h_1 = \frac{1}{0.97} \times 50,000 \times 45.45 \approx 2,343 \text{ kJ}$$

Car going upward—Steady State

The elevator car is cruising up with constant speed $v = 10.0$ m/s, and acceleration is equal to zero, $(a = 0.0)$. The duration time of the steady-state phase is calculated on the assumption that the car accelerates and decelerates with the same rate $|a| = 1.1$ m/s^2. Then, the deceleration time t_d as well as the corresponding distance h_d is the same as the acceleration time $t_a = t_1$ and acceleration distance h_a, i.e.,

$$t_d = t_3 - t_1 - t_s \approx 9.09 \text{ s}$$

where t_3 is the total time of traveling from start to stop. Then, the travel distance during the steady state is

$$h_s = h - 2h_a = 1000 - 2 \times 45.45 = 909.1 \text{ m}$$

The duration of the steady state

$$t_s = \frac{h_s}{v} = \frac{909.08}{10.0} \approx 90.91 \text{ s}$$

The steady-state thrust

$$F_{xs} = mg = 4583 \times 9.81 = 44,960 \text{ N}$$

The LSM steady-state input power is independent of time, i.e.,

$$P_{ins} = \frac{1}{\eta}F_{ds}v = \frac{1}{0.97} \times 44,960 \times 10.0 \approx 464.0 \text{ kW}$$

Energy delivered to the LSM

$$E_s = P_{ins}t_s = 464,000 \times 90.91 \approx 42,182 \text{ kJ}$$

Car going upward—Deceleration

The time instant when the car starts to decelerate, $t_2 = t_1 + t_s = 9.09 + 90.91 = 100.0$ s.

The thrust directed upward that is required to maintain the deceleration of 1.1 m/s^2

$$F_{xd} = m(g - a) = 4583 \times (9.81 - 1.1) = 39,918 \text{ N}$$

Energy delivered to the LSM during the up-deceleration phase

$$E_d = \frac{1}{\eta}F_{xd}h_d = \frac{1}{0.97} \times 39,918 \times 45.45 \approx 1,870 \text{ kJ}$$

The total energy consumed by the LSM during the travel up

$$E_{up} = E_a + E_s + E_d = 2,343 + 42,182 + 1,870 \approx 46,395 \text{ kJ}$$

Car going downward

To keep the acceleration and speed at acceptable levels, the LSM operates as a brake (plugging) during the entire period of downward travel. The LSM produces the force directed upward limiting the acceleration to 1.1 m/s^2. The regeneration starts when power developed by moving masses exceeds linear motor losses. Calculations conducted in the same way as for a car going up produced the following results:

Car going downward—Acceleration

- Time to reach contract speed $v = 10.0$ m/s, $t_a = 9.09$ s
- Car position at the end of the acceleration, $h(t_a) = 954.55$ m
- Distance traveled during acceleration, $h_a = 45.45$ m
- Thrust required to keep acceleration at 1.1 m/s^2, $F_{xa} = 39.918$ kN
- Time to start regeneration, $t_{ch} = 0.34$ s
- Generated power at the end of acceleration, $P_{ina} = 194.0$ kW
- Energy for downward acceleration, $E_a = 1,755$ kJ

Car going downward—Steady State

- Braking force in steady state, $F_{xs} = 44.96$ kN
- Power generated by the LSM at steady state, $P_{gs} = 436.0$ kW

- Generated energy, $E_s = 39,637$ kJ

Car going downward—Deceleration

- Upward thrust to maintain the deceleration of -1.1 m/s^2, $F_{xd} = 50$ kN
- Generated power at the beginning of deceleration, $P_{gd}(t_2) = 485.0$ kW
- Energy recovered during deceleration, $E_d = 2,198$ kJ

The total energy recovered when traveling downward

$$E_{dn} = E_a + E_s + E_d = 1,755 + 39,637 + 2,198 \approx 43,590 \text{ kJ}$$

One fully loaded car descending down generates 43,590 kJ energy, which is returned to the power system.

The energy balance for one fully loaded car for a round trip is

$$\Delta E = E_{up} + E_{dn} = 46,395 - 43,590 \approx 2,805.0 \text{ kJ}$$

It means that the fully loaded car (with 16 passengers) during one round trip consumes only 2,805.0 kJ energy.

The hoistway *cycling energy efficiency*, defined as the ratio of *energy recovered* during the descent of a fully loaded car to the *energy consumed* by this car during the ascend, is

$$\eta_{cy} = \left| \frac{E_{dn}}{E_{up}} \right| = \frac{41,551.0}{46,395.0} = 0.8956$$

For comparison, Fig. 9.19 shows the speed, power, and energy versus time curves measured for the roped elevator with propulsion system based on the tandem of LIMs. This experimental system was built and tested by Otis Elevator Company in the early 1990s. The elevator was driven by two flat linear induction motors installed in the counterweight running over long stationary reaction rails consisting of copper backed by a steel plate. The elevator was capable of carrying 16 passengers between 7 floors with 2.5 m/s rated speed. The peak 20,000 N thrust has been developed by the tandem of flat LIMs during the acceleration phase. The energy curve (Fig. 9.19b) shows that even for a fully loaded car going downward, the total energy at the end of the cycle is positive. This means that the energy supplied by the power system exceeds that generated during regenerative braking. This is due to the relatively low efficiency of LIMs. In the considered example, the peak LIM efficiency is 43%.

Example 9.2

Consider a 250 story, 1,000 m high hyper building, with population of 100,000 people. The elevator system consists of 17 ropeless hoistways with fifty 16-passenger cars each, plus two additional ropeless hoistways containing 20 cars

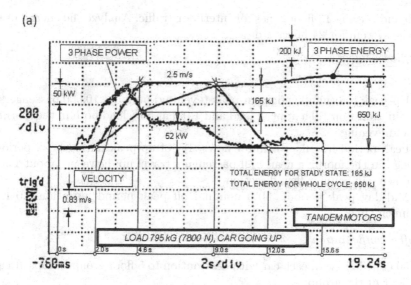

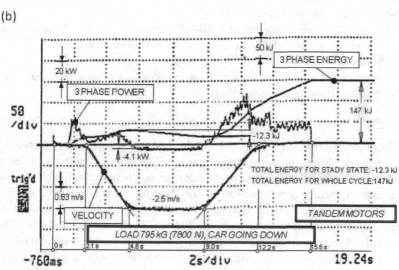

Fig. 9.19. Velocity, power, and energy for flat linear motor roped elevator system with counterweight: (a) ascending car, (b) descending car. Courtesy of Otis Elevator Company, Farmington, CT, USA.

each and serving 13 floor zones for interfloor traffic. Analyze the energy usage by the system for 24 hours a day.

Traffic scenario

- Up-peak (morning) and down-peak (evening) periods: 10,000 passengers/5 min, duration 1.0 h at peak load and 1.0 h at 50% load both in the morning and evening
- Peak period interfloor counter flow: 1% of up-peak during peak periods, e.g., in the morning nearly all passengers travel up; however, about 1% go down—in counter flow direction
- Mid-day peak period: 50% nominal off peak period, duration 2.0 h at lunch time

Traffic configuration

- 50 cars/hoistway, each car with destination to 5 floor group—stops at each floor of the group
- Nominal round trip time = 400 s
- Car size: 4583 kg, 16 passengers, (3483 kg empty car +1100 kg duty load)
- Motor efficiency $\eta = 97\%$ (propulsion and regeneration)
- Time for car to reach top floor = 110 s
- Time to service next 4 floors = 18 s per floor (4 s floor–to–floor run, 4 s door time, 10 second enter–exit)

Assumptions

(a) There are no friction losses (due to the noncontact guidance system), and no windage losses are taken into account. The power requirement for guidance, control, lighting, and ventilation is neglected. Similarly, the power to move in transverse direction in the loading areas is not considered.

(b) All cars in 17 fast traffic hoistways are assumed to be full when leaving the lobby and empty upon return. The up-peak energy analysis is based on fully loaded car operating at full propulsion power (calculated in the previous *Example 1*—car running to the highest floor) for the entire run time to the first stop at the lowest floor serviced by each car. The regenerated energy for the down run has been calculated on the basis of previous *Example 1* with the 3483 kg mass of empty car used in all equations.

(c) Energy usage by the two interfloor hoistways is calculated on the basis of the following: the average load of a car is 50%; the extra acceleration power is offset by the reduced deceleration power; cars are making an average of four 3-floor runs up plus another four 3-floor runs down in each round trip with three 7.0 s interfloor stops. The two interfloor hoistways together can carry 733 passengers up and another 733 passengers down in each 5 min period, at 50% capacity. If required, the nominal 50% capacity can be upgraded to the full capacity with higher number of passengers.

Solution

The ropeless elevators are estimated to consume approximately 255,000 kWh (255 MWh) of energy during a 24 h day. This number has been obtained from computer simulation using a dedicated software. The highest power is necessary for the 1 h up-peak period. During that time, the elevator system, i.e., all elevators, require approximately 49,500 kW power for the peak load.

It is interesting that this peak power consumption, considered together with typical linear motor efficiency, yields a thermal dissipation of approximately 11,500 kW (power losses in all LSMs) during that one peak hour. It should be mentioned that thermal dissipation is considerably less at all other times of the day. Most of the heat is generated in the armature windings of LSMs since PM excitation systems do not generate power losses. The estimated heat dissipation leads to the conclusion that an effective cooling system must be applied directly to the armature windings. Preferably, the cooling system should use a chilled liquid rather than the air flowing directly through cooling ducts in the armature. It is interesting to note that one person can generate during 1 h approximately 0.117 kWh of thermal energy in an office environment. The 100,000 occupants in the hyper building collectively generate about 11,700 kWh, or roughly the same amount as that of ropeless elevators during the 1 h peak period. Thus, the energy usage by the air-conditioning system is comparable with elevator energy converted into heat losses.

Industrial Automation Systems

10.1 Automation of Manufacturing Processes

Manufacturing processes can be classified as follows [49]:

- Casting, foundry or molding processes
- Forming or metalworking processes
- Machining (material removal) processes
- Joining (fastening, welding, fitting, bonding) and assembly
- Surface treatments and finishing (cleaning, tumbling, scribing, polishing, coating, plating)
- Heat treating
- Other, e.g., inspection, testing, packaging, storing, etc.

Manufacturing process automation frees the human operator from control functions, i.e., from the need to perform certain actions in a particular sequence to carry out an operation in accord with preset machining conditions.

Automation is the use of the energy of a nonliving system to control and carry out a process or process operation without direct human intervention [200]. The object of automation is to make the best use of available resources, materials, and machines. The human worker's function is limited to machine supervisory control and elimination of possible deviations from the prescribed process (corrective adjustment).

According to the *Yardstick for Automation* chart (Table 10.1) presented by G.H. Amber and P.S. Amber in 1962 [8], each level of automation is tied to a human attribute that is being replaced by the machine. Thus the A(0) level of automation, in which no human attribute was mechanized, covers Stone Age through Iron Age. So far, levels A(5), A(6), and A(7) have partially been implemented or are subject to intensive research. Levels A(8) and A(9) are still topics of science fiction.

In automatic systems, actuation components are adapted to moving various mechanisms of machine tools to execute the required step of control, e.g.,

Table 10.1. Yardstick for Automation [8].

Orders of Automation	Human attribute replaced	Examples
A(0)	*None*: Lever, screw, pulley, wedge	Hand tools, manual machines
A(1)	*Energy*: Muscles replaced	Powered machines and tools; Whitney's milling machine
A(2)	*Dexterity*: Self-feeding	Single-cycle automatics
A(3)	*Diligence*: No feedback	Repeats cycle; open-loop numerical control or automatic screw machine; transfer lines
A(4)	*Judgment*: Positional feedback	Closed-loop; numerical control; self-measuring and adjusting
A(5)	*Evaluation*: Adaptive control; deductive analysis; feedback from the process	Computer control, model of process required for analysis and optimization
A(6)	*Learning*: By experience	Limited self-programming; some artificial intelligence (AI); expert systems
A(7)	*Reasoning*: Exhibit intuition; relates causes and effects	Inductive reasoning; Advanced AI in control software
A(8)	*Creativeness*: Performs design unaided	Originality
A(9)	*Dominance*: Supermachine, commands others	Machine is master

change the mode of operation, release a workpiece, start or stop the machine, etc. One of the emerging technologies is the use of linear motors as electrical actuators.

10.2 Ball Lead Screws

10.2.1 Basic Parameters

Normally, the rotation of a rotary servo motor, usually PM brushless motor, is converted into linear motion in x, y, z, or w (e.g., tool) direction with the aid of *ball lead screws* (Fig. 10.1).

For a screw of mean diameter d, the helix angle Θ is given by the equation

$$tan(\Theta) = \frac{l}{\pi d} = \frac{2t_r}{d} \qquad (10.1)$$

where t_r is the transmission parameter (axial distance moved per one radian of screw revolution) of the lead screw. The *lead* l of the screw (helix) is the axial distance moved by the nut in one revolution of the screw:

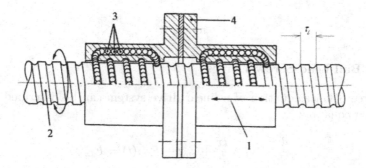

Fig. 10.1. Longitudinal section of a ball lead screw. 1 — table movement, 2 — recirculating ball lead screw, 3 — recirculating balls, 4 — ball nut attached to the table.

$$l = 2\pi t_r \qquad (10.2)$$

In general, the lead l is not the same as the pitch τ_l of the screw. The *pitch* τ_l is the axial distance between two adjacent threads (Fig. 10.1). For a screw with k independent threads on the screw shaft,

$$l = k\tau_l \qquad (10.3)$$

For a single start screw, $k = 1$ and $l = \tau_l$. If T_b is the torque provided by the screw, i.e., net torque after subtracting the inertia torque due to inertia of the motor rotor and lead screw, the force is

$$F = \frac{\eta_b}{t_r} T_b = \frac{2\pi\eta_b}{l} T_b \qquad (10.4)$$

where η_b is the efficiency of the ball screw. The frictional force

$$F_f = \mu_v \frac{T_b}{t_r} \qquad (10.5)$$

where the coefficient of friction $\mu_v = 0.2$ for steel (dry), $\mu_v = 0.15$ for steel (lubricated), $\mu_v = 0.1$ for bronze, and $\mu_v = 0.1$ for plastic. The torque required to overcome this frictional force is

$$T_f = F_f \frac{d}{2} = \mu_v d \frac{T_b}{2t_r} \qquad (10.6)$$

The efficiency of the ball screw

$$\eta_b = \frac{T_b - T_f}{T_b} = 1 - \mu_v \frac{d}{2t_r} = 1 - \frac{\mu_v}{\tan(\Theta)} \qquad (10.7)$$

The linear speed (travel rate) is the rotational speed n at which the screw or nut is rotating multiplied by the lead of the screw, i.e.,

$$v = n\tau_l \tag{10.8}$$

10.2.2 Ball Lead Screw Drives

The force balance equation of a linear drive system can be described by the following equation:

$$m\frac{d^2x}{dt^2} + D_v\frac{dx}{dt} + k_sx + F_{ext}(t) = F_{dx}(t) \tag{10.9}$$

where m is the total mass of the load including the table (thrust block), D_v is the viscous friction (damping) constant of the load, k_s is the spring constant, F_{dx} is electromagnetic force produced by the motor, F_{ext} represent all external (disturbance) forces acting on the system, and x is the linear displacement of the load.

The position displacement, velocity, and acceleration of a linear drive can be expressed as

$$x(t) = X_m \sin(\omega t) \tag{10.10}$$

$$\dot{x}(t) = \frac{dx(t)}{dt} = \omega X_m \cos(\omega t) \tag{10.11}$$

$$\ddot{x}(t) = \frac{d^2x(t)}{dt^2} = -\omega^2 X_m \sin(\omega t) \tag{10.12}$$

where X_m is the maximum position error. Assuming in eqn (10.9) $F_{dx} = 0$ (uncontrolled system) and $k_s = 0$, the external (disturbance) force is

$$F_{ext} = m\omega^2 X_m \sin(\omega t) + D_v\omega X_m \cos(\omega t) \tag{10.13}$$

Neglecting the viscous friction constant ($D_v \approx 0$), the external force becomes

$$F_{ext} \approx m\omega^2 X_m \sin(\omega t) \tag{10.14}$$

The *static stiffness* of a linear drive system is defined as the external *force-to-position displacement* ratio, i.e.,

$$K_s = \frac{F_{ext}}{x} \tag{10.15}$$

The *dynamic stiffness* of a linear drive system is defined as the external *force-to-system response* ratio, i.e.,

$$K_d = \frac{F_{ext}(t)}{x(t)} \tag{10.16}$$

Putting eqns (10.10) and (10.14) into eqn (10.16), the following simplified equation for dynamic stiffness can be written,

$$K_d \approx m\omega^2 \tag{10.17}$$

For a ball screw system with rotary motor and tooth or belt gear, the equation of motion is [151]

$$\left\{ m_t + \left(\frac{2\pi}{\tau_l} \right)^2 \left[J_{bs} + \left(\frac{N_2}{N_1} \right)^2 J_m \right] \right\} \ddot{x} \tag{10.18}$$

$$= \frac{2\pi}{\tau_l} \left[\frac{N_2}{N_1} \left(T_d - D_{vm} \dot{\Theta}_m - T_{extm} \right) \right] - \frac{2\pi}{\tau_l} \left(D_{vbs} \dot{\Theta}_{bs} + T_{extbs} \right) - D_{vt} \dot{x} - F_{ext}$$

where m_t is the mass of the moving table, τ_l is the screw pitch; J_{bs} and J_m are moments of inertia of the ball screw and motor, respectively; N_1 and N_2 are the number of teeth on the motor and ball screw gears, resectively; D_{vt}, D_{vm}, D_{vbs} are viscous friction coefficients of the table, motor and ball screw, respectively; F_{ext} is the external (disturbance) force on the table; T_d is the torque developed by the motor; T_{extm} and T_{extbs} are external torques acting on the motor shaft and ball screw, respectively; x is the linear displacement of the load; and Θ_m, Θ_{bs} are angular displacements of the motor and ball screw, respectively.

Assuming $T_d = 0$, $D_{vm} = 0$, $D_{vbs} = 0$, D_{vt}, $T_{extm} = 0$, $T_{extbs} = 0$ and denoting

$$J_e = J_{hs} + \left(\frac{N_2}{N_1} \right)^2 J_m \tag{10.19}$$

the external (disturbance) force of a ball screw system obtained from eqn (10.18) is

$$F_{ext} \approx \left[m_t + \left(\frac{2\pi}{\tau_l} \right)^2 J_e \right] \omega^2 X_m \sin(\omega t) \tag{10.20}$$

and dynamic stiffness

$$K_d \approx \left[m_t + \left(\frac{2\pi}{\tau_l} \right)^2 J_e \right] \omega^2 \quad \text{N/m} \tag{10.21}$$

where J_e is the equivalent moment of inertia of the balls screw and motor. The acceleration can be found from eqns (10.4), (10.12) and (10.20), i.e.,

$$a \approx \frac{F_{ext}}{m_t + m_t \left(\frac{2\pi}{\tau_l} \right)^2 J_e} = \frac{T \tau_l}{2\pi \left[J_e + m_t \left(\frac{\tau_l}{(2\pi)} \right)^2 \right]} \quad \text{m/s}^2 \tag{10.22}$$

It has been assumed in eqn (10.4) that $l = \tau_l$ $(k = 1)$ and $\eta_b = T_b/T$, where T is the torque developed by the rotary electric motor. Maximum acceleration a_{max} is for maximum torque T_{max} produced by the electric rotary motor.

In [103] the static servo stiffness of the ball screw feed drive system is expressed as

$$K_s = a_b K_p K_v k_{T_{max}} (1 + b_b K_i) \left(\frac{2\pi}{l}\right)^2 \tag{10.23}$$

where K_p is the proportional gain, K_i is the integral gain, K_v is the position loop gain, T_{max} is the peak torque of the servo motor, k_T is the torque constant of the servo motor, and a_b and b_b are constant parameters. For example, conventional machining centers use ball screws with leads $\tau_l = 10$ to 14 mm and servo motors with maximum speeds $n_{max} = 2000$ to 2500 rpm which gives maximum linear speed $v_{max} = 20$ to 35 m/min, maximum acceleration $a_{max} = 0.2$ g to 0.3 g, and $K_s \approx 23 \times 10^8$ N/m (x-axis of the table) [103].

10.2.3 Replacement of Ball Screws with LSMs

Linear motors can successfully replace ball lead screws. Fig. 10.2 shows two methods of obtaining high linear speed (feed rate) and high acceleration by using a ball lead screw and linear motor. Rotary servo motors can use either a rotary encoder or linear encoder with table-mounted scale, while linear motors use only linear encoders.

The tubular LSM (Fig.1.2) with NdFeB PMs in the thrust rod (reaction rail) offers an attractive alternative to ball screw, hydraulic, and pneumatic motion-control solutions. Comparison of tubular LSMs with ball screws is given in Table 10.2.

Sustainable thrust developed by a tubular LSM is a function of the motor ability to dissipate heat. Maximum force and velocity are controlled by the choice of the winding current density and cooling option (heat sink, forced air, water jacket). An extruded housing (Fig. 10.3) can serve as an armature (forcer) heat sink and increase the continuous thrust by a factor of 1.15. Modern tubular LSMs can deliver thrust density (thrust per mass) up to 500 N/kg. Thrust rod masses for 25 mm diameter rods are typically 3.5 kg/m. Examples of motion-control systems with tubular LSMs are shown in Figs 10.4 and 10.5. Prime targets for replacement of traditional motion control mechanisms with tubular LSMs include high-speed packaging, bottling and canning; high-speed printing and stamping; garment production (sewing, weaving and tufting); injection molding, vibratory part feeders and mixers, fastback conveyors. Tubular LSMs are also excellent actuators for systems that require varying controlled force or pressure (resistance welding, material testing, fluid pumping, high-speed crimping), clean operation and high force-to mass ratio (aerospace).

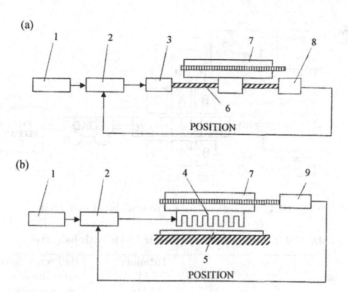

Fig. 10.2. Axis drive systems with (a) ball lead screw, (b) linear motor. 1 — interpolator, 2 — controller, 3 — low inertia servomotor, 4 — armature of a linear motor, 5 — reaction rail of a linear motor, 6 — ball lead screw, 7 — table (guide), 8 — rotary encoder or resolver, 9 — linear sensor.

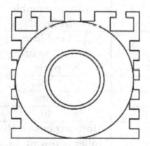

Fig. 10.3. Extruded heat sink — housing of tubular LSM.

Modern motion control systems with tubular LSMs require high-resolution feedback for position control and commutation feedback for controlling the frequency and phase of the three-phase motor signal (Fig. 10.4).

Although tubular PM LSM in comparison with ball or roller screw PM brushless motor linear actuators develop 10 times higher linear speed, provide fully programmable, zero-backlash controllability and emit much lower noise, the maximum force density is only 500 N/kg versus over 1200 N/kg for PM brushless motor linear actuators (Table 10.2).

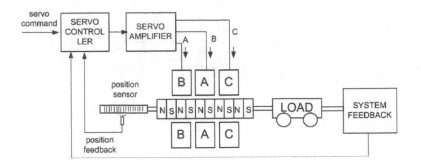

Fig. 10.4. Motion control system with tubular LSM.

Table 10.2. Comparison of tubular LSMs with ball screws

Performance	Tubular PM LSMs	Ball screw and roller screw linear actuators with rotary motors
Thrust density, N/kg	max. 500	over 1200
Linear speed, m/s	2.5	0.25
Position accuracy, mm	±0.025	±0.025
Stiffnes ability to hold a position	High	High
Controllability	Fully programmable; zero backlash	Backlash increases with time
Life-cycle cost	Very low; only 2 wearing parts	Moderate 4 to 200 wearing parts
Maximum temperature °C	≤ 125	≤ 125
Environmental impact	None	None
Noise level, db	40	80
Cost	More expensive than roller screw and ball screw actuators due to limited demand	Moderate

10.3 Linear Positioning Stages

Linear positioning stages have been partially discussed in Chapter 6, Section 6.6. Every motorized positioned stage comprises three essential components: (1) stage, (2) motor, and (3) controller. Positioning stages have travel range from a few micrometers to several meters.

Configurations of positioning stages are shown in Fig. 10.6 [119]. The simplest form of positioning stage is a *single-axis stage* (Figs 6.17 and 10.6a). It typically consists of a moving table (carriage), base, motor, encoder, limit

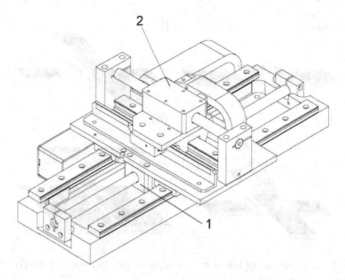

Fig. 10.5. Two-axis linear positioning stage as an example of replacement ball lead screw mechanisms with tubular LSMs. 1 — x-direction tubular LSM, 2 — y-direction tubular LSM. Courtesy of em ABTech, Swanzey, NH, USA.

switches and cable carriers. A *compound xy positioning stage* (Fig. 10.6b) provides the simplest form of 2 linear DOF of a positioning system where the base of the top axis is bolted to the moving table of the lower axis [119]. A *compound xyz stage* (Fig. 10.6c) provides the simplest form of 3 linear DOF of a positioning system with the smallest footprint. A *split xyz positioning stage* (Fig. 10.6d) provides typically higher precision and higher stiffness than a compound configuration of the same number of axes [119].

Positioning stages with linear motors are far simpler to design and assemble compared to a stage that is based on a rotary motor. The ball screw, coupling, gears, rack-and-pinion, or belts are all eliminated.

The single-axis positioning stage shown in Fig. 10.7 is driven by a three-phase, double-sided PM LBM with ironless core, commutated either sinusoidally or trapezoidally using Hall sensors. The ironless armature assembly has no electromagnetic attractive force to the stationary PM assembly, which reduces the load on and increases the life of the bearing system. The encapsulated armature coil assembly moves, and the multipole PM assembly is stationary. The lightweight coil assembly allows for higher acceleration of light payloads than heavier PM assembly. Linear guidance is achieved by using a single linear rail with one or two linear recirculating ball-bearing guides. The bearing is sealed with wipers to contain the lubrication and to keep out debris.

When neglecting the electrical dynamics of the LSM, the dynamics of a single-axis positioning stage can be described by the following equation:

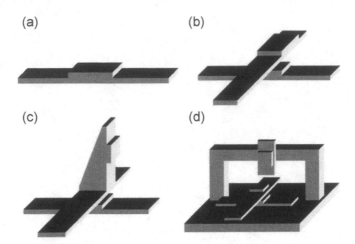

Fig. 10.6. Configurations of linear positioning stages: (a) single axis, (b) compound xy, (c) compound xyz, (d) split xyz [119].

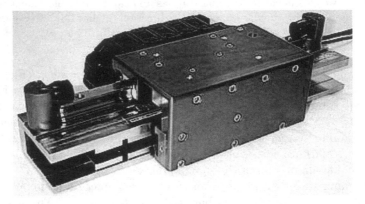

Fig. 10.7. Single-axis positioning stage with air-cored PM LSM. Photo courtesy of H2W Technologies, Santa Clarita, CA, USA.

$$m\frac{d^2x}{dt^2} + D_v\frac{dx}{dt} + k_s x + F_r + F_{ext} = k_F I_a \qquad (10.24)$$

where m is the mass of the moving table, D_v is the damping/friction constant; k_s is the spring constant; F_r is the external disturbance force, e.g., ripple force, F_{ext} is the external force during manufacturing operation; k_F is the thrust constant (6.6); and I_a is the armature current. Compare eqn (10.24) with eqns (1.25 and (10.9).

A linear precision positioning stage with two LSMs to obtain the $x - y$ motion is shown in Fig. 10.8. Multiaxis linear positioning stages provide a very compact platform for accurate positioning of delicate payloads in high-

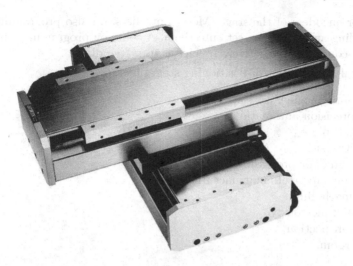

Fig. 10.8. Compound xy positioning stage with two PM LSMs. Photo courtesy of Hiwin Corporation, San Jose, CA, USA.

Fig. 10.9. Compact five-axis positioning stage with PM LSMs. Photo courtesy of H2W Technologies, Santa Clarita, CA, USA.

cycle applications. For example, the extremely smooth running five-axis linear stage shown in Fig. 10.9 enables reduction of the footprint of the machine and, at the same time, maximization of throughput. It is designed for high-speed assembly, test and measurements, grinding, polishing, etc.

Controllers are units that interface the user with the linear positioning stage. They have computer interfaces and interfaces to the linear motor of the stage. The controller can also be remotely operated. The controller has inputs for encoders of the stage. Most controllers are also programmable by downloading an instruction set onto them. A common programming platform for many controllers is ActiveX[1], which is user friendly.

Typical applications of positioning stages include

- precision machinery,
- high-precision automation,
- pick-and-place,
- parts transfer,
- semiconductor processing,
- optical components manufacturing,
- laser machining,
- precision metrology,
- vision inspection,
- clean room.

10.4 Gantry Robots

The term "gantry" defines a system with two motors controlling a single linear axis. Each motor/bearing system is separated a finite distance orthogonal to the direction of the axis. The most common mechanical systems use either linear motors or rotary motors with ballscrews (or belts). A typical gantry configuration is shown in Fig. 10.10.

Gantry robots are also called Cartesian or linear robots. They are usually large systems that perform material handling tasks (palletizing, unitizing, stacking, order picking, and machine loading) and drive-through wash systems for tracks and cars, but they can also be used in manufacturing processes, e.g., welding, removing paint from large aircraft, flat panel manufacture, and coordinate measuring.

A *gantry robot* consists of a manipulator mounted onto an overhead system that allows movement across a horizontal plane. Each of the motions is arranged to be perpendicular to the other, and are typically labeled x, y, and z. Motions in the x and y directions are located in the horizontal plane, while z is the vertical direction. Specifications in the x and y-axis require an absolute position accuracy of less than ± 5.0 mm and repeatability of ± 1.0 mm.

Gantry robot systems with PM LSMs provide the following advantages:

- Large work envelopes (not restricted by arm length)
- Less limited by floor space constraints than other robots
- Better suited for multiple machines and conveyor lines
- Better handling of large or awkward payloads

---

[1] ActiveX is a loosely defined set of technologies developed by *Microsoft* for sharing information among different applications.

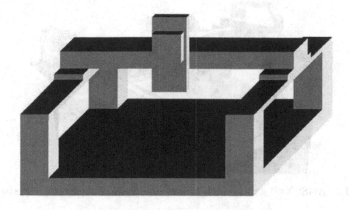

Fig. 10.10. Gantry configuration [119]

- Very good positioning accuracy[2]
- Better acceleration
- Improved efficiency and versatility
- Easy programming with respect to motion, because gantry robots work with *xyz* coordinate system
- Superiority of optimum schedule

Fig. 10.11. *Hercules* series gantry *x-y* stages. Photo courtesy Anorad Rockwell Automation, Shirley, NY, USA.

---

[2] Position accuracy is the ability of the robot to place a part correctly.

Fig. 10.12. AGS20000 gantry with LSMs. Courtesy of Aerotech, Pittsburgh, PA, USA.

The *Hercules* family of gantrie (Anorad, Shirley, NY, USA) are designed to address a multitude of performance requirements for inspection, pick-and-place, assembly, or dispensing applications. The *Hercules* gantry stages (Fig. 10.11) are based on a single platform design, where many linear servo motor selections, linear encoder options, and several travel lengths allow for uniquely supporting a variety of applications with a cost-effective solution. The standard model *Hercules* gantries feature iron-cored LSMs that are ideally suited to meet the rapid point-to-point motion common in the electronics assembly industry. High force produced by LSMs, combined with a low mass x-axis crossbeam, allows high acceleration to maximize throughput.

Aerotech, Pittsburgh, PA, USA has introduced the AGS20000 gantries with LSMs, which are believed to be the most powerful and accurate Cartesian gantries in the world (Fig. 10.12). Dual PM linear brushless servomotors and dual linear encoders offer [5]:

- high velocity up to 3 m/s and high acceleration up to 5g;
- lower-axis continuous force 1644 N (air cooling), peak force 3288 N;
- upper-axis continuous force 276 N (air cooling), peak force 1106 N;
- accuracy ±3.0 mm;
- repeatability ±1.0 mm;
- resolution 0.02 to 1.0 mm;
- optimized mechanical structure for high servo bandwidth.

10.5 Material Handling

10.5.1 Monorail Material Handling System

Fig. 10.13 shows an *overhead monorail* system for material handling with two HLSMs to obtain a linear motion control in the x and y direction. Such a monorail system can be computer controlled and installed in automated assembly lines or material transfer lines where high precision of positioning or clean atmosphere is required.

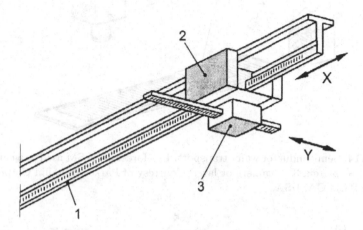

Fig. 10.13. Automated monorail system with HLSMs. 1 — overhead monorail, 2 — x-direction forcer of an HLSM, 3 — y-direction forcer of a HLSM.

10.5.2 Semiconductor Wafer Transport

Manual handling and manual *semiconductor wafer transport* are not an acceptable methods for advanced manufacturing processes. HLSMs simplify the process of semiconductor wafer transport as shown in Fig. 10.14 [40]. The HLSM offers increased throughput and gentle handling of the wafer.

Magnetically levitated wafer transport systems are also used for the semiconductor fabrication process to get rid of the particle and oil contaminations that normally exist in conventional transport systems.

10.5.3 Capsule-Filling Machine

To dispense radioactive fluid into a capsule, an HLSM driven machine can be used. Such a solution has the following advantages [40]:

- increased throughput,
- no spilling of radioactive fluid,
- automation in two axes,
- smooth, repeatable motion,
- cost-effective solution.

In a *capsule-filling machine* shown in Fig. 10.15, the forcer of an HLSM moves a tray with empty capsules along a horizontal axis [40]. The filling head driven

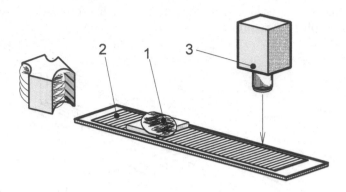

Fig. 10.14. Semiconductor wafer transport. 1 — forcer of HLSM used as a carrier for wafers, 2 — platen, 3 — camera or laser. Courtesy of Parker Hannifin Corporation, Rohnert Park, CA, USA.

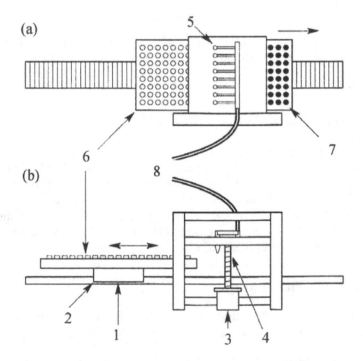

Fig. 10.15. Capsule-filling machine with an HLSM. 1 — forcer, 2 — platen, 3 — rotary microstepping motor, 4 — lead screw, 5 — filling heads, 6 — tray of empty capsules, 7 — full capsules, 8 — hose. Courtesy of Parker Hannifin Corporation, Rohnert Park, CA, USA.

by a rotary stepping motor and ball lead screw, is raised and lowered in the vertical axis. A linear motor in the vertical axis has also been considered, but with loss of power, the fill head will drop onto the tray [40]. The simple mechanical construction guarantees a long maintenance-free life.

In addition to the foregoing examples of material-handling systems, HLSMs offer solutions to a variety of factory automation systems that include [40]:

- printed circuit board assembly,
- industrial sewing machines,
- light assembly automation,
- automatic inspection,
- wire harness making,
- automotive manufacturing,
- gauging,
- packaging,
- medical applications,
- parts transfer,
- pick-and-place,
- laser cut and trim systems,
- flying cutters,
- semiconductor technology,
- water jet cutting,
- print heads,
- fiber optics manufacture,
- x-y plotters.

Speed, distance, and acceleration are easily programmed in a highly repeatable fashion.

10.6 Machining Processes

The seven basic machining processes are *shaping*, *drilling*, *turning*, *milling*, *sawing*, *broaching*, and *abrasive machining*. To accomplish the basic machining processes, eight basic types of cutting *machine tools* have been developed [49]:

- shapers and planers,
- drill presses,
- lathes,
- boring machines,
- milling machines,
- saws,
- broaches,
- grinders.

For example, constructions of milling machines are shown in Fig. 10.16.

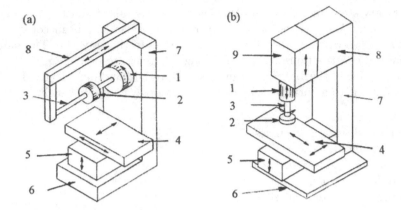

Fig. 10.16. Major components of knee type milling machines with (a) horizontal spindle, (b) vertical spindle. 1 — spindle, 2 — cutter, 3 — arbor, 4 — table, 5 — knee, 6 — base, 7 — column, 8 — overarm, 9 — head.

10.6.1 Machining Centers

Most of machine tools are capable of performing more than one of the basic machining processes. This advantage has led recently to the development of machining centers. A *machining center* is a specifically designed and numerically controlled (NC) single machine tool with a single workpiece set up to permit several of the basic processes, plus other related processes [49]. Thus, a machining center can perform a variety of processes and change tools automatically while under programmable control. Numerical control is a method of controlling the motion of machine components by means of coded instructions generated by microprocessors or computers.

A part to be machined is fixed to the x-y table of the machining center, which must provide a movement both in the x and y directions (Fig. 10.17). The vertical z movement is provided by the head, e.g., to control the depth of drilled holes. This can also be achieved by moving the cutting tool in the w direction (Fig. 10.17). To obtain the full flexibility of machining, the vertical cutting tool should rotate around its horizontal axis (α direction). Closed-loop position control is used in each of five axes.

In the early 1990s, high-speed modern machining centers have been developed (Makino Milling Company and Comau Company). Specifications of high-speed machining centers are given in Table 10.3 [103]. The maximum speed $v_{max} = 60$ m/min is at least twice higher, and the maximum acceleration $a_{max} \approx 1$ g is at least three times higher than those of conventional machining centers. On the other hand, the static servo stiffness of high speed machining centers decreases about 18 times in comparison with conventional machining centers. This is mainly caused by the increase of the pitch τ_l of the screw and decrease of the rotor moment of inertia J_m of the servo motor.

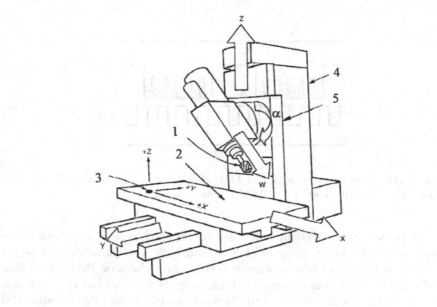

Fig. 10.17. Five-axis vertical spindle machining center. 1 — cutting tool, 2 — table, 3 — machine zero point, 4 — column, 5 — head.

Table 10.3. Specifications of a modern high speed machining center with ball lead screws [103]

Parameter	x-axis	y-axis	z-axis
Stroke, m	0.56	0.41	0.41
Maximum linear speed (feed rate), m/min	60		
Maximum acceleration	1.0 g	1.2 g	0.9 g
Mass of slider, kg	235	500	425
Ball lead screw			
• Diameter, mm	36		
• Pitch τ_l, mm	20		
• Type of support	Single		
Type of guideway	Sliding		
Control	Semi-closed loop		
• Resolution of positioning, mm	0.001		
Spindle			
• Diameter, mm	65		
• Rated output power, kW	30		
• Rotational speed, rpm	20,000		

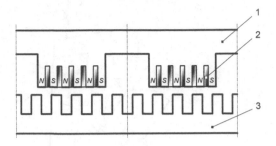

Fig. 10.18. High thrust density PM LBM developed by Shinko Electric Co. Ltd, Japan. 1 — armature core, 2 — PM, 3 — reaction rail.

LIMs [63] are not recommended motors for machine tool applications because they emit large amounts of heat and have low efficiency and power factor. PM LSMs or LBMs are much better motors because they are smaller and can provide efficiency over 75%, high power factor, and fast response. According to Shinko Electric Co. Ltd, Takegahana Ise, Japan, the so-called PM *high-thrust density linear motor* (HDL) has efficiency over 90% at low speed and very high acceleration [105, 155]. An HDL is similar to HLSM, i.e., both have forcer windings and toothed reaction rail; however, the HDL has a different arrangement of PMs (Fig. 10.18). Table 10.4 compares the performance of three different linear motors, i.e., LIM, LSM and HDL for applications to machine tools [103, 155]. The HDL is sometimes called *flux reversal* PM LSM [38].

The most important variable that describes the behavior of a position control loop for computer numerically controlled (CNC) machine tool driven by a linear motor servo drive is position loop gain K_v [164]. This is the ratio of the command velocity (feed rate) v to the position control deviation (following error, tracking error, lag) Δx, i.e.,

$$K_v = \frac{v}{\Delta x} \; 1/\text{s} \tag{10.25}$$

In general, position loop gain K_v should be high for faster system response and higher accuracy, but the maximum gains allowable are limited due to undesirable oscillatory responses at high gains and low damping factor [164]. Usually, K_v factor is experimentally tuned on the already assembled machine tool. To obtain the required position control loop damping, the position loop gain K_v should be calculated with the following equation [164]

$$K_v = \frac{1}{4\zeta^2 \left(\frac{2D_v}{\omega} + \frac{T_s}{2} \right)} \; 1/\text{s} \tag{10.26}$$

where ω is the angular frequency, D_v is the damping constant, ζ is the position control loop damping ($0 < \zeta < 1$), and T_s is the sampling period. In

Table 10.4. Comparison of linear motors used in servo drives of machine tools

Parameter	LIM	LSM	HLD
Dimensions L×W×H, mm	222×50×49.5	160×50×27	290×79×40
Mass of armature, kg	4.29	1.89	6.47
Air gap, mm	1.0	0.5	0.2
Resistance per phase, Ω	1.72	26.8	0.193
Inductance per phase, mH	14.4	54.7	5.62
Pole pitch τ, mm	54	33	10
Construction of armature	Winding	Winding	Winding + PMs
Construction of reaction rail	Cu+Fe	PMs+Fe	Fe
Cogging and detent force	Nonexisting	Exists	Exists
Rated current, A	4.0	1.03	16.9
Rated thrust, N	48	139	880
Thrust constant k_F, N/A	12	135	52
Maximum efficiency, η	0.18	0.77	0.92
Efficiency at rated thrust and $v = 1$ m/s	0.125	0.60	0.62
Thermal resistance, ^{0}C/W	0.87	0.686	0.291

practice, the K_v factor must be decreased up to 40% due to the presence of nonlinearities, i.e.,

$$K_v = \frac{0.6}{4\zeta^2 \left(\frac{2D_v}{\omega} + \frac{T_s}{2}\right)} \ 1/s \qquad (10.27)$$

Using eqn (10.27), the position loop gain K_v of a CNC machine tool driven by linear motor servo drive can be estimated without performing any experiments.

First machining centers with linear motor propulsion were built by Ingersoll Milling Machine Company, Rockford, IL, USA and LMT Consortium, Japan [103] in the mid 1990s. In machining centers built by Ingersoll, series LF LSMs manufactured by Anorad [12] have been installed. Table 10.5 shows specifications of the LMT96 machining center manufactured by LMT Consortium [103].

Mazak Corporation, uses linear motors in the F3-660L horizontal machining center shown in Fig. 10.19, which is designed for automotive applications, especially for die-cast aluminum such as transmission casings [53]. This machining center uses LSMs for the x, y and z axes. With a rapid traverse of 120 m/min in the x-axis and 50 m/min on the y and z axes, the table is capable of accelerating at 0.5 g in all axes. Machining centers with LSMs are mainly used for machining processes that require high *contouring* accuracy, e.g., manufacturing of dies or molds. Contouring permits two or three axes to be controlled simultaneously in two or three dimensions. The price of machin-

Fig. 10.19. Horizontal machining center F3-660L with LSMs for all three axes. Photo courtesy of Mazak Corporation, Florence, KY, USA.

ing centers with linear motors is almost twice of that with ball lead screws.

10.6.2 Aircraft Machining

Most wing ribs and floor spars on aircraft are made of stamped sheet metal cut to a contour [131]. Extrusion stock is then riveted to one or both faces to add strength. A small wing rib could be $50 \times 250 \times 1350$ mm in dimension, while floor spars can be up to 0.9 m wide and more than 6 m in length. Since a typical commercial jetliner has around 100 wing ribs and another 100 floor spars, the time in labor, as well as the cumulative weight of the components, tends to add up [131].

Monolithic parts, i.e., structural components hogged out of single billets of metal, usually aluminum, can save time and cost in manufacturing. Making wing ribs, floor spars, fuselage frames, and other parts with pockets and honeycomb structures from single aluminum pieces would use about the same amount of material as sheet metal assembly. The real savings would be in the elimination of hundreds of fasteners, increased production speed, elimination of tooling, and the production of those tools [131]. The main problem has been how to machine these large parts with high speed, efficiency and repeatable accuracy.

The HyperMach program (Boeing, McDonnel Douglas, United Technologies Corporation, and US Air Force as primary partners) has solved this problem with the aid of LSMs. The HyperMach™ aerospace vertical profiler (Fig. 10.20) with *Kollmorgen* and *Anorad* LSMs can make contours at feed rates of over 100 m/min accelerating as fast as 2 g. Machine range is over 0.914 m in the x-axis, 1.5 m in the y-axis, and 0.75 m in the z-axis. This machine

Table 10.5. Specifications of machining center LMT96 manufactured by LMT Consortium, Japan

Mass of machining center, kg	16,000
Dimensions length×width×height, m	4.4×4.4×2.9
Stroke, m	
• x axis	0.8
• y axis	0.5
• z axis	0.5
Maximum feed rate v_{max}, m/min	
• x axis	80
• y axis	80
• z axis	80
Maximum acceleration a_{max}	
• x axis	1.0 g
• y axis	2.0 g
• z axis	1.0 g
Spindle	
• Diameter, mm	65
• Rated output power, kW	15/22
• Rotational speed, rpm	30,000
Tools	
• Number of tools	12
• Tool selection	Random
• Maximum diameter of tool, mm	100
• Maximum length of tool, mm	250
• Time required to change tool-to-tool	1.5 s
• Time required to change chip-to-chip	3.5 s

Fig. 10.20. HyperMach^{TM}vertical profiler with PM LSMs in the x and y axes. Photo courtesy of MAG, Hebron, KY, USA.

Fig. 10.21. IEV flexible vertical grinding center with LSMs. Photo courtesy of Danobat Group, Elgoibar, Spain.

platform offers high-efficiency machining for both thin and thick plate processing of a wide variety of large aluminum structural components such as ribs, bulkheads, plate, frames, stringers, and spars to meet the demands of next-generation aircraft parts. The x and y axes are driven by PM LSMs.

A high precision vertical grinding centre for complete machining of a wide range of aerospace components such as nozzles, ballscrews, landing-gear flaps, etc., is shown in Fig. 10.21. This machining center is designed using LSM technology and precision linear slides.

10.7 Welding and Thermal Cutting

10.7.1 Friction Welding

The heat required for *friction welding* is produced as a result of mechanical friction between two pieces of metal to be joined [49]. Those two pieces are held together while one rotates and the other is stationary. The rotating part is held in a motor-driven collet, while the stationary part is held against it under controlled pressure produced by a hydraulic or pneumatic actuator (Fig. 10.22a). The frictional heat is a function of the rotational speed and applied pressure (linear force). The rotational speed and force depend on the size of the piece of metal to be welded. For example, for welding a 3 mm mild steel stud, a force about 900 N at 20,000 rpm is required [92].

In manufacturing plants where clean atmosphere is required, hydraulic or pneumatic actuator must be replaced by electric linear actuator (Fig. 10.22b). PM linear actuators [92] or PM LSMs can successfully be used.

10.7.2 Welding Robots

HLSMs provide a simple solution to high-precision motion control of a torch or electrode of *welding robots*. Fig. 10.23 shows a robot for the resistance spot

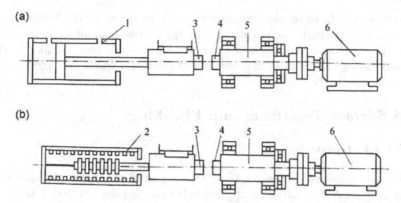

(a)

(b)

Fig. 10.22. Equipment used for friction welding: (a) with hydraulic or pneumatic actuator, (b) with electric linear actuator. 1 — mechanical actuator, 2 — PM actuator or tubular LSM, 3 — chuck for stationary part, 4 — chuck for rotating part, 5 — spindle, 6 — rotary electric motor.

welding in which the electrode is moved in the x-y plane with the aid of two HLSMs [240]. The robot is microprocessor controlled.

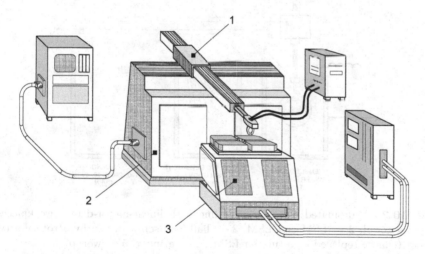

Fig. 10.23. Linear motor driven welding robot. 1 — arm driven by two HLSMs in x-y plane, 2 — base, 3 — table.

10.7.3 Thermal Cutting

Most of all *thermal cutting* is done by *oxyfuel gas cutting* where acetylene is used as a fuel [49]. The tip of the torch contains a circular array of small holes through which the oxygen–acetylene mixture is supplied for the heating flame. In many manufacturing applications, cutting torches cannot be manipulated manually, and electrically driven carriages are used to hold cutting torches. LSMs or LIMs can be used as direct linear drives for torch carriages.

10.8 Surface Treatment and Finishing

10.8.1 Electrocoating

In the *electrocoating* process, a workpiece is placed in a tank with paint and water solvent. A d.c. voltage is applied between the tank (cathode) and workpiece to be coated (anode). The paint particles are attracted to the workpiece and deposited on it creating a uniform thin coating (0.02 to 0.04 mm) [49]. Then the workpiece is removed from the dip tank, rinsed, and baked at about 195^0C for 10 to 20 min. This process is especially suitable for complex metal structures such as automobile bodies [49].

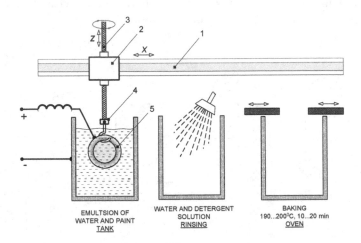

EMULTSION OF
WATER AND PAINT
TANK

WATER AND DETERGENT
SOLUTION
RINSING

BAKING
$190...200^0$C, $10...20$ min
OVEN

Fig. 10.24. Automated electrocoating line with linear-motor-driven workpieces. 1 — monorail, 2 — LSM or HLSM, 3 — ball lead screw driven by a rotary servo motor (can be replaced by a tubular LSM), 4 — gripper, 5 — workpiece.

Fig. 10.24 shows an automated electrocoating line with linear-motor-driven overhead monorail system for horizontal transfer of bulk workpieces. Depending on the required thrust, mass of the workpiece, and travel distance in the x direction, either LSMs or HLSMs can be used. To raise or lower a heavy workpiece vertically, ball lead screws driven by rotary servo motors are better than linear motors.

10.8.2 Laser Scribing Systems

Aircraft aluminum skin panels are usually scribed manually through templates prior to chemical milling. This costly process can be automated by the use of laser scribers and LSMs [55, 110]. An automatic *laser maskant scriber* with LSMs was built in 1993 for Boeing, Wichita, KS, USA. [55]. In the chemical milling process, most aircraft component parts are covered with a coating called maskant[3]. In this etched machining process, the scribed metal surface is coated with polymer material. The laser process is kept smoke free, and the scribed line is kept clean through the use of the smoke extraction hood installed on the optical laser wrist. The polymer is scribed to a desired pattern and then removed from the surface. The metal is then placed in an acid solution that etches out the exposed surfaces.

Two parallel LSMs, (series LF, Anorad[12]) produce the peak thrust up to 9 kN at speed up to 5 m/s [110]. LSMs rapidly and precisely position a huge gantry with the CO_2 laser maskant scribing system. Two laser systems have been used: one for scribing and another for position feedback. The use of a laser interferometer for positioning feedback with LSM drive mechanisms results in high accuracy and repeatability. Large over 30 m long 5-axis gantry for laser maskant scriber driven by *Anorad*'s LSMs is shown in Fig. 10.25.

Fig. 10.25. Large 5-axis gantry for laser maskant scriber driven by Anorad's LSMs. Photo courtesy of G. Hagiz [76].

The advantages of the new LSM driven laser maskant scriber include mass, scrap and inventory reduction, noncontact design, no lubrication and adjustments, elimination of the stress channel, and minimizing back bending and repetitive motions. According to Boeing, this is the largest automatic laser

---

[3] Maskant is a material that protects a metal surface during the etching process.

(a)   (b)

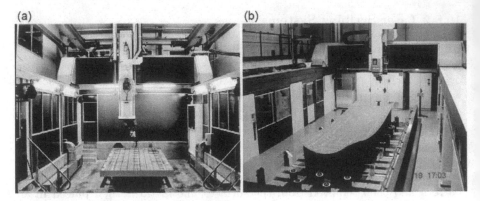

Fig. 10.26. UVS 5-axis machining center with Siemens SIMODRIVE LSM 840D CNC for aircraft industry: (a) overhead gantry (b) Airbus A320 external skin around front landing box to be laser processed. Photo courtesy of Le Creneau Industriel, Annecy le Vieux, France.

scriber in the world (33 m long, 3 m wide, and 5.1 m tall) that can hold two Boeing 747 wings simultaneously, scribing one while positioning the other and scribe up to 2.54 m/s or complete up to four skin panels in 1 h [110].

Le Creneau Industriel, Annecy le Vieux, France, is a manufacturer of 5 axis machining centers (Fig. 10.26) for routing and drilling aluminum and large-dimension composite components for the aircraft industry. The UGV 5 axis machine with overhead moving gantry, Coherent 100 W CO_2 laser, and optical laser wrist, have been designed for aircraft manufacturers for laser-scribing maskant prior to chemical machining [75].

The optical laser wrist is designed with a large 9 m $\times$4.5 m $\times$1.5 m work envelope moving gantry. It incorporates a Siemens *Simodrive* 840D CNC with 1FN LSMs to provide high accuracy with speeds up to 60 m/min [75]. The curved component to be laser processed is placed on a 99-peg support fixture where each peg has its own independent numerical axis to form the precise 3D shape required. The optical laser wrist incorporates a measurement probe to confirm the actual shape of the component prior to being laser processed. The result is compared to the theoretical shape within the system's software for quality control. The complex pattern is scribed up to 30 m/min with an accuracy of ± 0.025 mm.

10.8.3 Application of Flux-Switching PM Linear Motors

Large gantry systems and machining centers require powerful linear motors, preferably with PM-free reaction rail. Siemens 1FN6 PM LSMs with a magnet-free reaction rail belong to the group of the so called *flux-switching PM machines* [90, 176]. The armature system is air cooled, degree of protection IP23, class of insulation F, line voltage from 400 to 480 V, rated thrust from 235 to 2110 N, maximum velocity at rated thrust from 170 to 540 m/min (Table 10.6), overload capacity 3.8 of rated thrust, modular type construction [198]. These LSMs operate with Siemens *Sinamics* or *Simodrive* solid-state converters and external encoders. According to Siemens [198], these new LSMs (Fig. 10.27) produce thrust forces and velocities equivalent to competitive classical models for light-duty machine tool, machine accessory, and material handling applications.

Table 10.6. Specifications of 1FN6 PM LSMs manufactured by Siemens, Erlangen, Germany [198]

Armature unit	Rated thrust N	Max. thrust N	Max. speed at rated thrust m/min	Max. speed at max. thrust m/min	Rated current A	Max. current A
1FN6008-1LC17	235–350	900	263	103	1.7–2.6	9.0
1FN6008-1LC37	235–350	900	541	224	3.5–5.3	18.0
1FN6016-1LC30	470–710	1800	419	176	5.4–8.0	28.0
1FN6016-1LC17	935–1400	3590	263	101	7.0–10.5	36.0
1FN6024-1LC12	705–1060	2690	176	69	3.5–5.3	18.0
1FN6024-1LC20	705–1060	2690	277	114	5.4–8.0	28.0
1FN6024-1LG10	2110–3170	8080	172	62	10.5–16.0	54.0
1FN6024-1LG17	2110–3170	8080	270	102	16.2–24.3	84.0

The magnet-free reaction rail is easy to install and does not require the safety considerations of standard PM reaction rails. Without PMs, there is no problem with ferrous chips and other debris being attracted to these sections. Maintenance becomes a simple matter of installing a wiper or brush on the moving part of the slide.

The 1FN6 flux-switching LSMs comprises an armature section that is equipped with coils and PMs as well as a nonmagnetic, toothed reaction rail section (Fig. 10.28). The key design innovation is an LSM in which PMs are integrated directly into the lamination of the armature core along with the individual windings for each phase. Both magnitudes and polarities of the linkage flux in the armature winding vary periodically along with the reaction rail movement. The magnetic flux between the armature core and steel reaction rail is controlled by switching the three-phase armature currents according to a designated algorithm [172]. The passive reaction rail consists of milled steel with poles (teeth) and is much simpler to manufacture.

The relationships between the pole pitches and number of poles of the armature and reaction rail are

$$\tau_2 = \frac{\tau_1}{1 \pm \frac{k}{m_1}} \tag{10.28}$$

Fig. 10.27. Novel 1FN6 flux-switching LSM with PM-free reaction rail. Photo courtesy of Siemens AG, Erlangen, Germany.

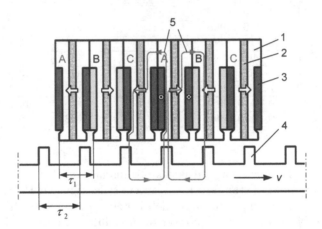

Fig. 10.28. Construction of flux-switching LSM with PM-free reaction rail. 1 — laminated armature core, 2 — PM, 3 — armature coil (phase C), 4 — toothed passive steel reaction rail, 5 — linkage magnetic flux (phase A is on).

$$P_2 = P_1 \frac{\tau_1}{\tau_2} \qquad (10.29)$$

where τ_1, τ_2 are pole pitches of the armature and reaction rail, respectively, P_1 and P_2 are the numbers of the armature and reaction rail poles, respectively, and $k = 1, 2, 3, \ldots$ is integer. For example, if $\tau_1 = 42$ mm, $m_1 = 3$ and $k = 1$, the reaction rail pole pitch $\tau_2 = 42/[1 \pm (1/6)] = 36$ mm or 50.4 mm. Assuming $P_1 = 12$, the possible number of reaction rail poles is $P_2 = 12 \times (42/36) = 14$ or $P_2 = 12 \times (42/50.4) = 10$.

As far as the authors are aware, the first paper on flux-switching PM brushless rotary machines was published in 1955 [176]. There are numerous papers on the analysis of this type of machines, e.g., [21, 37, 90, 94, 251]. However, it is impossible to find a convincing analysis published so far on how flux-switching PM brushless motors compare to standard PM brushless motors in terms of thrust density, efficiency, and power factor. Some constructions of flux-switching PM LSMs have been patented [P212, P213, P215].

10.9 2D Orientation of Plastic Films

High-quality plastic films can be obtained as a result of *simultaneous orientation technology* in which the film is stretched in two directions [32]. Simultaneous orientation is a process where the distances between clips, i.e., gripping points of the film, are continuously increased as a result of moving clips apart lengthwise and across.

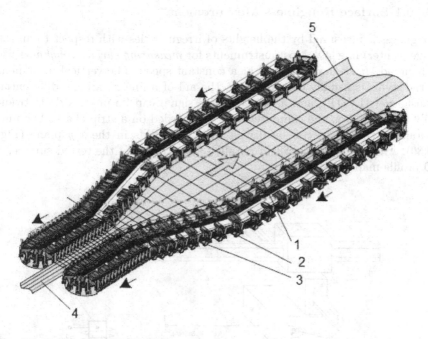

Fig. 10.29. Application of LSMs to simultaneous film orientation technology. 1 — LSM, 2 — driven clips, 3 — idle clips, 4 — cast film, 5 — oriented film. Courtesy of Brueckner Maschinenbau GmbH, Siegsdorf, Germany.

LSMs can be used to move clips forward with adjustable speeds [32]. Driven carriages (clips) with built-in PMs are arranged in a closed circuit to form a roller rail (Fig. 10.29). Armature systems of LSMs are stationary. Additional gripping points for the film are provided by idle carriages that move between driven carriages. Each carriage can move freely along the rail without any

mechanical limitations. As many as 900 LSMs with rated thrust of 900 N
each can be employed, and linear velocities up to 7.5 m/s (450 m/min) can
be achieved [32]. The working width of foil is 7.5 m, and the length of the
track is 2×126 m. High speeds and flexibility of speed sequences of individual
carriages allow for high productivity and adjustable product features.

The electromagnetic thrust of each LSM is proportional to the load (power)
angle δ, i.e., $F_{dx} \approx (m_1/v_s)(V_1 E_f/X_{sd}) \sin \delta$. Variation of external forces
causes variation of the angle δ and, as a consequence, oscillations are generated. To damp oscillations effectively, an active damping control system is
implemented in addition to the damper. An observer estimates the load angle
δ from the known currents and sends a feedback to the controller [32].

10.10 Testing

10.10.1 Surface Roughness Measurement

Roughness is measured by the heights of irregularities with respect to an average (center) line [49]. Most instruments for *measuring surface roughness* use
a diamond *stylus* that is moved at a constant speed. The vertical movement
of the stylus is usually detected with the aid of a linear variable differential
transformer (LVDT), and as an electrical signal can be processed electronically and stored on the computer disk or recorded on a strip chart. The unit
containing the stylus can be driven by two HLSMs in the x-y plane (Fig.
10.30). After making a series of parallel offset traces on the tested surface, a
2D profile map is obtained.

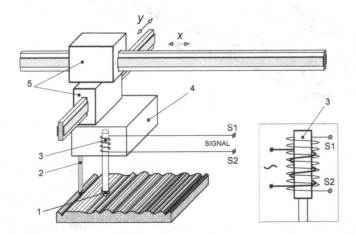

Fig. 10.30. Stylus profile device for measuring surface roughness and profile. 1 —
diamond stylus, 2 — rider, 3 — LVDT, 4 — head, 5 — HLSM.

10.10.2 Generator of Vibration

Analysis and simulation of dynamic behavior of buildings and large-scale constructions requires long-stroke and high-speed *generators of vibration*. Typical parameters of such a device are: 7.7 kN maximum thrust, 6.5 kN rated thrust, 2 m/s maximum speed, 0.5 m effective stroke, and 2000 kg maximum load mass [105]. Oil hydraulic devices are large, need complex maintenance, and have low efficiency and nonlinear characteristics. Linear electric actuators or motors are smaller, provide high speed and acceleration, and do not need maintenance. Shinko Electric Co. Ltd has built a prototype generator of vibration with HDL [105]. A good response, i.e., thrust versus sinusoidal current signal, has been obtained in the vibration domain less than 5 Hz.

10.11 Industrial Laser Applications

Laser systems driven by LSM x-y positioning stages have been sucessfully applied to *diamond processing*, including sawing, cutting, kerfing, and shaping [55]. Advantages of applications of LSMs versus ball lead screws include better-quality finished surface, insensitivity of LSMs to diamond dust, and positioning accuracy of 0.2 μm over travel distances 300 by 100 mm. A multitask *diamond processing laser system* has been built by Or-Ziv, Rehovot, Israel [55].

A linear motor gantry for industrial *laser-cutting systems* has been built e.g., by CBLT, Starnberg, Germany [55]. High-beam-quality CO_2 laser together with LBM positioning stages provide smoother cuts, more vertical surfaces and cleaner edges.

Large *laser scribing-systems* with LSMs have been described in Section 10.8.2.

Examples

Example 10.1

A linear ball screw system driven by a rotary PM brushless motor has the following parameters: maximum rotational speed of the motor $n = 2400$ rpm, maximum stator current $I_{amax} = 8.0$ A, torque constant $k_T = 2.255$ Nm/A, mass of the rotor $m_r = 5.2$ kg, diameter of the rotor $D_r = 48$ mm, diameter of ball screw $d_b = 32$ mm, mass of ball screw $m_b = 4.4$ kg, pitch of screw $\tau_l = 11$ mm, number of threads $k = 1$ (single start screw), number of teeth on the motor wheel $N_1 = 54$, number of teeth on the ball screw wheel $N_2 = 18$, and mass of the table $m_t = 4.6$ kg. Find the maximum linear acceleration and dynamic stiffness of the ball screw drive.

Solution

The maximum rotational speed of the ball screw

$$n_b = \frac{N_2}{N_1}n = \frac{54}{18} \times 2400 = 800 \text{ rpm} = 13.33 \text{ rev/s}$$

For $k = 1$ (single start screw) according to eqn (10.3), the lead is equal to the screw pitch, i.e., $l = 1 \times 11 = 11$ mm, and maximum linear speed according to eqn (10.8) is $v_{max} = 13.33 \times 0.011 = 0.147$ m/s $= 8.8$ m/min. The various moments of inertia are as follows

- Moment of inertia of the motor cylindrical rotor

$$J_m = \frac{1}{2}m_r\frac{D_r^2}{4} = \frac{1}{2}5.2\frac{0.048^2}{4} = 1.498 \times 10^{-3} \text{ kgm}^2$$

- Moment of inertia of the cylindrical ball screw

$$J_{bs} = \frac{1}{2}m_b\frac{d_b^2}{4} = \frac{1}{2}4.4\frac{0.032^2}{4} = 0.563 \times 10^{-3} \text{ kgm}^2$$

- Equivalent moment of inertia of the motor–ball screw system according to eqn (10.19)

$$J_e = 0.563 \times 10^{-3} + \left(\frac{54}{18}\right)^2 \times 1.498 \times 10^{-3} = 0.014 \text{ kgm}^2$$

Maximum torque produced by the rotary PM brushless motor

$$T_{max} = k_T I_{amax} = 2.255 \times 8.0 = 18.04 \text{ Nm}$$

Maximum output power of the motor

$$P_{out} = 2\pi\frac{2400}{60} \times 18.04 = 4534 \text{ W}$$

Linear acceleration according to eqn (10.22)

$$a_{max} \approx \frac{18.04 \times 0.011}{2\pi\left[0.014 + 4.6\left(\frac{0.011}{2\pi}\right)^2\right]} = 2.247 \text{ m/s}^2$$

Angular frequency of ball screw

$$\omega = 2\pi\frac{800}{60} = 83.78 \text{ rad/s}$$

Dynamic stiffness according to eqn (10.21)

$$K_d \approx \left[4.6 + \left(\frac{2\pi}{0.011}\right)^2 \times 0.014\right] \times 83.78^2 = 32.19 \times 10^6 \text{ N/m}$$

Example 10.2

A gantry is driven by a flat PM LSM. The steady-state speed is $v_{const} = 4.0$ m/s, overall time of operation $t = 1.2$ s, acceleration time $t_1 = 0.2$ and deceleration time $t_3 = 0.25$ s. For the speed profile given in Fig. 1.22a, find the maximum acceleration, maximum deceleration, and estimate how the thrust will change if (a) the overall time is extended to $1.2t$, and (b) the acceleration time t_1 is reduced twice. In both cases, assume that all other parameters are the same.

Solution

Total distance

$$s = \frac{1}{2}vt_1 + vt_2 + \frac{1}{2}vt_3 = \frac{1}{2}v(t_1 + t_3) + v(t - t_1 - t_3)$$

$$= vt - \frac{1}{2}v(t_1 + t_3) = 4.0 \times 1.2 - \frac{1}{2}4.0 \times (0.2 + 0.25) = 3.9 \text{ m}$$

Maximum acceleration

$$a_{max} = \frac{v}{t_1} = \frac{s}{t_1(t - 0.5t_1 - 0.5t_3)} = \frac{3.9}{0.2(1.2 - 0.5 \times 0.2 - 0.5 \times 0.25)} = 20 \text{ m/s}^2$$

Maximum deceleration

$$d_{max} = \frac{v}{t_3} = \frac{s}{t_3(t - 0.5t_1 - 0.5t_3)} = \frac{3.9}{0.25(1.2 - 0.5 \times 0.2 - 0.5 \times 0.25)} = 16 \text{ m/s}^2$$

How will the thrust change, if the overall time is extended to $1.2t$ at the same speed, acceleration, and deceleration time? Since the thrust is proportional to the acceleration, the thrust will decrease by the factor

$$\frac{t - 0.5t_1 - 0.5t_3}{1.2t - 0.5t_1 - 0.5t_3} = \frac{1.2 - 0.5 \times 0.2 - 0.5 \times 0.25}{1.2 \times 1.2 - 0.5 \times 0.2 - 0.5 \times 0.25} = 0.802$$

The required thrust will decrease. If the existing LSM is rated, say, at 2000 N, for longer overall time, a smaller LSM rated at $0.8 \times 2000 = 1600$ N is needed.

How will the thrust change if the acceleration time is reduced to $0.5t_1$ at the same speed, acceleration, and deceleration time? Again, the thrust is proportional to the acceleration, i.e.,

$$\frac{t_1}{0.5t_1} = 2$$

If the existing LSM is rated at 2000 N, for faster acceleration, an LSM with doubled thrust is necessary.

Example 10.3

The model of an LSM-driven positioning stage with mass m_p, spring constant k_{sp}, and viscous friction coefficient D_{vp} is shown in Fig. 10.31. The mass m_p includes the mass of LSM and positioning stage. The mass of external structure (load) is m, spring constant k_s, and coefficient of viscous friction D_v. The thrust developed by a linear motor is $F_{dx}(t)$ and the external (load) force is $F_{ext}(t)$. Find the equations of motion using:

(a) Free-body diagram;
(b) Euler–Lagrange equation (1.43).

Solution

(a) Free-body diagram

This is a two-DOF system subjected to external forces (Fig. 10.31a). A free-body diagram is shown in Fig. 10.31b. According to eqn (1.28), the mechanical balance equations are

$$F_{dx}(t) - D_{vp}\dot{x}_1 - k_{sp}x_1 + D_v(\dot{x}_2 - \dot{x}_1) + k_s(x_2 - x_1) = m_p\ddot{x}_1$$

$$F_{ext}(t) - D_v(\dot{x}_2 - \dot{x}_1) - k_s(x_2 - x_1) = m\ddot{x}_2$$

or

$$m_p\ddot{x}_1 + (D_{vp} + D_v)\dot{x}_1 - D_v\dot{x}_2 + (k_{sp} + k_s)x_1 - k_sx_2 = F_{dx}(t) \qquad (10.30)$$

$$m\ddot{x}_2 + D_v\dot{x}_2 - D_v\dot{x}_1 + k_sx_2 - k_sx_1 = F_{ext}(t) \qquad (10.31)$$

It is more convenient to write eqns (10.30) and (10.31) in matrix–vector form, i.e.,

$$\begin{bmatrix} m_p & 0 \\ 0 & m \end{bmatrix} \begin{bmatrix} \ddot{x}_1 \\ \ddot{x}_2 \end{bmatrix} + \begin{bmatrix} D_{vp} + D_v & -D_v \\ D_v & D_v \end{bmatrix} \begin{bmatrix} \dot{x}_1 \\ \dot{x}_2 \end{bmatrix}$$

$$+ \begin{bmatrix} k_{sp} + k_s & -k_s \\ -k_s & k_s \end{bmatrix} \begin{bmatrix} x_1 \\ x_2 \end{bmatrix} = \begin{bmatrix} F_{dx}(t) \\ F_{ext}(t) \end{bmatrix} \qquad (10.32)$$

The short form of the matrix–vector notation

$$[\mathbf{m}][\ddot{\mathbf{x}}] + [\mathbf{D_v}][\dot{\mathbf{x}}] + [\mathbf{k_s}][\mathbf{x}] = [\mathbf{F(t)}] \qquad (10.33)$$

The mass matrix in eqn (10.32) is diagonal, so the system is uncoupled inertially. The damping and stiffness matrices in eqn (10.32) are coupled.

(a)

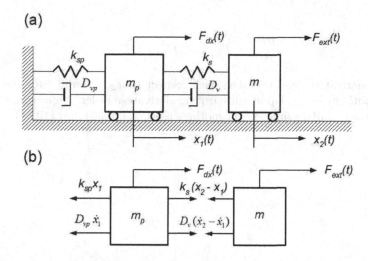

(b)

Fig. 10.31. Mathematical model of a single-axis positioning stage driven by a linear motor: (a) two-DOF system subjected to forces $F_{dx}(t)$ and $F_x(t)$; (b) free-body diagrams.

(b) Euler–Lagrange equation

Kinetic energy, potential energy, and Rayleigh dissipation function (1.44) for $\xi = x$ are, respectively,

$$E_k = \frac{1}{2}m_p\dot{x}_1{}^2 + \frac{1}{2}m\dot{x}_2{}^2 \tag{10.34}$$

$$E_p = \frac{1}{2}k_{sp}x_1^2 + \frac{1}{2}k_s(x_2 - x_1)^2 = \frac{1}{2}k_{sp}x_1^2 + \frac{1}{2}k_sx_2^2 - k_sx_1x_2 + \frac{1}{2}k_sx_1^2 \tag{10.35}$$

$$Ra = \frac{1}{2}D_{vp}\dot{x}_1{}^2 + \frac{1}{2}D_v(\dot{x}_2 - \dot{x}_1)^2 = \frac{1}{2}D_{vp}\dot{x}_1{}^2 + \frac{1}{2}D_v\dot{x}_2^2 - D_v\dot{x}_1\dot{x}_2 + \frac{1}{2}D_v\dot{x}_1^2 \tag{10.36}$$

Derivatives with respect to $\dot{x}_1$, x_1, and t for the positiong stage with mass m_p taken according to the definion of Lagrangian (1.33) and Euler–Lagrange equation (1.43), are

$$\frac{\partial \mathcal{L}}{\partial \dot{x}_1} = \frac{\partial(E_k - E_p)}{\partial \dot{x}_1} = m_p\dot{x}_1; \qquad \frac{d}{dt}\frac{\partial \mathcal{L}}{\partial \dot{x}_1} = m_p\ddot{x}_1$$

$$\frac{\partial \mathcal{L}}{\partial x_1} = \frac{\partial(E_k - E_p)}{\partial x_1} = -k_{sp}x_1 + k_sx_2 - k_sx_1$$

$$\frac{\partial Ra}{\partial \dot{x}_1} = D_{vp}\dot{x}_1 - D_v\ddot{x}_2 + D_v\dot{x}_1$$

Similar derivatives can be taken with respect to $\dot{x}_2$, x_2, and t for the linear motor with mass m. Upon substituting derivates, Euler–Lagrange equation (1.43) gives similar equations of motions as eqns (10.30) and (10.31).

Appendix A

Magnetic Circuits with Permanent Magnets

A.1 Approximation of Demagnetization Curve and Recoil Line

The most widely used approximation of the demagnetization curve is the approximation using a hyperbola, i.e.,

$$B = B_r \frac{H_c - H}{H_c - a_0 H} \tag{A.1}$$

where B and H are coordinates, and a_0 is the constant coefficient that can be evaluated as [19]

$$a_0 = \frac{B_r}{B_{sat}} = \frac{2\sqrt{\gamma} - 1}{\gamma} \tag{A.2}$$

or

$$a_0 = \frac{1}{n} \sum_{i=1}^{n} \left(\frac{H_c}{H_i} + \frac{B_r}{B_i} - \frac{H_c}{H_i} \frac{B_r}{B_i} \right) \tag{A.3}$$

where (B_i, H_i) are coordinates of points $i = 1, 2, \ldots n$, on the demagnetization curve, arbitrarily chosen, and n is the number of points on the demagnetization curve.

The recoil magnetic permeability is assumed to be constant and equal to [19]

$$\mu_{rec} = \frac{B_r}{H_c}(1 - a_0) \tag{A.4}$$

The above equations give a good accuracy between calculated and measured demagnetization curves for Alnicos and isotropic ferrites with low magnetic energy. Application to anisotropic ferrites with high coercivity can in some cases cause errors.

For rare-earth PMs, the approximation is simple due to their practically linear demagnetization curves, i.e.,

$$B = B_r \left(1 - \frac{H}{H_c} \right) \tag{A.5}$$

This means that putting $a_0 = 0$ or $\gamma = 0.25$, eqn (A.1) gets the form of eqn (A.5).

A.2 Operating Diagram

A.2.1 Construction of the Operating Diagram

The energy of a PM in the external space only exists if the reluctance of the external magnetic circuit is higher than zero. If a previously magnetized PM is placed inside a closed ideal ferromagnetic circuit, i.e., toroid, this PM does not show any magnetic properties in the external space in spite of the fact that there is magnetic flux Φ_r corresponding to the remanent flux density B_r inside the PM.

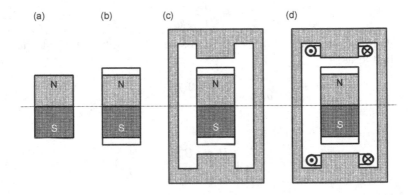

Fig. A.1. Stabilization of a PM: (a) PM alone, (b) PM with pole shoes, (c) PM inside an external magnetic circuit, (d) PM with a complete external armature system.

A PM previously magnetized and placed alone in an open space, as in Fig. A.1a, generates a magnetic field. To sustain a magnetic flux in the external open space, an MMF developed by the magnet is necessary. The state of the PM is characterized by the point K on the demagnetization curve (Fig. A.2). The location of the point K is at the intersection of the demagnetization curve with a straight line representing the permeance of the external magnetic circuit (open space), i.e.,

$$G_{ext} = \frac{\Phi_K}{\mathcal{F}_K}, \qquad \tan\alpha_{ext} = \frac{\Phi_K/\Phi_r}{\mathcal{F}_K/F_c} = G_{ext}\frac{F_c}{\Phi_r} \qquad (A.6)$$

The permeance G_{ext} corresponds to a flux Φ–MMF coordinate system and is referred to as MMF at the ends of the PM. The magnetic energy per unit produced by the PM in the external space is $w_K = B_K H_K/2$. This energy is proportional to the rectangle limited by the coordinate system and lines perpendicular to the Φ and $\mathcal{F}$ coordinates projected from the point K. It is obvious that the maximum magnetic energy is for $B_K = B_{max}$ and $H_K = H_{max}$.

If the poles are furnished with pole shoes (Fig. A.1b), the permeance of the external space increases. The point that characterizes a new state of the PM in Fig. A.2 moves along the recoil line from the point K to the point A. The recoil line KG_M is the same as the internal permeance of the PM, i.e.,

$$G_M = \mu_{rec}\frac{w_M l_M}{h_M} = \mu_{rec}\frac{S_M}{h_M} \qquad (A.7)$$

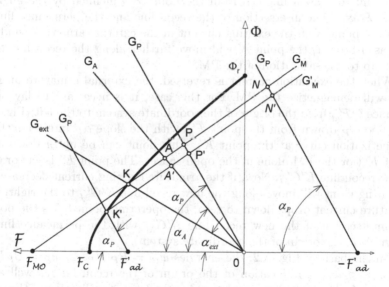

Fig. A.2. Diagram of a PM for finding the origin of the recoil line and operating point.

The point A is the intersection of the recoil line KG_M and the straight line OG_A representing the leakage permeance of the PM with pole shoes, i.e.,

$$G_A = \frac{\Phi_A}{\mathcal{F}_A}, \qquad \tan\alpha_A = G_A\frac{F_c}{\Phi_r} \qquad (A.8)$$

The energy produced by the PM in the external space decreases as compared with the previous case, i.e., $w_A = B_A H_A / 2$.

The next stage is to place the PM in an external ferromagnetic circuit as shown in Fig. A.1c. The resultant permeance of this system is

$$G_P = \frac{\Phi_P}{\mathcal{F}_P}, \qquad \tan \alpha_P = G_P \frac{\mathcal{F}_c}{\Phi_r} \qquad (A.9)$$

which meets the condition $G_P > G_A > G_{ext}$. For an external magnetic circuit without any electric circuit carrying the armature current, the magnetic state of the PM is characterized by the point P (Fig. A.2), i.e., the intersection of the recoil line KG_M and the permeance line OG_P.

When the external magnetic circuit is furnished with an armature winding and when this winding is fed with a current that produces an MMF magnetizing the PM (Fig. A.1d), the magnetic flux in the PM increases to the value Φ_N. The d-axis MMF $\mathcal{F}'_{ad}$ of the external (armature) field acting directly on the PM corresponds to Φ_N. The magnetic state of the PM is described by the point N located on the recoil line on the right-hand side of the origin of the coordinate system. To obtain this point, it is necessary to lay off the distance $O\mathcal{F}'_{ad}$ and to draw a line G_P from the point $\mathcal{F}'_{ad}$ inclined by the angle α_P to the $\mathcal{F}$-axis. The intersection of the recoil line and the permeance line G_P gives the point N. If the exciting current in the external armature winding is increased further, the point N will move further along the recoil line to the right, up to the saturation of the PM.

When the excitation current is reversed, the external armature magnetic field will demagnetize the PM. For this case, it is necessary to lay off the distance $O\mathcal{F}'_{ad}$ from the origin of the coordinate system to the left (Fig. A.2). The line G_P drawn from the point $\mathcal{F}'_{ad}$ with the slope α_P intersects the demagnetization curve at the point K'. This point can be up or down of the point K (for the PM alone in the open space). The point K' is the origin of a new recoil line $K'G'_M$. Now, if the armature-exciting current decreases, the operating point will move along the new recoil line $K'G'_M$ to the right. If the armature current drops down to zero, the operating point takes the position P' (intersection of the new recoil line $K'G'_M$ with the permeance line G_P drawn from the origin of the coordinate system).

On the basis of Fig. A.2, the energies $w_{P'} = B_{P'} H_{P'} / 2$, $w_P = B_P H_P / 2$, and $w_{P'} < w_P$. The location of the origin of the recoil line as well as the location of the operating point determine the level of utilization of the energy produced by the PM. A PM behaves in a different way than a d.c. electromagnet: the energy of a PM is not constant if the permeance and exciting current of the external armature changes.

The location of the origin of the recoil line is determined by the minimum value of the permeance of the external armature or the demagnetization action of the external field.

To improve the properties of PMs independent of the external fields, PMs are stabilized. In magnetic circuits with stabilized PMs, the operating point describing the state of the PM is located on the recoil line. *Stabilization* means the PM is demagnetized up to a value that is slightly higher than the most dangerous demagnetization field during the operation of a system where the PM is installed.

A.2.2 Magnetization without Armature

In most practical applications, the PM is magnetized without the armature and is then placed in the armature system with an air gap. In Fig. A.3 the demagnetization curve is plotted in flux Φ–MMF coordinate system. The origin of the recoil line is determined by the leakage permeance G_{ext} of the PM alone located in open space (Fig. A.3).

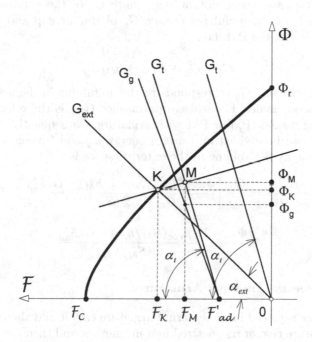

Fig. A.3. Location of the operating point for magnetization without the armature.

For rare-earth PMs, the recoil permeability μ_{rec} and the equation of the recoil line is the same as that for the demagnetization line.

The armature field usually demagnetizes the PM so that the line of the resultant magnetic permeance, G_t, intersects the recoil line between the point K and the magnetic flux axis.

The magnetic flux in the PM is $\Phi_M = G_t(\mathcal{F}_M - \mathcal{F}'_{ad})$. Using the coefficient of leakage flux (2.14), the useful flux density in the air gap can be found as [70]

$$B_g = \frac{\Phi_M}{S_g \sigma_{lM}} = \frac{G_t(\mathcal{F}_M - \mathcal{F}'_{ad})}{S_g \sigma_{lM}}$$

$$= \frac{G_t}{S_g \sigma_{lM}} \left[\frac{\Phi_K + \mathcal{F}_K \mu_{rec}(S_M/h_M) + G_t \mathcal{F}'_{ad}}{G_t + \mu_{rec}(S_M/h_M)} - \mathcal{F}'_{ad} \right] \qquad (A.10)$$

where S_g is the surface of the air gap. With the fringing effect being neglected, the corresponding magnetic field intensity is

$$H_g = H_M = \frac{\mathcal{F}_M}{h_M} = \frac{\Phi_K + \mathcal{F}_K \mu_{rec}(S_M/h_M) + G_t \mathcal{F}'_{ad}}{h_M[G_t + \mu_{rec}(S_M/h_M)]} \qquad (A.11)$$

In the general case, the resultant permeance G_t of the external magnetic circuit consists of the useful permeance G_g of the air gap and the leakage permeance G_{lM} of the PM, i.e.,

$$G_t = G_g + G_{lM} = \sigma_{lM} G_g \qquad (A.12)$$

The useful permeance G_g corresponds to the useful flux in the active portion of the magnetic circuit. The leakage permeance G_{lM} is the referred leakage permeance of a single PM or PM with armature. Consequently, the external energy w_{ext} can be divided into useful energy w_g and leakage energy w_{lM}. The useful energy per volume in the external space is

$$w_g = \frac{B_g H_g}{2} = \frac{G_g}{V_M} \left[\frac{\Phi_K + \mathcal{F}_K \mu_{rec}(S_M/h_M) + G_t \mathcal{F}'_{ad}}{G_t + \mu_{rec}(S_M/h_M)} - \mathcal{F}'_{ad} \right]$$

$$\times \frac{\Phi_K + \mathcal{F}_K \mu_{rec}(S_M/h_M) + G_t \mathcal{F}'_{ad}}{G_t + \mu_{rec}(S_M/h_M)} \qquad (A.13)$$

A.2.3 Magnetization with Armature

If the magnet is placed in the external armature circuit and then magnetized by the armature field or magnetized in a magnetizer and then the poles of the magnetizer are in a continuous way replaced by the poles of the armature, the origin K of the recoil line is determined by the resultant magnetic permeance G_t drawn from the point $\mathcal{F}'_{admax}$ at the $\mathcal{F}$-coordinate. The MMF $\mathcal{F}'_{admax}$ corresponds to the maximum demagnetizing d-axis field acting directly on the magnet that can appear during the machine operation. In Fig. A.4, this is the intersection point K of the demagnetization curve and the line G_t:

$$G_t = \frac{\Phi_K}{\mathcal{F}_K - \mathcal{F}'_{admax}} \qquad (A.14)$$

The maximum armature demagnetizing MMF F'_{admax} can be determined for the reversal or locked-rotor condition.

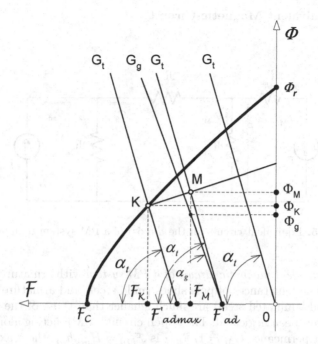

Fig. A.4. Location of the operating point for magnetization with the armature.

The origin of the recoil line is determined by the resultant permeance G_t of the PM mounted in the armature (Fig. A.4).

The rest of the construction is similar to that shown in Fig. A.3 for the demagnetization action of the armature winding (point M). The beginning of the recoil line is determined by the resultant permeance G_t of the PM mounted in the armature (Fig. A.4).

If $a_0 > 0$, the MMF corresponding to the point K is [70]:

$$\mathcal{F}_k = b_0 \pm \sqrt{b_0^2 - c_0} \tag{A.15}$$

where

$$b_0 = 0.5 \left(\frac{\mathcal{F}_c}{a_0} + \mathcal{F}'_{admax} + \frac{\Phi_r}{a_0 G_t} \right) \quad \text{and} \quad c_0 = \frac{(G_t \mathcal{F}'_{admax} + \Phi_r)\mathcal{F}_c}{a_0 G_t}$$

If $a_0 = 0$ (for rare-earth PMs), the MMF $\mathcal{F}_K$ is [70]:

$$\mathcal{F}_K = \frac{\Phi_r + G_t \mathcal{F}'_{admax}}{G_t + \Phi_r/\mathcal{F}_c} \tag{A.16}$$

The magnetic flux Φ_K can be found on the basis of eqn (A.14).

The rest of the construction is similar to that shown in Fig. A.3 for the demagnetization action of the armature winding (point M).

A.2.4 Equivalent Magnetic Circuit

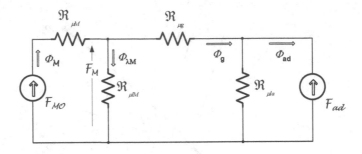

Fig. A.5. Equivalent circuit (in the d-axis) of a PM system with armature.

The *equivalent magnetic circuit of a PM system* with armature is shown in Fig. A.5. The reluctances of pole shoes (mild steel) and armature stack (electrotechnical laminated steel) are much smaller than those of the air gap and PM and have been neglected. The "open circuit" MMF acting along the internal magnet permeance $G_M = 1/\Re_{\mu M}$ is $\mathcal{F}_{M0} = H_{M0}h_M$, the d-axis armature reaction MMF is $\mathcal{F}_{ad}$, the total magnetic flux of the permanent magnet is Φ_M, the leakage flux of the PM is Φ_{lM}, the useful air gap magnetic flux is Φ_g, the leakage flux of the external armature system is Φ_{la}, the flux produced by the armature is Φ_{ad} (demagnetizing or magnetizing), the reluctance for the PM leakage flux is $\Re_{\mu lM} = 1/G_{lM}$, the ai rgap reluctance is $\Re_{\mu g} = 1/G_g$, and the external armature leakage reactance is $\Re_{\mu la} = 1/G_{gla}$. The following Kirchhoff's equations can be written on the basis of the equivalent circuit shown in Fig. A.5

$$\Phi_M = \Phi_{lM} + \Phi_g$$

$$\Phi_{la} = \frac{\pm \mathcal{F}_{ad}}{\Re_{\mu la}}$$

$$\mathcal{F}_{M0} - \Phi_M \Re_{\mu M} - \Phi_{lM} \Re_{\mu lM} = 0$$

$$\Phi_{lM} \Re_{lM} - \Phi_g \Re_{\mu g} \mp \mathcal{F}_{ad} = 0$$

The solution to the above equation system gives the air gap magnetic flux:

$$\Phi_g = \left[\mathcal{F}_{M0} \mp \mathcal{F}_{ad} \frac{G_g}{G_g + G_{lM}} \frac{(G_g + G_{lM})(G_M + G_{lM})}{G_g G_M} \right] \frac{G_g G_M}{G_g + G_{lM} + G_M}$$

or

$$\Phi_g = \left[\mathcal{F}_{M0} \mp \mathcal{F}'_{ad} \frac{G_t(G_M + G_{lM})}{G_g G_M} \right] \frac{G_g G_M}{G_t + G_M} \qquad (A.17)$$

where the total resultant permeance G_t for the flux of the PM is according to eqn (A.12) and the direct-axis armature MMF acting directly on the PM is

$$\mathcal{F}'_{ad} = \mathcal{F}_{ad} \frac{G_g}{G_g + G_{lM}} = \mathcal{F}_{ad} \left(1 + \frac{G_{lM}}{G_g} \right)^{-1} = \frac{\mathcal{F}_{ad}}{\sigma_{lM}} \qquad (A.18)$$

The upper sign in eqn (A.17) is for the demagnetizing armature flux and the lower sign is for the magnetizing armature flux.

The coefficient of the PM leakage flux can also be expressed in terms of permeances, i.e.,

$$\sigma_{lM} = 1 + \frac{\Phi_{lM}}{\Phi_g} = 1 + \frac{G_{lM}}{G_g} \qquad (A.19)$$

Appendix B

Calculations of Permeances

B.1 Field Plotting

The procedure to be followed in *field plotting* is simple. On a diagram of the magnetic circuit, several equipotential lines are drawn. Flux lines connecting the surfaces of opposite polarity are then added in such a manner so as to fulfill the following requirements:

- All flux lines and equipotential lines must be mutually perpendicular at each point of intersection.
- Each figure bonded by two adjacent flux lines and two adjacent equipotential lines must be a curvilinear square.
- The ratio of the average width to average height of each square should be equal to unity.

When the full plot has been completed (Fig. B.1), the magnetic permeance can be found by dividing the number of curvilinear squares between any two adjacent equipotential lines, designated as n_e, by the number of curvilinear squares between any two adjacent flux lines, n_Φ, and multiplying by the length l_M of the field perpendicular to the plane of the flux plot, i.e.,

$$G = \mu_0 \frac{n_e}{n_\Phi} l_M \qquad (B.1)$$

Permeances of air gaps between poles of different configurations are expressed by the following formulae:

(a) Rectangular poles neglecting fringing flux paths (Fig. B.2a)

$$G = \mu_0 \frac{w_M l_M}{g} \qquad (B.2)$$

where $g/w_M < 0.1$ and $g/l_M < 0.1$

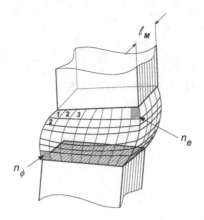

Fig. B.1. Permeance evaluation by flux plotting.

(b) Halfspace and a rectangular pole (Fig. B.2b)

$$G = \mu_0 \frac{1}{g}(w_M + 0.614g/\pi)(l_M + 0.614g/\pi) \tag{B.3}$$

(c) Fringe paths originating on lateral flat surfaces (Fig. B.2c)

$$G = \mu_0 \frac{w_M x}{0.17g + 0.4x} \tag{B.4}$$

or

$$G = \mu_0 \frac{w_M}{\pi} \ln\left[1 + 2\sqrt{\frac{x + (x^2 + xg)}{g}}\right] \tag{B.5}$$

(d) Cylindrical poles neglecting fringing flux (Fig. B.2d)

$$G = \mu_0 \frac{\pi d_M^2}{4g} \tag{B.6}$$

A more accurate formula for $g/d_M < 0.2$ is

$$G = \mu_0 d_M [\frac{\pi d_M}{4g} + \frac{0.36 d_M}{2.4 d_M + g} + 0.48] \tag{B.7}$$

For fringe paths originating on lateral cylindrical surfaces

$$G = \mu_0 \frac{x d_M}{0.22g + 0.4x} \tag{B.8}$$

(e) Between identical rectangles lying on the same surface (Fig. B.2e)

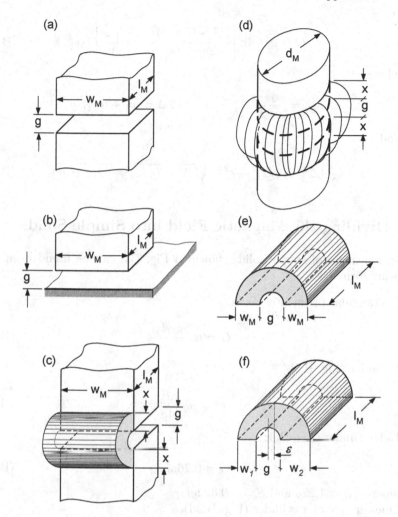

Fig. B.2. Configurations of poles and air gaps between them: (a) rectangular poles, (b) halfspace and a rectangular pole, (c) fringe paths originating on lateral flat surfaces, (d) cylindrical poles, (e) identical rectangles lying on the same surface, (f) two rectangles of different area lying in the same plane.

$$G = \mu_0 \frac{1}{2\pi} \ln[2m^2 - 1 + 2m\sqrt{m^2 - 1}]l_M \qquad (B.9)$$

or

$$G = \mu_0 \frac{1}{\pi} \ln\left(1 + \frac{2w_M}{g}\right) l_M \qquad (B.10)$$

(f) Between two rectangles of different area lying in the same plane (Fig. B.2f)

$$G = \mu_0 \frac{1}{\pi} \ln \left[\frac{\Delta^2 - (\epsilon + x)^2}{\Delta(g - x)} - \frac{\epsilon + x}{\Delta} \right] l_M \qquad (B.11)$$

where

$$\epsilon = \frac{w_2 - w_1}{2}, \qquad 2\Delta = w_1 + w_2 + g$$

and

$$x = \frac{1}{2\epsilon}(\Delta^2 - (g/2)^2 - \epsilon^2 - \sqrt{\Delta^2 - (g/2)^2 - \epsilon^2 - \epsilon^2 g^2}) \qquad (B.12)$$

B.2 Dividing the Magnetic Field into Simple Solids

The permeances of simple solids shown in Fig. B.3 can be found using the following formulae:

(a) Rectangular prism (Fig. B.3a)

$$G = \mu_0 \frac{w_M l_M}{g} \qquad (B.13)$$

(b) Cylinder (Fig. B.3b)

$$G = \mu_0 \frac{\pi d_M^2}{4g} \qquad (B.14)$$

(c) Half-cylinder (Fig. B.3c)

$$G = 0.26 \mu_0 l_M \qquad (B.15)$$

where $g_{av} = 1.22g$ and $S_{av} = 0.322 g l_M$
(d) One-quarter of a cylinder (Fig. B.3d)

$$G = 0.52 \mu_0 l_M \qquad (B.16)$$

(e) Half-ring (Fig. B.3e)

$$G = \mu_0 \frac{2 l_M}{\pi(g/w_M + 1)} \qquad (B.17)$$

For $g < 3 w_M$,

$$G = \mu_0 \frac{l_M}{\pi} \ln \left(1 + \frac{2 w_M}{g} \right) \qquad (B.18)$$

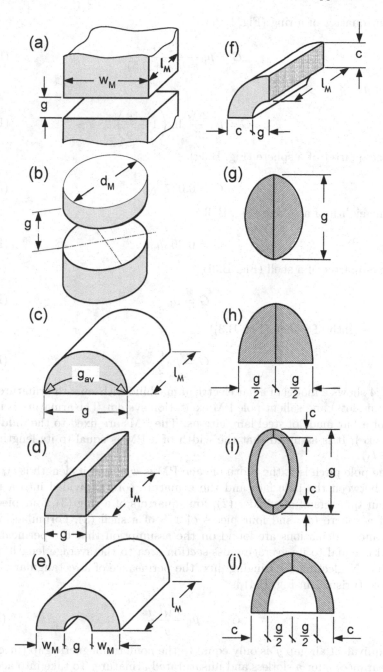

Fig. B.3. Simple solids: (a) rectangular prism, (b) cylinder, (c) half-cylinder, (d) one-quarter of a cylinder, (e) half-ring, (f) one-quarter of a ring, (g) one-quarter of a sphere, (h) one-eighth of a sphere, (i) one-quarter of a shell, (j) one-eighth of a shell.

(f) One-quarter of a ring (Fig. B.3f)

$$G = \mu_0 \frac{2l_M}{\pi(g/c + 0.5)} \tag{B.19}$$

For $g < 3c$,

$$G = \mu_0 \frac{2l_M}{\pi} \ln\left(1 + \frac{c}{g}\right) \tag{B.20}$$

(g) One-quarter of a sphere (Fig. B.3g)

$$G = 0.077\mu_0 g \tag{B.21}$$

(h) One-eighth of a sphere (Fig. B.3h)

$$G = 0.308\mu_0 g \tag{B.22}$$

(i) One-quarter of a shell (Fig. B.3i)

$$G = \mu_0 \frac{c}{4} \tag{B.23}$$

(j) One-eighth of a shell (Fig. B.3j)

$$G = \mu_0 \frac{c}{2} \tag{B.24}$$

Fig. B.4 shows a model of a flat electrical machine with smooth armature core (without slots) and salient-pole PM excitation system. The armature is in the form of a bar made of steel laminations. The PMs are fixed to the mild steel rail (yoke). It is assumed that the width of a PM is equal to its length, i.e., $w_M = l_M$.

The pole pitch is τ, the width of each PM is w_M, and its length is l_M. The space between the pole face and the armature core is divided into a prism (1), four quarters of a cylinder (2), four quarters of a ring (3), four pieces of 1/8 of a sphere (4), and four pieces of 1/8 of a shell (5). Formulae for the permeance calculations are found on the assumption that the permeance of a solid is equal to its average cross-section area to the average length of the flux line. Neglecting the fringing flux, the permeance of a rectangular air gap per pole (prism 1 in Fig. B.4) is

$$G_{g1} = \mu_0 \frac{w_M l_M}{g'} \tag{B.25}$$

The equivalent air gap g' is only equal to the nonferromagnetic gap (mechanical clearance) g for a slotless and unsaturated armature. To take into account slots (if they exist) and magnetic saturation, the air gap g is increased to $g' = g k_C k_{sat}$, where $k_C > 1$ is Carter's coefficient taking into account slots, and $k_{sat} > 1$ is the saturation factor of the magnetic circuit defined as the

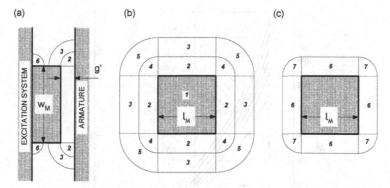

Fig. B.4. Electric machine with flat slotless armature and flat PM excitation system — division of the space occupied by the magnetic field into simple solids: (a) longitudinal section, (b) air gap field, (c) leakage field (between the PM and steel yoke). The width of the PM w_M is equal to its length l_M.

ratio of the MMF per pole pair to the air gap magnetic voltage drop taken twice, i.e.,

$$k_{sat} = 1 + \frac{2(V_{1t} + V_{2t}) + V_{1c} + V_{2c}}{2V_g} \qquad (B.26)$$

where V_g is the magnetic voltage drop (MVD) across the air gap, V_{1t} is the MVD along the armature teeth (if they exist), V_{2t} is the MVD along the PM pole shoe teeth (if there is a pole shoe and cage winding), V_{1c} is the MVD along the armature core (yoke), and V_{2c} is the MVD along the excitation system core (yoke).

To take into account the fringing flux it is necessary to include all paths for the magnetic flux coming from the excitation system through the air gap to the armature system (Fig. B.4), i.e.,

$$G_g = G_{g1} + 4(G_{g2} + G_{g3} + G_{g4} + G_{g5}) \qquad (B.27)$$

where G_{g1} is the air gap permeance according to eqn (B.25) and G_{g2} to G_{g5} are the air gap permeances for fringing fluxes. The permeances G_{g2} to G_{g5} can be found using eqns (B.16), (B.19), (B.22), and (B.24).

In a similar way, the resultant permeance for the leakage flux of the PM can be found, i.e.,

$$G_{lM} = 4(G_{l6} + G_{l7}) \qquad (B.28)$$

where G_{l6} (one-quarter of a cylinder) and G_{l7} (one-eight of a sphere) are the permeances for leakage fluxes between the PM and rotor yoke according to Fig. B.4c — eqns (B.16) and (B.22).

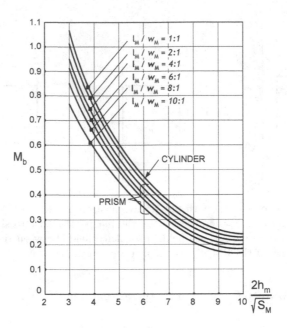

Fig. B.5. Ballistic coefficient of demagnetization M_b for cylinders and prisms of different ratios of l_M/w_M (experimental curves).

B.3 Prisms and Cylinders Located in an Open Space

In the case of simple-shaped PMs, the permeance for leakage fluxes of a PM alone can be found as:

$$G_{ext} = \mu_0 \frac{2\pi}{M_b} \frac{S_M}{h_M} \tag{B.29}$$

where M_b is the ballistic coefficient of demagnetization. This coefficient can be estimated with the aid of graphs as shown in Fig. B.5 [19]. The cross-section area is $S_M = \pi d_M^2/4$ for a cylindrical PM, and $S_M = w_M l_M$ for a rectangular PM. In the case of hollow cylinders (rings), the coefficient M_b is practically the same as that for solid cylinders. For cylindrical PMs with small h_M and large cross sections $\pi d_M^2/4$ (button-shaped PMs), the leakage permeance can be calculated using the following equation [19]:

$$G_{ext} \approx 0.716 \mu_o \frac{d_M^2}{h_M} \tag{B.30}$$

Eqns (B.29) and (B.30) can be used for finding the origin K of the recoil line for PMs magnetized without an armature (Fig. A.3).

Appendix C

Performance Calculations for PM LSMs

The *Case Study 3.1* in Chapter 3 deals with the performance calculations for a single-sided LSM with surface PMs. Let us make similar calculation for the same armature system and equivalent reaction rail with buried (embedded) configuration of PMs. The line *voltage-to-frequency* ratio should be $V_{1L-L}/f = 10$ ($V_{1L-L} = 200$ V at 20 Hz) and calculations should be done for $f = 20$ Hz and $f = 5$ Hz.

All specifications are the same as for PM surface configuration except the air gap, which in the d-axis, is equal to that in the q-axis, i.e., $g = g_q = 2.5$ mm, the depth of the reaction rail slot for the PM $h_2 = 21$ mm, and the width of the reaction rail slot for the PM $b_2 = 8$ mm. The axial length of the PM $l_M = 84$ mm, and pole pitch $\tau = 56$ mm.

The height of a surface PM per pole is $h_M = 4$ mm, and its width is $w_M = 42$ mm. The equivalent height of a buried PM is $2h_M$ and its width is $0.5w_M$ (Fig. C.1). In both cases, the volume V_M of PM materials remains the same, i.e.,

- for surface magnets

$$V_M = p \times 2h_M \times w_M \times l_M = 4 \times (2 \times 4.0) \times 42.0 \times 84.0 = 112,896 \text{ mm}^2$$

- for buried magnets

$$V_M = p \times 2h_M \times 2\frac{w_M}{2} \times l_M = 4 \times (2 \times 4.0)2 \times \frac{42.0}{2} \times 84.0 = 112,896 \text{ mm}^2$$

The width of the reaction rail slot for a buried PM is $b_2 = 2h_M$, and its depth is $h_2 = 0.5w_M$ or $w_M = 2h_2$.

The width of the pole shoe (distance between neighboring PMs) of the reaction rail with buried PMs

$$b_p = \tau - 2h_M = 56.0 - 2 \times 4.0 = 48 \text{ mm}$$

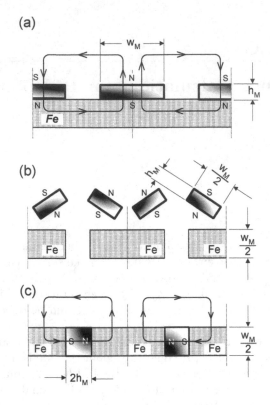

Fig. C.1. Comparison between surface and buried (embedded) magnets configurations: (a) reaction rail with surface PMs, (b) surface PMs are cut in half, (c) reaction rail with buried PMs of the same volume.

For $f = 20$ Hz, the line-to-line voltage $V_{1L-L} = 200$ V, and synchronous speed $v_s = 2.24$ m/s. For $f = 5$ Hz, the line-to-line voltage $V_{1L-L} = 50$ V, and synchronous speed $v_s = 0.56$ m/s.

Parameters independent of the frequency, voltage, and load angle are as follows:

- number of slots per pole per phase $q_1 = 1$
- winding factor $k_{w1} = 1.0$
- *pole shoe to pole pitch* ratio $\alpha = 0.8571$
- pole pitch in slots = 3
- coil pitch in slots = 3
- coil pitch in millimeters $w_c = 56$ mm
- armature slot pitch $t_1 = 18.7$ mm
- width of the armature slot $b_{11} = b_{12} = 10.3$ mm
- number of conductors in each slot $N_{sl} = 280$
- conductors area to slot area $k_{fill} = 0.2252$

- Carter's coefficient $k_C = 1.46703$
- form factor of the excitation field $k_f = 1.2413$
- form factor of the d-axis armature reaction $k_{fd} = 0.9126$
- form factor of the q-axis armature reaction $k_{fq} = 0.7190$
- reaction factor in the d-axis $k_{ad} = 0.7352$
- reaction factor in the q-axis $k_{aq} = 0.5793$
- coefficient of leakage flux $\sigma_l = 1.2469$
- permeance of the air gap $G_g = 0.1633 \times 10^{-5}$ H
- permeance of the PM $G_M = 0.1213 \times 10^{-5}$ H
- permeance for leakage fluxes $G_{lM} = 0.4032 \times 10^{-6}$ H
- magnetic flux corresponding to remanent magnetic flux density $\Phi_r = 0.3881 \times 10^{-2}$ Wb
- relative recoil magnetic permeability $\mu_{rrec} = 1.094$
- PM edge line current density $A_M = 800,000.00$ A/m
- mass of the armature yoke $m_{y1} = 5.56$ kg
- mass of the armature teeth $m_{t1} = 4.92$ kg
- mass of the armature wires $m_{Cu} = 15.55$ kg
- friction force $F_r = 1.542$ N

Resistances and reactances independent of magnetic saturation are

- armature winding resistance $R_1 = 2.5643$ Ω at 75°C
- armature winding leakage reactance $X_1 = 4.6532$ at $f = 20$ Hz
- armature winding leakage reactance $X_1 = 1.1633$ at $f = 5$ Hz
- d-axis armature reaction reactance $X_{ad} = 8.8093$ at $f = 20$ Hz
- d-axis armature reaction reactance $X_{ad} = 2.2090$ at $f = 5$ Hz
- q-axis armature reaction reactance $X_{aq} = 6.9411$ at $f = 20$ Hz
- q-axis armature reaction reactance $X_{aq} = 1.7353$ at $f = 5$ Hz
- specific slot leakage permeance $\lambda_{1s} = 1.3918$
- specific leakage permeance of end connections $\lambda_{1e} = 0.2192$
- specific tooth-top leakage permeance $\lambda_{t1} = 0.1786$
- specific differential leakage permeance $\lambda_{1d} = 0.4477$
- coefficient of differential leakage $\tau_{d1} = 0.0965$

The steady-state performance characteristics (Table C.1) have been calculated as functions of the load angle δ and the corresponding value of the angle Ψ between the phasor of the armature current $\mathbf{I}_a$ and the q-axis (or the phasor of the EMF $\mathbf{E}_f$). Magnetic saturation due to the main flux and leakage fluxes has been included. The maximum efficiency corresponds to the angle $\Psi \approx 0$ at which the current in the d-axis $I_{ad} \approx 0$ (theoretically, $\Psi = 0^0$ and $I_{ad} = 0$). An LSM should operate with maximum efficiency, and its rated parameters are usually for $\Psi \approx 0$. This gives a linear thrust F_x versus the armature current I_a.

Calculation results for $\Psi \approx 0^0$ (approximately maximum efficiency) with magnetic saturation taken into account are given in Table C.2. The electric and magnetic loadings for $\Psi \approx 0$ and $I_{ad} \approx 0$ are rather low, i.e., $A_{m1} = 41,460$

Table C.1. Steady state characteristics of a flat three-phase four-pole LSM with $\tau = 56$ mm and buried PM configuration

δ	Ψ	P_{out}	F_x	F_z	I_a	η	$\cos\phi$
deg	deg	W	N	N	A	–	–
$f = 20$ Hz, $V_{1L-L} = 200$ V, $v_s = 2.24$ m/s							
−20	34.01	−830.1	−371.6	1301	3.59	0.8796	0.5877
−10	58.63	−345.6	−159.7	1293	2.44	0.8921	0.3644
−5	77.98	−105.4	−54.6	1298	2.05	0.8219	0.1222
1	72.30	141.1	69.8	1320	1.88	0.7555	0.2873
5	52.31	320.4	151.5	1340	2.00	0.8581	0.5402
10	31.78	540.7	251.9	1373	2.36	0.8853	0.7457
15	16.73	756.1	350.2	1414	2.89	0.8873	0.8505
18	9.69	882.5	407.9	1443	3.26	0.8838	0.8854
20	5.62	965.5	445.7	1464	3.51	0.8804	0.9017
22	1.94	1047.0	483.1	1486	3.77	0.8763	0.9140
23	**0.21**	1088.0	501.6	1499	3.91	**0.8741**	0.9190
24	−1.42	1128.0	519.9	1511	4.05	0.8718	0.9234
25	−2.98	1168.0	538.1	1522	4.18	0.8694	0.9271
30	−10.01	1362.0	626.7	1592	4.89	0.8560	0.9397
35	−15.95	1546.0	710.9	1666	5.61	0.8411	0.9452
40	−21.17	1719.0	790.0	1750	6.36	0.8251	0.9465
45	−25.87	1880.0	863.4	1841	7.11	0.8081	0.9448
60	−38.02	2271.0	1043.0	2167	9.40	0.7523	0.9273
80	−51.63	2525.0	1161.0	2694	12.46	0.6650	0.8799
$f = 5$ Hz, $V_{1L-L} = 50$ V, $v_s = 0.56$ m/s							
−20	63.99	−82.8	−151.2	1254	2.76	0.3026	0.1046
−10	87.20	−3.6	−12.8	1313	1.85	0.1791	0.1254
−5	73.84	26.3	51.5	1352	1.52	0.5554	0.3609
1	42.91	65.9	123.8	1406	1.35	0.7832	0.7205
5	21.42	90.5	168.6	1447	1.43	0.8148	0.8956
6	16.52	96.4	179.4	1457	1.48	0.8164	0.9237
7	11.89	102.2	190.0	1468	1.53	0.8162	0.9461
8	7.54	107.9	200.3	1479	1.59	0.8146	0.9635
9	3.47	113.4	210.5	1490	1.65	0.8117	0.9764
9.8	**0.42**	117.8	218.5	1498	1.71	**0.8086**	0.9841
10	−0.33	118.9	220.5	1503	1.72	0.8077	0.9858
15	−15.70	144.5	267.3	1562	2.15	0.7768	0.9999
20	−26.67	167.1	308.6	1627	2.64	0.7347	0.9932
25	−34.98	186.4	343.8	1697	3.18	0.6873	0.9849
30	−41.64	202.1	372.7	1771	3.74	0.6376	0.9794
35	−47.24	214.1	394.6	1849	4.31	0.5870	0.9773
40	−52.12	222.0	409.3	1931	4.89	0.5364	0.9777
45	−56.51	225.8	416.3	2015	5.47	0.4862	0.9799
60	−67.79	210.2	388.6	2280	7.22	0.3393	0.9908
80	−80.48	123.6	231.8	2632	9.46	0.1509	0.9999

Table C.2. Calculation results for minimum angle Ψ ($I_{ad} \approx 0$) with magnetic saturation taken into account

Quantity	$f = 20$ Hz	$f = 5$ Hz
Load angle δ	23^0	9.8^0
Angle between the armature current I_a and q axis Ψ	0.21^0	0.42^0
Output power P_{out}, W	1088	118
Output power-to-armature mass, W/kg	41.80	4.53
Input power P_{in}, W	1245	146
Electromagnetic thrust F_{dx}, N	503.1	220.1
Thrust F_x, N	501.6	218.5
Normal force F_z, N	1499	1498
Electromagnetic power P_g, W	1127	123
Efficiency η	0.8741	0.8086
Power factor $\cos\phi$	0.9190	0.9841
Armature current I_a, A	3.91	1.71
d-axis armature current I_{ad}, A	0.01	0.01
q-axis armature current I_{aq}, A	3.91	1.71
Armature line current density, peak value, A_{m1}, A/m	41,460	18,140
Current density in the armature winding j_a, A/mm^2	2.411	1.054
Air gap magnetic flux density, maximum value B_{mg}, T	0.4833	0.4832
Per phase EMF excited by PMs E_f, V	96.07	24.02
Magnetic flux in the air gap Φ_g, Wb	0.1950×10^{-2}	0.1950×10^{-2}
Armature winding loss ΔP_{1w}, W	117.6	22.5
Armature core loss ΔP_{1Fe}, W	13.88	2.19
Mechanical losses ΔP_m, W	3.45	0.86
Additional losses ΔP_{ad}, W	21.8	2.4
Armature leakage reactance X_1, Ω	4.6103	1.1572
d-axis synchronous reactance X_{sd}, Ω	13.326	3.336
q-axis synchronous reactance X_{sq}, Ω	11.551	2.892
Magnetic flux density in the armature tooth, B_{1t}, T	1.1188	1.1185
Magnetic flux density in the armature yoke B_{1y}, T	0.6044	0.604
Saturation factor of the magnetic circuit k_{sat}	1.0107	1.0107

A/m, $j_a = 2.411$ A/mm^2, $B_{mg} = 0.4833$ T for 20 Hz and $A_{m1} = 18,140$ A/m, $j_a = 1.054$ A/mm^2, $B_{mg} = 0.4832$ T for 5 Hz. Those values are higher for $|\Psi| > 0^0$ ($\delta > 0^0$). Because the armature winding current density j_a is low, the LSM can operate as a continuous duty motor. The air gap magnetic flux density B_{mg} is independent of the input frequency f because, for $V_1/f = const$, the air gap magnetic flux $\Phi_g = const$.

The thrust constant k_F is practically independent of the input frequency, i.e.,

- for 20 Hz

$$k_F = \frac{F_x}{I_a} = \frac{501.6}{3.91} = 128.29 \text{ N/A}$$

- for 5 Hz

$$k_F = \frac{F_x}{I_a} = \frac{218.5}{1.71} = 127.8 \text{ N/A}$$

For $I_{ad} \approx 0$, the buried PM LSM has lower thrust and higher efficiency than its surface PM counterpart (Chapter 3, Table 3.3). It also draws less current from the power supply so that the temperature rise of the stator winding will be lower. By increasing the input voltage to obtain similar electromagnetic loadings, the thrust can be close to that developed by the surface PM LSM.

Appendix D

Field-Network Simulation of Dynamic Characteristics of PM LSMs

In order to understand the behavior of an electromagnetic system, it is important to comprehend the dynamics of its components.

There are several basic methods of formulating the equations of motion and thereby obtaining the characteristics of the electromagnetic energy conversion devices. These are through the applications of [145, 238]

- electromagnetic field theory;
- principle of virtual work;
- variational principles.

Each of the above approaches has its own merits and, in some cases, one cannot be conveniently substituted by the other [209, 225].

Mechanical and electrical parameters of the system are necessary to formulate the equations of motion. Electrical parameters can be evaluated on the basis of electromagnetic field theory. In most cases, such as tubular PM LSMs operating at power frequencies, the electric field distribution is not of importance, and only the magnetic field must be analyzed.

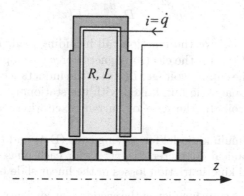

Fig. D.1. A segment of a tubular PM LSM for transient analysis.

In [218] the so-called *field-network model* has been obtained on the basis of the Lagrange function [238]. The tubular PM LSM has been divided into segments (modules). The symmetry axes of neighboring teeth are remote from each other by the distance equal to one pole pitch. Since each segment is magnetically separated, it can be modeled independently. Generalized coordinates ξ for one segment (Fig. D.1) are the position z of the reaction rail and the electric charge q. The Lagrangian (1.33) of a mechanical system can also be expressed with the aid of kinetic coenergy[1] E_k' and potential energy E_p, i.e.,

$$\mathcal{L} = E_k' - E_p \tag{D.1}$$

Assuming that the net potential energy $E_p = 0$, the kinetic coenergy E_k' of the system is

$$E_k' = \frac{1}{2}m\dot{z}^2 + \int_0^{\dot{q}} \Psi(\dot{\tilde{q}}, z)d\dot{\tilde{q}} \tag{D.2}$$

Using Hamilton's principle [145, 238], the Lagrangian (D.1) of the system is [210, 208]

$$\mathcal{L} = E_k' - 0 = \frac{1}{2}m\dot{z}^2 + \int_0^{\dot{q}} \Psi(\dot{\tilde{q}}, z)d\dot{\tilde{q}} \tag{D.3}$$

Euler–Lagrange differential equation (1.36) gives two differential equations, one for electrical part and another for mechanical part of the system, i.e.,

- electrical balance equation

$$v(t) = L(z, i)\frac{\partial i}{\partial t} + \frac{\partial \Psi(z, i)}{\partial z}\frac{\partial z}{\partial t} + Ri \tag{D.4}$$

- mechanical balance equation

$$m\frac{d^2z}{dt^2} + D_v\frac{dz}{dt} = F(z, i) \tag{D.5}$$

in which m is the mass of the reaction rail including load, D_v is the viscous friction constant, $F(z, i)$ is the electromagnetic force developed by the tubular PM LSM, $v(t)$ is the supply voltage, $L(z, i)$ is the inductance of the stationary coil, $\Psi(z, i)$ is the magnetic flux linked with the stationary coil, and R is the resistance of the coil. In the case of current excitation, eqn (D.4) can be neglected.

The Matlab/Simulink model for solving eqns (D.4) and (D.5) is presented in Fig. D.2. Mechanical characteristic $f(z, i)$ for each phase has been implemented. Static and kinetic friction losses in the linear slide bearings have also

---

[1] Coenergy (a second state function of the energy) is an auxiliary function necessary for calculations of the force or torque at constant current.

been taken into account. The magnetization curve B–H has been included in eqn (D.4).

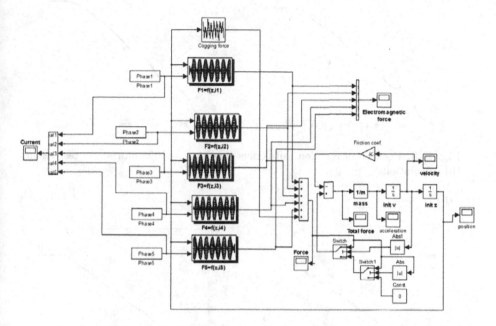

Fig. D.2. Circuit diagram for Matlab/Simulink simulation.

The field-network model performs the calculation of the dynamics of the tubular PM LSM (Fig. 5.14) fed from a sinusoidal source. Since the magnetic circuit is nonlinear, a direct solution to eqn (D.4) is almost impossible. Thus, after finding the integral parameters of the electromagnetic field, the combined mechanical balance and electric field equations have been solved simultaneously. Some calculated and measured values of the position of the reaction rail versus time are plotted in Figs. D.3 to D.5.

In computer simulations, the position of the reaction rail varies from $z(t) = 0$ to $|z(t)| = 0.1$ m, mass of reaction rail $m = 1.3$ kg, $\mu_v = 50$ Ns/m, $R = 1.6\,\Omega$, and $I_a = 8$ A. No-load operating conditions have been assumed. The average velocity of the reaction rail $v = 1000$ mm/s.

Experimental verification of the reaction rail movement has also been performed for the tubular PM LSM under load. Simulation results of the reaction rail movement have been verified experimentally for loaded tubular PM LSM. The load force has been created by an additional mass rigidly connected and moving with the reaction rail. For the mass $m = 3.2$ kg, the position of the reaction rail versus time is shown in Fig. D.4 and for the mass $m = 19.2$ kg, the same plot is shown in Fig. D.5. Test results (Figs D.4 and D.5) are in good agreement with the calculations.

(a) (b)

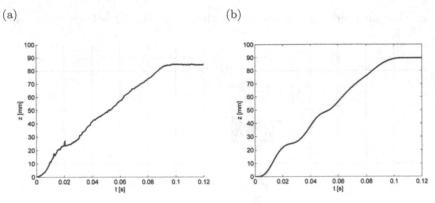

Fig. D.3. Position of reaction rail versus time at $v = 1000$ mm/s: (a) test results, (b) calculations.

(a) (b)

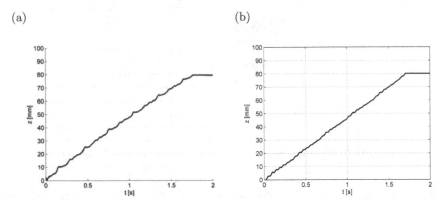

Fig. D.4. Position of reaction rail versus time at average velocity $v = 50$ mm/s and inertial mass $m = 3.2$ kg: (a) test results, (b) calculations.

(a) (b)

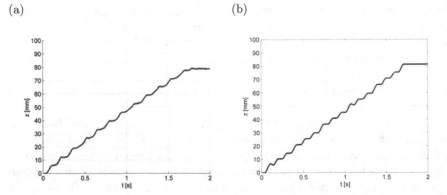

Fig. D.5. Position of reaction rail versus time at average velocity $v = 50$ mm/s and inertial mass $m = 19.2$ kg: (a) test results, (b) calculations.

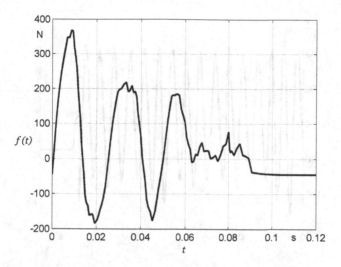

Fig. D.6. Computed electromagnetic force versus time at no load and $v = 1000$ mm/s.

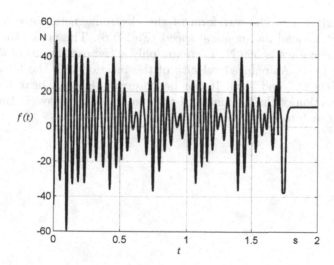

Fig. D.7. Electromagnetic force versus time at average velocity $v = 50$ mm/s and inertial mass $m = 3.2$ kg.

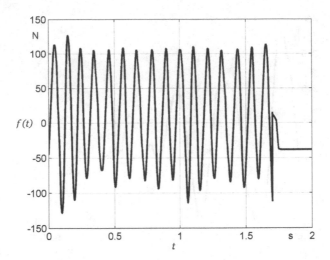

Fig. D.8. Electromagnetic force versus time at average velocity $v = 50$ mm/s and inertial mass $m = 19.2$ kg.

The simulation of the waveform of the electromagnetic force has been performed at no-load and nominal speed (Fig. D.6). The peak value of the electromagnetic force is 360 N. There are only a few oscillations of the electromagnetic force. At reduced velocity of the reaction rail, the force ripple increases (Figs D.7 and D.8). It can be observed that the peak force is a nonlinear function of the load mass. If the load mass increases six times, the electromagnetic force increases only twice.

Symbols and Abbreviations

A	magnetic vector potential
A	line current density; projection of magnetic vector potential; surface
$[A]$	magnetic vector potential matrix
a	instantaneous value of the line current density; number of parallel current paths of the a.c. armature winding
B	vector magnetic flux density
B	magnetic flux density
b	width of slot
b_p	pole shoe width
$[C]$	conductivity matrix that relates to the reaction rail
c	tooth width
c_E	EMF constant, 1/m
c_F	thrust constant, N/(VAs)
D	diameter; damping
D_v	viscous damping constant
d	thickness; diameter
E	vector electric field intensity
E	EMF, rms value; electric field intensity
E_f	EMF per phase induced by the excitation system without armature reaction
E_k	kinetic energy
E_k'	kinetic coenergy (a second state function of kinetic energy)
E_i	internal EMF per phase
E_p	potential energy
e	instantaneous EMF
F	force
F_x	thrust
F_{dx}	electromagnetic thrust developed by the motor
f_r	thrust ripple
$\mathcal{F}$	MMF
$\mathcal{F}_{exc}$	MMF of the rotor excitation system

$\mathcal{F}_a$ armature reaction MMF

G permeance

$[G]$ conductance matrix that relates to the region with a multiturn winding

g,   air gap (mechanical clearance)

g' equivalent airgap $g' = gk_Ck_{sat}$

$\mathbf{H}$ vector magnetic field intensity

H magnetic field intensity

h height

h_c height of armature coil

h_M height of PM per pole (in the direction of magnetization)

h_s height of module

I current

$[I]$ current matrix

I_a armature current

i instantaneous value of current or HLSM current

J electric current density

J_a current density in the armature winding

$[K]$ matrix of coefficients

k coefficient, general symbol

k_{1R} skin effect coefficient for armature resistance

k_{1X} skin effect coefficient for armature leakage reactance

k_C Carter's coefficient

k_{ad} reaction factor in d-axis

k_{aq} reaction factor in q-axis

k_{d1} distribution factor

k_E EMF constant, Vs/m

k_{end} end effect coefficient

k_F thrust constant, N/A

k_f form factor of the field excitation $k_f = B_{mg1}/B_{mg}$

k_{fill} slot fill factor (cross section area of all conductors to the area of slot)

k_i stacking factor of laminations

k_r coefficient of force (thrust) ripple

k_{p1} pitch factor

k_s spring constant

k_{sat} saturation factor of the magnetic circuit due to the main (linkage) magnetic flux

k_{w1} winding factor $k_{w1} = k_{d1}k_{p1}$

l integration path

L inductance; length

$[L]$ inductance matrix

$\mathcal{L}$ Lagrangian; Laplace transform

L_i armature stack effective length

l_{1e} length of the one-sided end connection

l_M axial length of PM

M_b ballistic coefficient of demagnetization

m number of phases; mass
N number of turns per phase
P active power
P_{elm} electromagnetic power
p number of pole pairs (number of poles is $2p$)
$\mathbf{Q}_I$ dimensionless matrix that relates to currents
$\mathbf{Q}_M$ matrix that relates to PMs, A
q_1 number of slots per pole per phase of armature winding
$|\mathbf{Q}|$ determinant of Jacobian matrix
R resistance
$[R]$ resistance matrix
R_1 armature winding resistance of a.c. motors
$\Re$ reluctance
$\Re_{\mu M}$ permanent magnet reluctance
$\Re_{\mu g}$ airgap reluctance
$\Re_{\mu la}$ external armature leakage reluctance
r radius
r, ϕ, z cylindrical (polar) coordinates (tubular motors)
S apparent power; surface
S_M cross-section area of PM $S_M = w_M L_M$ or $S_M = b_p L_M$
s displacement
s_1 number of armature teeth or slots
$[\mathbf{T}]$ Maxwell stress tensor matrix
t time; slot pitch
V electric voltage; volume
v instantaneous value of electric voltage; linear velocity; weighted function
W energy, J
w energy per volume, J/m^3; width
w_c coil pitch (span)
w_M width of PM
w_p axial width of ferromagnetic core between neighboring PMs
w_s axial width of space for coil (tubular motors)
w_{ss} axial width of module (tubular motors)
w_t axial thickness of module
X reactance
X_1 armature leakage reactance
X_{ad} d-axis armature reaction (mutual) reactance
X_{aq} q-axis armature reaction (mutual) reactance
X_{sd} d-axis synchronous reactance
X_{sq} q-axis synchronous reactance
x, y, z rectangular (Cartesian) coordinates (flat motors)
$\mathbf{Z}$ impedance $\mathbf{Z} = R + jX$; $|\mathbf{Z}| = Z = \sqrt{R^2 + X^2}$
α_i effective pole arc coefficient $\alpha_i = b_p/\tau$
β propagation constant ($\beta = \pi/\tau$)

χ	magnetic susceptibility
Γ	boundary surface of element; closed surface embracing a region (FEM analysis)
γ	form factor of demagnetization curve of PM material
ΔP	active power losses
$\Delta p_{1/50}$	specific core loss in W/kg at 1T and 50 Hz
∇	operator nabla (electromagnetic field theory)
δ	power (load) angle, equivalent depth of penetration
η	efficiency
ϵ	force angle (angle between phasors of the excitation flux Φ_f in the d-axis and the armature current $\mathbf{I}_a$)
ϑ	temperature
λ	coefficient of leakage permeance (specific leakage permeance)
μ_0	magnetic permeability of free space $\mu_0 = 0.4\pi \times 10^{-6}$ H/m
μ_r	relative magnetic permeability
μ_{rec}	recoil magnetic permeability
μ_{rrec}	relative recoil permeability $\mu_{rrec} = \mu_{rec}/\mu_0$
μ_v	coefficient of viscous friction
ξ	generalized coordinate (Euler–Lagrange equation)
ρ	specific mass density
σ	electric conductivity
τ	pole pitch
Φ	magnetic flux
Φ_f	excitation magnetic flux
Φ_l	leakage flux
ϕ	power factor angle; magnetic scalar potential; reduced magnetic scalar potential
Ψ	flux linkage $\Psi = N\Phi$; total magnetic scalar potential; angle between $\mathbf{I}_a$ and $\mathbf{E}_f$
Ψ_{sd}	total flux linkage in the d-axis
Ψ_{sq}	total flux linkage in the q-axis
Ω	region (in the FEM analysis)
ω	angular frequency $\omega = 2\pi f$

Subscripts

a	armature
av	average
c	counterweight; core (yoke); coercive; coil
Cu	copper
d	direct axis; differential; developed; dynamic
e	end connection; eddy current
elm	electromagnetic
eq	equivalent
exc	excitation

ext	external
Fe	ferromagnetic
f	field
g	air gap
h	hysteresis
$hoist$	hoist
in	inner
l	leakage
M	magnet
m	peak value (amplitude)
n, t	normal and tangential components
o	outside, outer
out	output, outer
q	quadrature axis
r	rated; remanent
rec	recoil
rel	reluctance; relative
s	synchronous; static; slot
sat	saturation
sh	shaft
st	starting
syn	synchronous or synchronizing
t	tooth
x, y, z	cartesian coordinate system
1	armature; stator; fundamental harmonic
2	reaction rail

Superscripts

(e)	element (FEM analysis)
$fine$	fine mesh (FEM analysis)

Abbreviations

A/D	analog-to-digital
AMG	active magnetic guidance
AMSC	American Superconductors (MA, USA)
a.c.	alternating current
BCD	binary code decimal
BSCCO	first generation HTS BiSrCaCuO
CAD	computer aided design
CGS	conjugate gradient solver
CNC	computer numerical control
DARPA	Defense Advanced Research Projects Agency (USA)
DCRM	distance-coded reference mark

DoE	Department of Energy (USA)
DOF	degree of freedom
DSP	digital signal processor
DTC	direct thrust control
d.c.	direct current
EDL	electrodynamic levitation
EML	electromagnetic levitation
EMF	electromotive force
EMI	electromagnetic interference
ENNA	Engineering Advancement Association of Japan
FEM	finite element method
HDL	high-thrust density linear motor
HLSM	hybrid linear stepping motor
HTS	high-temperature superconductor
IBAD	ion-beam-assisted deposition
IC	integrated circuit
IPT	inductive power transfer
LBM	linear brushless motor
LCX	leaky coaxial cable
LED	light-emitting diode
LIM	linear induction motor
LOA	linear oscillatory actuator
LRM	linear reluctance motor
LSM	linear synchronous motor
LTS	low-temperature superconductor
LVDT	linear variable differential transformer
MMF	magnetomotive force
MR	magnetoresistive
MRI	magnetic resonance imaging
MVD	magnetic voltage drop
NBC	natural binary code
NC	numerically controlled
NMR	nuclear magnetic resonance
PCG	preconditioned conjugate gradient
PDE	partial differential equation
PLC	programmable logic controller
PM	permanent magnet
PVD	physical vapor deposition
PWM	pulse width modulation
RABITS	rolling-assisted biaxially textured substrates
RFI	radio frequency interference
RNA	reluctance network approach
SC	superconductor
STO	= $SrTiO_3$
TTL	transistor–transistor logic

TVE	Emsland Transrapid test facility (Germany)
VSI	voltage source inverter
VVVF	variable voltage variable frequency
YBCO	second-generation HTS YBaCuO

References

1. *Accucore Components*, information brochure, TSC Ferrite International, Wadsworth, IL, USA, 2001, www.tscinternational.com
2. Afonin, A., *The Application of Metal Powder Technology for Linear Motors*, 1st Int. Symp. on Linear Drives for Ind. Appl. LDIA'95, Nagasaki, Japan, 1995, pp. 271–274.
3. Afonin, A., Szymczak, P. and Bobako, S., *Linear Drives with Controlled Current Layer*, 1st Int. Symp. on Linear Drives for Ind. Appl. LDIA'95, Nagasaki, Japan, 1995, pp. 275–278.
4. Adamiak, K., Barlow, D., Choudhury, C.P., Cusack, P.M., Dawson, G.E., Eastham, A.R., Grady, B., Ho, E., Yuan Hongpin, Pattison, L. and Welch, J., *The Switched Reluctance Motor as a Low-Speed Linear Drive*, Int. Conf. on Maglev and Linear Drives, Las Vegas, USA, 1987, pp. 39–43.
5. *AGS20000 Linear Motor Gantries*, Aerotech, Inc., Pittsburgh, PA, USA, 2002, www.aerotechinc.com
6. Akmeşe, R. and Eastham, J.F., *Design of Permanent Magnet Flat Linear Motors for Standstill Applications*, IEEE Trans. on MAG, Vol. 28, 1992, No. 5, pp. 3042–3044.
7. Albicini, F., Andriollo, M., Martinelli, G. and Morini, A., *General Expressions of Propulsion Force in EDS-Maglev Transport Systems with Superconducting Coils*, IEEE Trans. on AS, Vol. 3, 1993, No. 1, pp 425–429.
8. Amber, G.H. and Amber, P.S., *Anatomy of Automation*, Prentice-Hall, Englewood Cliffs, NJ, 1962.
9. Amoros, J.G. and Andrada, P., *Sensitivity Analysis of Geometrical Parameters on a Double-Sided Linear Switched Reluctance Motor* IEEE Trans. on IE, Vol. 57, 2010, No. 1, pp. 311–319.
10. *American Superconductors (AMSC)*, Westborough, MA, USA, www.amsuper.com
11. Anders, M., Binder, A. and Suess, M., *A Spherical Linear Motor as Direct Drive of an Airborne Optical Infrared Telescope*, IEEJ Trans. on IA, Vol. 126, 2006, No. 10, pp. 1363–1367.
12. *Anorad Linear Motors*, information brochure, Anorad, Hauppauge, NY, USA, 2007, www.anorad.com
13. *Ansys Manual*, Ansys, Inc., Southpointe, PA, USA, www.ansys.com

14. Atherton, D.L. et al, *Design, Analysis and Test Results for a Superconducting Linear Synchronous Motor*, Proc. IEE, Vol. 124, 1977, No. 4, pp. 363–372.

15. Atzpodien, H.C., *Magnetic Levitation System on Route from Berlin to Hamburg—Planning, Financing, State of Project*, 14th Int. Conf. on Magnetically Levitated Systems Maglev'95, Bremen, Germany, 1995, pp. 25–29.

16. Ayoma, H., Araki, H., Yoshida, T., Mukai, R. and Takedoni, S., *Linear Motor System for High Speed and High Accuracy Position Seek*, 1st Int. Symp. on Linear Drives for Ind. Appl. LDIA'95, Nagasaki, Japan, 1995, pp. 461–464.

17. Azukizawa, T., *Optimum Linear Synchronous Motor Design for High Speed Ground Transportation*, IEEE Trans. on PAS, Vol. 102, 1983, No 10, pp. 3306–3314.

18. Baker, N.J., Mueller, M.A., Tavner, P.J. and Li Ran, *Prototype Development of Direct-Drive Linear Electrical Machines for Marine Energy Converters*, World Renewable Energy Congress (WREC), 2005, Elsevier, pp. 271-276.

19. Balagurov, V.A., Galtieev, F.F. and Larionov, A.N., *Permanent Magnet Electrical Machines* (in Russian), Energia, Moscow, 1964.

20. Bardeen, J., Cooper, L.N., and Schrieffer, J.R., *Theory of Superconductivity*, Phys. Review, Vol. 108, 1957, pp. 1175–1204.

21. Bart, L., Gysen, J., Ilhan, E., Meessen, K.J., Paulides, J.J.H. and Lomonova, E.A., *Modeling of Flux Switching Permanent Magnet Machines with Fourier Analysis*, IEEE Trans. on MAG, Vol. 46, 2010, No. 6, pp. 1499–1502.

22. Beakley, B., *Linear Motors for Precision Positioning*, Motion Control, October 1991.

23. Bednorz, J.G. and Mueller, K.A., *Possible High T_c Superconductivity in the Ba-La-Cu-O System*, Zeitschrift für Physics B - Condensed Matter, Vol. 64, 1986, pp. 189–193.

24. Bianchi N., *Analytical Field Computation of a Tubular Permanent-Magnet Linear Motor*, IEEE Trans. on MAG, Vol. 36, No. 5, 2000, pp. 3798–3801.

25. Binns, K.J., Lawrenson, P.J. and Trowbridge, C.W., *The Analytical and Numerical Solution of Electric and Magnetic Fields*, John Wiley & Sons, New York, 1992.

26. Bladel, J. Van, *Electromagnetic Fields*, 2nd ed., John Wiley & Sons, Wiley-Interscience, IEEE Press, Hoboken, 2007.

27. Blakley, J.J., *A Linear Oscillating Ferroresonant Machine*, IEEE Trans. on MAG, Vol. 19, 1983, No. 4, pp. 1574–1579.

28. Blaugher, R.D., *Low-Calorie, High-Energy Generators and Motors*, IEEE Spectrum, Vol. 34, 1997, No. 7, pp. 36–42.

29. Boçarov, V.I. and Nagorsky, V.D., *High-Speed Ground Transport with Linear Propulsion and Magnetic Suspension System* (in Russian), Transport, Moscow, 1985.

30. Boldea, I. and Nasar, S.A., *Linear Electric Actuators and Generators*, Cambridge University Press, New York, 2005.

31. Boldea, I. and Nasar, S.A., *Linear Motion Electromagnetic Systems*, John Wiley & Sons, New York, 1985.

32. Breil, J., Oedl, G. and Sieber, B., *Synchronous Linear Drives for many Secondaries with Open Loop Control*, 2nd Int. Symp. on Linear Drives for Ind. Appl. LDIA'98, Tokyo, Japan, 1998, pp. 142–146.

33. *CEDRAT Software for Field Calculation*, CEDRAT Group, Meylan Cedex, France, www.cedrat.com/en/software-solutions/flux.html

34. Ceramawire, Elizabeth City, NC, USA, http://www.ceramawire.com/msds.html
35. Concordia, C., *Synchronous Machines: Theory and Performance*, J Wiley & Sons, New York, 1951.
36. Chai, H.D., Permeance Model and Reluctance Force between Toothed Structures, in *Theory and Applications of Step Motors*, ed. Kuo, B.C., West Publishing, 1974, pp.141–153.
37. Chen, A., Nilssen, R. and Nysveen, A., *Analytical Design of High-Torque Flux-Switching Permanent Magnet Machine by a Simplified Lumped Parameter Magnetic Circuit*, Int. Conf. on Electr. Machines ICEM'10, Rome, Italy, Available on CD.
38. Chung, S.U., Lee, H.J., Hong, D.K., Lee, J.Y., Woo, B.C. and Koo D.H., *Development of Flux Reversal Linear Synchronous Motor for Precision Position Control*, Int. Journal of Precision Engineering and Manufacturing, Vol. 12, 2011, No. 3, pp. 443–450.
39. Compter, J., *Towards Planar Drives for Lithography*, Keynote Address, Int. Symp. on Linear Drives for Ind. Appl. LDIA'07, Lille, France, 2007, available on CD.
40. *Compumotor Digiplan: Positioning Control Systems and Drives*, Parker Hannifin Corporation, Rohnert Park, CA, USA, 2011.
41. *Computational Magnetics*, ed. Sykulski, J.K., Chapman & Hall, London, 1995.
42. Coris, N., Coleman, R. and Piaget, D., *Status and New Development of Linear Drives and Subsystems*, Keynote Address, Int. Symp. on Linear Drives for Ind. Appl. LDIA'07, Lille, France, 2007, available on CD.
43. Cruise, R.J. and Landy, C.F., *Design Considerations of Linear Motor Hoists for Underground Mining Operations*, 7th South African Universities Power Engineering Conf. SAUPEC'98, Stellenbosch, RSA, 1998, pp. 65–68.
44. Cruise, R.J. and Landy, C.F., *Linear Synchronous Motor Propelled Hoist for Mining Applications*, 31st IEEE IA Conf., San Diego, CA, 1996.
45. *Concise Encyclopedia of Traffic and Transportation Systems*, ed. Papageorgiu, M., Pergamon Press, 1991, pp. 36–49.
46. Coulomb, J. and Meunier, G., *Finite Element Implementation of Virtual Work Principle for Magnetic and Electric Force and Torque Computation*, IEEE Trans. on MAG, Vol. 20, 1984, No. 5, pp. 1894–1896.
47. Dabrowski, M., *Magnetic Field and Circuits of Electrical Machines* (in Polish), WNT, Warsaw, Poland, 1971.
48. Dawson, G.E., Eastham, A.R., Gieras, J.F., Ong, R. and Ananthasivam K., *Design of Linear Induction Drives by Field Analysis and Finite-Element Techniques*, IEEE Trans. on IA, 1986, Vol. 22, No.5, pp. 865–873.
49. DeGarmo, E.P., Black, J.T. and Kohser, R.A., *Materials and Processes in Manufacturing*, Macmillan, New York, 1988.
50. Demenko, A., *Equivalent RC Networks with Mutual Capacitances for Electromagnetic Field Simulation of Electrical Machine Transients*, IEEE Trans. on MAG, Vol. 28, 1992, No. 2, pp. 1406–1409.
51. Demenko, A., *Time-Stepping FE Analysis of Electric Motor Drives with Semiconductor Converters*, IEEE Trans. on MAG, Vol. 30, 1994, No. 5, pp. 3264–3267.
52. Deng, Z., Boldea, I. and Nasar, S.A., *Forces and Parameters of Permanent Magnet Linear Synchronous Machines*, IEEE Trans. on MAG, Vol. 23, No. 1, 1987, pp. 305–309.
53. Donahue, B., *The Line on Linear*, Today's Machining World, June 2010, pp. 16–20.

54. Edwards J.D., *An Introduction to MagNet for Static 2D Modeling, Case Studies: Rotational Geometry*, Infolytica, Montreal, Canada, 2007.
55. Eidelberg, B., *Linear Motors Drive Advances in Indsutrial Laser Applications*, Industrial Laser Review, 1995, No. 1, pp. 15–18.
56. Eidelberg, B., *Simulation of Linear Motor Machines*, 2nd Int. Symp. on Linear Drives for Ind. Appl. LDIA'98, Tokyo, Japan, 1998, pp. 30–33.
57. Elgerd, O.I., *Electric Energy Systems Theory: Introduction*, McGraw-Hill, New York, 1971.
58. Ellerthorpe, S. and J. Blaney, J., *Force Estimation for Linear Step Motor with Variable Airgap*, 25th Annual Symp. on Incremental Motion Control Systems and Devices, San Jose, CA, USA, 1996, pp. 327–335.
59. Everes, W., Henneberger, G., Wunderlich, H. and Selig, A., *A Linear Homopolar Motor for a Transportation System*, 2nd Int. Symp. on Linear Drives for Ind. Appl. LDIA'98, Tokyo, Japan, 1998, pp. 46-49.
60. Fitzgerald, A.E. and Kingsley, C., *Electric Machinery*, 2nd edition, McGraw-Hill, New York, 1961.
61. Gieras, J.F., *Electrodynamic Forces in Electromagnetic Levitation Systems*, Acta Technica ČSAV, 1982, No. 5, pp. 532–535.
62. Gieras, J.F. and Miszewski, M., *Performance Characteristics of the Air-Core Linear Synchronous Motor* (in Polish), Rozprawy Elektrot. PAN, Warszawa, Poland, Vol. 29, 1983, No. 4, pp. 1101–1124.
63. Gieras, J.F., *Linear Induction Drives*, Clarendon Press, Oxford, UK, 1994.
64. Gieras, J.F., Spannenberg, A.,Wing, M. and Yamada, H., *Analysis of a Linear Synchronous Motor with Buried Permanent Magnets*, 1st Int. Symp. on Linear Drives for Ind. Appl. LDIA'95, Nagasaki, Japan, 1995, pp. 323–326.
65. Gieras, J.F. and Wang, R., *Calculation of Forces in a Hybrid Linear Stepping Motor for Machine Tools*, 7th European Conf. on Power Electronics and Appl. EPE'97, Trondheim, Norway, 1997, Vol. 3, pp. 591–595.
66. Gieras, J.F., Santini, E. and Wing.M., *Calculation of Synchronous Reactances of Small Permanent-Magnet Alternating-Current Motors: Comparison of Analytical Approach and Finite Element Method with Measurements*, IEEE Trans. on MAG, Vol. 34, 1998, No. 5, pp. 3712–3720.
67. Gieras, J.F., *Status of Linear Motors in the United States*, Int. Symp. on Linear Drives for Ind. Appl. LDIA'03, Birmingham, UK, 2003.
68. Gieras J.F., *Advancements in Electrical Machines*, Springer Verlag, Berlin, 2008.
69. Gieras, J.F., *Linear Electric Motors in Aircraft Technology: an Overview*, Int. Symp. on Linear Drives for Ind. Appl. LDIA'09, Seoul, S. Korea, 2009, available on CD.
70. Gieras J.F., *Permanent Magnet Motor Technology: Design and Applications*, 3rd edition, Taylor & Francis, CRC Press, Boca Raton, FL, 2010.
71. Goncavales, G. J., *Modeling of Two-Dimensional Electromagnetic Field in both Linear and Tubular Actuators*, Int. Conf. Electrical Machines ICEM 2004, Cracow, Poland, 2004, Paper 407, available on CD.
72. Gordon, S. and Hillery, M.T., *Development of a High-Speed CNC Cutting Machine Using Linear Motors*, Journal of Materials Processing Technology, 2005, No. 166, pp. 321–329.
73. Guderjahn, C.A., Wipf, S.I., Fink, H.J., Boom, R.W., MacKenzie, K.E., Williams, D. and Downey, T., *Magnetic Suspension and Guidance for High Speed Rockets by Superconducting Magnets*, Journal of Applied Physics, Vol. 40, 1998, No. 5, pp. 3519–3521.

74. Gurol, S., Baldi, R. and Post R., *General Atomics Urban Maglev Program Status*, 20th Int. Conf. on Magnetically Levitated Systems and Linear Drives MAGLEV2008, San Diego, CA, USA, 2008, Paper No. 99, available on CD.

75. Haas, G.J., *Industrial Laser Solutions*, 2003, No 11, pp. 17-19, www.industriallaser.com

76. Hagiz, G., *My Love Affair with Computers and CNC*, 2007, http://numeryx.com

77. Hague, B., *The Principles of Electromagnetism Applied to Electrical Machines*, Dover Publications, New York, 1929.

78. Hakala, H., *Application of Linear Motors in Elevator Hoisting Machines*, Ph.D., Publication No. 157, Tampere University of Technology, Finland, 1995.

79. Halbach, K., *Design of Permanent Multipole Magnets with Oriented Rare Earth Cobalt Material*, Nuclear Instruments and Methods, Vol. 169, 1980, pp. 1–10.

80. Halbach, K., *Application of Permanent Magnets in Accelerators and Electron Storage Rings*, Journal of Applied Physics, vol. 57, 1985, pp. 109–117.

81. Hamler, A., Trlep, M. and Hribernik, B., *Optimal Secondary Segment Shapes of Linear Reluctance Motors Using Stochastic Searching*, IEEE Trans. on MAG, Vol. 34, 1998, No. 5, pp. 3519–3521.

82. Hannakam, L., *Wirbelströme in dünnen leitenden Platten infolge bewegter stromdurchflossener Leiter*, etz-a, Vol. 86, 1965, No. 13, pp. 427–431.

83. Hashimoto, M., Kitano, J., Inden, K., Tanitsu, H., Kawaguchi, I., Kaga, S., Nakashima, T., Koike, S., Migiya, Y. and Sogihara, H., *Driving Control Characteristic Using the Inverter System at Yamanashi Maglev Test Line*, 15th Int. Conf. on Magnetically Levitated Systems and Linear Drives Maglev'98, Mt. Fuji, Yamanashi, Japan, 1998, pp. 287–291.

84. *Heidenhein General Catalogue: Linear Encoders*, Heidenhain, GmbH, Traunreut, Germany, 1998, www.heidenhain.com

85. *Heidenhain Catalogue: Linear Encoders for Numerical Controlled Machine Tools*, Heidenhain GmbH, Traunreut, Germany, September 2005, www.heidenhain.de

86. Heller, B. and Hamata, V., *Harmonic Field Effects in Induction Motors*, Academia, Prague, Czech Republics, 1977.

87. Henneberger, G. and Reuber, C., *A Linear Synchronous Motor for a Clean Room Conveyance System*, 1st Int. Symp. on Linear Drives for Ind. Appl. LDIA'95, Nagasaki, Japan, 1995, pp. 227–230.

88. Hinds, W. and Nocito, B., The Sawyer Linear Motor, in *Theory and Applications of Step Motors* ed. Kuo, B.C., West Publishing, 1974, pp. 327–340.

89. Hippner, M. and Piech, Z., *Ripple Free Linear Synchronous Motor*, Int. Conf. on Electr. Machines ICEM'98, Istanbul, Turkey, 1998, pp. 845–850.

90. Hoang, E., Ahmed, A.H.B. and Lucidarme, J., *Switching Flux Permanent Magnet Polyphase Synchronous Machines*, 7th European Conf. on Power Electronics and Appl. EPE'97, Vol. 3, Trondheim, Norway, 1977, pp. 903–908.

91. Hor, P.J., Zhu, Z.Q., Churn, P.M., Howe, D. and Rees-Jones, J., *Design and Analysis of a Novel Long-Stroke Tubular Permanent Magnet Linear Motor*, 2nd Int. Symp. on Linear Drives for Ind. Appl. LDIA'98, Tokyo, Japan, 1998, pp. 34–37.

92. Howe, D. and Zhu, Z.Q., *Status of Linear Permanent Magnet and Reluctance Motor Drives in Europe*, 2nd Int. Symp. on Linear Drives for Ind. Appl. LDIA'98, Tokyo, Japan, 1998, pp. 1–8.

93. Huth, E., Canders, W.R. and Mosebach, H., *Linear Motor Transfer Technology (LMTT) for Container Terminals*, 2nd Int. Symp. on Linear Drives for Ind. Appl. LDIA'98, Tokyo, Japan, 1998, pp. 38–41.

94. Ilhan, E., Gysen, B.L.J., Paulides, J.J.H. and Lomonova, E.A., *Analytical Hybrid Model for Flux Switching Permanent Magnet Machines*, IEEE Trans. on MAG, Vol. 46, 2010, No. 6, pp. 1762–1765.

95. Ishii, T., *Elevators for Skyscrapers*, IEEE Spectrum, 1994, No. 9, pp. 42–46.

96. Jahns, T.M., Kliman, G.B. and Neumann, T.W., *Interior PM Synchronous Motors for Adjustable-Speed Drives*, IEEE Trans. on IA, Vol. 22, 1986, No. 4, pp. 738–747.

97. Jahns, T.M., *Motion Control with Permanent-Magnet a.c. Machines*, Proc. IEEE, Vol. 82, 1994, No. 8, pp. 1241–1252.

98. Jahns, T.M., Variable Frequency Permanent Magnet a.c. Machine Drives, in *Power Electronics and Variable Frequency Drives*, ed. Bose, B.K., IEEE Press, New York, 1997.

99. Jansen, J.W., van Lierop, C.M.M., de Boeij, J., Lomonowa, E.A., Duarte, J.L. and Vandenput, A.J.A., *Moving Magnet Multi-DOF Planar Actuator Technology with Contactless Energy and Data Transfer*, Keynote Address, Int. Symp. for Linear Drives for Ind. Appl. LDIA'07, Lille, France, 2007, available on CD.

100. Jung, I.S., Yoon, S.B., Shim, J.H. and Hyun, D.S., *Analysis of Forces in a Short Primary Type and a Short Secondary Type Permanent Magnet Linear Synchronous Motor*, IEEE Trans. on EC, Vol. 14, No. 4, pp. 1265–1269.

101. Jung I.S., Hyun D.S., *Dynamic Characteristics of PM Linear Synchronous Motor Driven by PWM Inverter by Finite Element Method*, IEEE Trans. on MAG, vol. 35, 1999, No. 5, pp. 3697–3699.

102. Kajioka, M., Torii, S. and Ebihara, D., *A Comparison of Linear Motor Performance Supported by Air bearings*, 2nd Int. Symp. on Linear Drives for Ind. Appl., LDIA'98, Tokyo, Japan, 1998, pp. 252–255.

103. Kakino, Y., *Tools for High Speed and High Acceleration Feed Drive System of NC Machine Tools*, 2nd Int. Symp. on Linear Drives for Ind. Appl., LDIA'98, Tokyo, Japan, 1998, pp. 15–21.

104. Kaminski, G., *Electric Motors with Multi-DOF Motion* (in Polish), OWPW, Warsaw, Poland, 1994.

105. Karita, M., Nakagawa, H. and Maeda, M., *High Thrust Density Linear Motor and its Applications*, 1st Int. Symp. on Linear Drives for Ind. Appl. LDIA'95, Nagasaki, Japan, 1995, pp. 183–186.

106. Kawanishi, T., *Linear Motor Application for Architecture*, 1st Int. Symp. on Linear Drives for Ind. Appl., LDIA'95, Nagasaki, Japan, 1995, pp. 239–242.

107. Khan, S.A. and Ivanov, A.A., *Methods of Calculation of Magnetic Fields and Static Characteristics of Linear Step Motor For Control Rod Drives of Nuclear Reactors*, IEEE Trans. on MAG, Vol. 28, 1992, No. 5, pp. 2277–2279.

108. Khan, S.H. and Ivanov, A.A., *An Analytical Method for the Calculation of Static Characteristics of Linear Step Motor For Control Rod Drives in Nuclear Reactors*, IEEE Trans. on MAG, Vol. 31, 1995, No. 3, pp. 2324–2330.

109. Kim, W.J. and Murphy, B.C., *Development of a Novel Direct-Drive Tubular Linear Brushless Permanent-Magnet Motor*, Int. Journal of Control, Automation and Systems, Vol. 2, 2004, No. 3, pp. 279–288.

110. King, R., *Precision Laser Scribes Aircraft Skins*, Design News, 1994, No. 5-9, pp. 66–67.

111. Kitamori, T., Inoue, A., Yoshimura, M., Matsudaira, Y. and Hosaka, S., *Outline of the Second Train Set for the Yamanashi Maglev Test Line*, 15th Int. Conf. on Magnetically Levitated Systems and Linear Drives Maglev'98, Mt. Fuji, Yamanashi, Japan, 1998, pp. 220–224.

112. *Kollmorgen Linear Motors Aim to Cut Cost of Semiconductors and Electronics Manufacture*, Kollmorgen, Radford, VA, USA, 1997.

113. Kostenko, M. and Piotrovsky, L., *Electrical Machines, Vol. 2: Alternating Current Machines*, Mir Publishers, Moscow, 1974.

114. Kurobe, H., Kaminishi, K., Miyamoto, S. and Seki, A., *Current Test Status of the Superconducting Maglev System on the Yamanashi Test Line*, 15th Int. Conf. on Magnetically Levitated Systems and Linear Drives Maglev'98, Mt. Fuji, Yamanashi, Japan, 1998, pp. 56–61.

115. Kwon, B.I., Woo, K.I., Rhyu, S.H. and Park, S.C., *Analysis of Direct Thrust Control in Permanent Magnet Type Linear Synchronous Motor by FEM*, 2nd Int. Symp. on Linear Drives for Ind. Appl., LDIA'98, Tokyo, Japan, 1998, pp. 404–407.

116. Kyutoku, S., Shinya, T., *Development of Linear Synchronous Motor for Air Suspension Table*, 1st Int. Symp. on Linear Drives for Ind. Appl. LDIA'95, Nagasaki, Japan, 1995, pp. 223–226.

117. Laithwaite, E.R., *A History of Linear Electric Motors*, Macmillan, London, 1987.

118. Laugis, J. and Lehtla, T, *Control of Special Purpose Linear Drives*, 7th European Conf. on Power Electronics and Appl. EPE'97, Trondheim, Norway, Vol. 3, pp. 541–546.

119. *Linear and Rotary Positioning Stages Engineering Reference*, Parker Hannifin, Rohnert Park, CA, USA, 2005, www.parkermotion.com/engineeringcorner

120. *Linear Motor Conveyance System for Hospitals*, Shinko Electric Co. Ltd., Tokyo, Japan, 1999, www.shinko-elec.co.jp

121. *Linear-Motor-Driven Vertical Transportation System*, Elevator Word, September, 1996, pp. 66–72, www.elevator-world.com

122. *Linear Step Motor*, information brochure, Tokyo Aircraft Instrument Co., Ltd., Tokyo, 1998.

123. *Linear Synchronous Motors*, MagneMotion, 1999, www.magnemotion.com/linear.html

124. Lingaya, S. and Parsch, C.P., Characteristics of the Force Components on an Air-Core Linear Synchronous Motor with Superconducting Excitation Magnets, in *Electric Machines and Electromechanics*, Hemisphere Publishing, 1979, No. 4, pp. 113–123.

125. *LinMot Design Manual*, Sulzer Electronics, Ltd, Zürich, Switzerland, 1999.

126. Liwinski W., *Transverse induction heaters* (in Polish), WNT, Warsaw, Poland, 1968.

127. Locci, N. and Marongiu, I., *Modelling and Testing a New Linear Reluctance Motor*, Int. Conf. on Electr. Machines ICEM'92, vol. 2, Manchester, UK, pp. 706–710.

128. Luukko, J., Kaukonen, J., Niemelaä, M., Pyrhönen, O., Tiitinen, P. and Väänänen, J., *Permanent Magnet Synchronous Motor Drive Based on Direct Flux Linkage Control*, 7th European Conf. on Power Electronics and Appl. EPE'97, Trondheim, Norway, 1997, Vol. 3, pp. 683–688.

129. *Macro Sensors Technical Bulletin 1101: Hermetically Sealed Frictionless Position Sensors*, Macro Sensors, Pennsauken, NJ, USA, 2008, www.macrosensors.com

130. *Magnetic Sensors*, BEI Sensors & System Company, Goleta, CA, USA, 1998.
131. Mandel, R., *Linear Motors Drive Aerospace Machining to Higher Speeds*, 2000, www.manufacturingcenter.com/dfx/archives/1100
132. Mangan, J. and Warner, A., *Magnet Wire Bonding*, Joyal Product, Inc., Linden, NJ, USA, 1998, www.joyalusa.com
133. Marshall, S.V., Skitek, G.G.,*Electromagnetic Concepts and Applications*, Prentice–Hall, Englewood Cliffs, NJ, 1987.
134. Masada, E., *Linear Drives for Industry Applications in Japan—History, Existing State and Future Prospects*, 1st Int. Symp. on Linear Drives for Ind. Appl. LDIA'95, Nagasaki, Japan, 1995, pp. 9–12.
135. Masada, E.,*Development of Maglev and Linear Drive Technology for Transportation in Japan*, 14th Int. Conf. on Magnetically Levitated Systems Maglev'95, Bremen, Germany, 1995, pp. 11–16.
136. Masada, E., *High Power Converters and their Future Applications*, 7th Int. Power Electronics and Motion Control Conf. PEMC'96, Budapest, Hungary, Vol. 3, 1996, pp. K1–K4.
137. Matlack, J., *Smart Linear Position Sensors, Motion System Design*, 2010, pp. 32–33, www.motionsystemdesign.com
138. Matsui, N., Nakamura. M. and Kosaka, T., *Instantaneous Torque Analysis of Hybrid Stepping Motors*, IEEE Trans. on IA, Vol.32, 1996, No. 5, pp. 1176–1182.
139. Matsuoka, K. and Kondou, K., *Development of Permanent Magnet Linear Motor for the Next Generation High Speed Railways*, Symp. on Power Electronics, Electr. Drives, Adv. Electr. Machines SPEEDAM'94, Taormina, Italy, 1994, pp. 237–242.
140. Menden, W., Mayer, W.J. and Rogg, D., *State of Development and Future Prospects on the Maglev System Transrapid, M-Bahn and Starlim*, Int. Conf. Maglev'89, Yokohama, Japan, 1989, pp. 11–18.
141. Mendrela, E.A., *Comparision of the Performance of a Linear Reluctance Oscillating Motor Operating Under AC Supply with One Under DC Supply*, IEEE Trans. on EC, Vol. 14, 1999, No. 3. pp. 328-332.
142. Mendrela, E. and Song, T., *A Performance of Switched-Reluctance Linear Oscillating Motor Operating under Different Switching Circuits*, 3rd Int. Conf. on Unconventional Electromech. and Electr. Systems UEES'97, Alushta, The Crimea, Ukraine, 1997, pp. 349–354.
143. *Metglas®Amorphous Magnetic Alloys*, information brochure, AlliedSignal Inc., Morristown, NJ, USA, 1999, www6.alliedsignal.com/metglas/magnetic/
144. Meeker, D., *FEMM 4.0, User's Manual*, University of Virginia, Charlottesville, VA, USA, 2004.
145. Meisel, J., *Principles of Electromechanical Energy Conversion*, R.E. Krieger Publishing Company, Malabar, FL, 1984.
146. Miller, L., *Superspeed Maglev System Transrapid: System Description*, 14th Int. Conf. on Magnetically Levitated Systems Maglev'95, Bremen, Germany, 1995, pp. 37–43.
147. Miller, L. and Löser, F., *System Characteristics of the Transrapid Superspeed Maglev System*, 15th Int. Conf. on Magnetically Levitated Systems and Linear Drives Maglev'98, Mt. Fuji, Yamanashi, Japan, 1998, pp. 19–24.
148. Mishler, W.R., *Test Results on a Low Loss Amorphous Iron Induction Motor*, IEEE Trans. on PAS, Vol. 100, 1981, No. 6, pp. 860–866.

149. Miyatake, M., Koseki, T. and Sone, S., *Design and Traffic Control of Multiple Cars for an Elevator System Driven by Linear Synchronous Motors*, 2nd Int. Symp. on Linear Drives for Ind. Appl. LDIA'98, Tokyo, Japan, 1998, pp. 94–97.

150. Mizuno, T. and Yamada, H., *Magnetic Circuit Analysis of a Linear Synchronous Motor with Permanent Magnets*, IEEE Trans. on MAG, Vol. 28, 1992, No. 5, pp. 3027–3029.

151. Moscrop, J., Cook, C. and Naghdy, F., *Development and Performance Analysis of a Single-Axis Linear Motor Test-Bed*, Australasian Universities Power Eng. Conf. AUPEC'01, Perth, Australia, 2001, pp. 607–612

152. Mosebach, H., *Direct Two-Dimensional Analytical Thrust Calculation of Permanent Magnet Excited Linear Synchronous Machines*, 2nd Int. Symp. on Linear Drives for Ind. Appl. LDIA'98, Tokyo, Japan, 1998, pp. 396–399.

153. Mosebach, H. and Canders, W.R., *Average Thrust of Permanent Magnet Excited by Linear Synchronous Motors for Different Stator Current Waveforms*, Int. Conf. on Electr. Machines ICEM'98, Istanbul, Turkey, 1998, Vol. 2, pp. 851–865.

154. *Motor and Magnet Wire Industry Bulletin*,DuPont High Performance Films, DuPontTM^{TM}Kapton®, 2006, Circleville, OH, USA, www.kapton.dupont.com

155. Muraguchi, Y., Karita, M., Nakagawa, H., Shinya, T. and Maeda, M.,*Method of Measuring Dynamic Characteristics for Linear Servo Motor and Comparison of their Performance*, 2nd Int. Symp. on Linear Drives for Ind. Appl. LDIA'98, Tokyo, Japan, 1998, pp. 204–207.

156. Nai K., Forsythe W., Goodall R.M., *Improving Ride Quality in High-Speed Elevators*, Elevator World, 1997, No. 6, pp. 80–93.

157. Nakao, H., Takahashi, M., Sanada, Y., Yamashita, T., Yamaji, M., Miura, A., Terai, M., Igarashi, M., Kurihara, T. and Tomioka, K., *Development of the New Type On-board GM Refrigeration System for the Superconducting Magnet in Maglev Use*, 15th Int. Conf. on Magnetically Levitated Systems and Linear Drives Maglev'98, Mt. Fuji, Yamanashi, Japan, 1998, pp. 250–255.

158. Nakashima, H. and Seki, A., *The Status of the Technical Development for the Yamanashi Maglev Test Line*, 14th Int. Conf. on Magnetically Levitated Systems MAGLEV'95, Bremen, Germany, 1995, pp 31–35.

159. Nakashima, H. and Isoura, K., *Superconducting Maglev Development in Japan*, 15th Int. Conf. on Magnetically Levitated Systems and Linear Drives Maglev'98, Mt. Fuji, Yamanashi, Japan, 1998, pp. 25–28.

160. Nasar, S.A. and Boldea, I., *Linear Electric Motors*, Prentice-Hall, Englewood Cliffs, NJ, 1987.

161. *Northern Magnetics Linear Motors Technology*, Normag (Baldor Electric Company), Santa Clarita, CA, USA, 1998.

162. *OPERA 3-D User Guide*, Vector Fields Ltd, Oxford, UK, 1999.

163. Osada, Y., Gotou, H., Sawada, K. and Okumura, F., *Outline of Yamanashi Maglev Test Line and Test Schedule*, 15th Int. Conf. on Magnetically Levitated Systems and Linear Drives Maglev'98, Mt. Fuji, Yamanashi, Japan, 1998, pp. 50–55.

164. Pandilov, Z., *Analytical Determination of the Position Loop Gain for Linear Motor CNC Machine Tool*, Journal AMME, Vol. 26, 2008, No 2, pp. 171–174, www.journalamme.org

165. Park, R.H., *Two-Reaction Theory of Synchronous mMchines: Generalized Method of Analysis—Part 1*, Trans AIEE, July 1929, pp. 716–730.

166. Parker, R.J., *Advances in Permanent Magnetism*, John Wiley & Sons, New York, 1990.

167. *Partial Differential Equations*, Toolbox for use with Matlab, The MathWorks Inc., 1999.

168. *Platinum^{TL} DDL Direct Drive Linear Motors*, Kollmorgen, Radford, VA, USA, 1998, www.kollmorgen.com

169. Pepperl+Fuchs Group, *Identification Systems Catalogue* 2009, edition 2009-03-01, part no. 33152, Pepperl+Fuchs GmbH, Mannheim, Germany, 2009, www.pepperl-fuchs.com.

170. Post, R.F.: *Inductrack Demonstration Model*, Report No UCRL-ID-129664, Lawrence Livemore National Laboratory, Livermore, CA, USA, 1998, www.askmar.com/Inductrack/

171. Post, R.F., *The Inductrack: A Home-Grown Maglev System for our Nation*, Lockheed Martin Colloquium, Advanced Technology Center, Palo Alto, CA, USA, 2004, www.askmar.com/Inductrack/

172. Presher, A., *New Class of Linear Motors*, Design News, November 16, www.designnews.com/article/388467

173. Rais, V.R., Turowski, J. and Turowski, M., Reluctance Network Analysis of Coupled Fields in Reversible Electromagnetic Motor, Chapter 6.7 in *Electromagnetic Fields in Electrical Engineering*, Plenum Press, New York–London, 1988, pp. 279–283.

174. Rajagopal, K.,Krishnaswamy, M., Singh, B. and Singh, B.P., *High Thrust Density Linear Motor and its Applications*, 1st Int. Symp. on Linear Drives for Ind. Appl. LDIA'95, Nagasaki, Japan, 1995, pp. 183–186.

175. Raschbichler, H.G. and Miller, L., *Readiness for Application of the Transrapid Maglev System*, RTR Railway Technical Review, vol. 33, 1991/92, pp. 3–7.

176. Rauch, S.E. and Johnson, L.J., *Design Principles of Flux-Switch Alternators*, AIEE Trans., Part III, Vol. 74, 1955, No. 12, pp. 1261–1269.

177. Rees, J.J., *Dynamic Consideration and Candidacy Requirements for Linear Servo-Driven Motors in Factory Automation*, 1st Int. Symp. on Linear Drives for Ind. Appl. LDIA'95, Nagasaki, Japan, 1995, pp. 255–258.

178. *RG2 Linear Encoder System*, Renishaw plc, New Mills, Gloucestershire, UK, 1998, www.renishaw.com

179. *RLS Data Sheet LM10D17 01: DCRM Distance Coded Reference Mark System*, RLS Merilna Tehnika D.O.O., Ljubljana-Dobrunje, Slovenia, 2009, No 1, www.ris.si

180. *RLS Data Sheet LM10D01 10: LM Linear Magnetic Encoder System*, RLS Merilna Tehnika D.O.O., Ljubljana-Dobrunje, Slovenia, 2010, No 10, www.ris.si

181. Rosenmayr, M., Casat, Glavitsch, A. and Stemmler, H., *Swissmetro — Power Supply for a High-Power-Propulsion System with Short Stator Linear Motors*, 15th Int. Conf. on Magnetically Levitated Systems and Linear Drives Maglev'98, Mount Fuji, Yamanashi, Japan, 1998, pp. 280–286.

182. Sanada, M., Morimoto, S. and Takeda, Y., *Interior Permanent Magnet Linear Synchronous Motor for High-Performance Drives*, IEEE Trans. on IA, Vol. 33, 1997, No. 4, pp. 966–972.

183. Sanada, M., Morimoto, S. and Takeda, Y., *Reluctance Equalization Design of Multi Flux Barrier Construction for Linear Synchronous Reluctance Motors*, 2nd Int. Symp. on Linear Drives for Ind. Appl. LDIA'98, Tokyo, Japan, 1998, pp. 259–262.

184. Seki, K., Watada, Torii, S. and Ebihara, D., *Experimental Device of Long Stator LSM with Discontinuous Arrangement and Result*, 7th European Conf. on Power Electronics and Appl. EPE'97, Trondheim, Norway, 1997, Vol. 3, pp. 532–536.

185. Seki, K., Oka, K., Watada, M., Torii, S. and Ebihara, D., *Synchronization of Discontinuously Arranged Linear Synchronous Motor for Transportation System*, 2nd Int. Symp. on Linear Drives for Ind. Appl. LDIA'98, Tokyo, Japan, 1998, pp. 82–85.

186. Seok-Myeong, J. and Sang-Sub, J., *Design and Analysis of the Linear Homopolar Synchronous Motor for Integrated Magnetic Propulsion and Suspension*, 2nd Int. Symp. on Linear Drives for Ind. Appl. LDIA'98, Tokyo, Japan, 1998, pp. 74–77.

187. Setbacken, R., *Feedback Devices in Motion Control Systems*, RENCO Encoders, Fiftian Press, Santa Barbara, CA, USA, 1997.

188. Shiraki, M., Song, R., Itoh, A., Mizuno, T. and Yamada, H., *High Speed High Accuracy Positioning System for Industrial Printer by LDM*, 2nd Int. Symp. on Linear Drives for Ind. Appl. LDIA'98, Tokyo, Japan, 1998, pp. 98–101

189. Silvester, P.P. and Ferrari, R.L., *Finite Elements for Electrical Engineers*, Cambridge University Press, Cambridge, 1990.

190. Skalski, C.A., *The Air-Core Linear Synchronous Motor: An Assessment of Current Development*, MITRE Technical Report, McLean, VA, USA, 1975.

191. Smith, A.C., *Magnetic Forces on a Misaligned Rotor of a PM Linear Actuator*, Int. Conf. on Electr. Machines ICEM'90, Boston, MA, USA, 1990, pp. 1076–1081.

192. Smith, B., *Going Direct*, California Linear Drives, Carslbad, CA, USA, 2001, www.calinear.com

193. *Software Developed by International Center for Numerical Methods in Engineering*, EMANT, Barcelona, Spain, www.cimne.upc.es/emant/

194. Stec, T.F., *Amorphous Magnetic Materials Metglass 2605S-2 and 2605TCA in Application to Rotating Electrical Machines*, NATO ASI Modern Electrical Drives, Antalya, Turkey, 1994.

195. Stec T.F., *Electric Motors from Amorphous Magnetic Materials*, Int. Symp. on Nonlinear Electromagnetic Systems, Cardiff, UK, 1995, www.ammtechnologies.com

196. *SuperPower*, Schenectady, NY, USA, www.superpower-inc.com

197. Suwa, H., Turuga, H., Iida, T., Tujimoto, S., Kobayashi, Y. and Itabashi, Y., *Features of Ground Coils for Yumanashi Maglev Test Line*, 15th Int. Conf. on Magnetically Levitated Systems and Linear Drives Maglev'98, Mt. Fuji, Yamanashi, Japan, 1998, pp. 292–296.

198. *Synchronous Linear Motor 1FN6: The Electrical Gear Rack*, Siemens AG Industry Sector, Drive Technologies. Motion Control, Erlangen, Germany, 2008.

199. Takahashi, Y., Yoshihiro, J., Hidenari, A., Motoaki, T. Motohiro, I. and Masatoshi, S., *Vibration Characteristics and Mechanical Heat Load of Superconducting Magnets of Maglev Trains*, 15th Int. Conf. on Magnetically Levitated Systems and Linear Drives Maglev'98, Mt. Fuji, Yamanashi, Japan, 1998, pp. 244–249.

200. Tergan, V., Andreev, I. and Liberman, B., *Fundamentals of Industrial Automation*, Mir Publishers, Moscow, 1986.

201. *The Contactless Power Supply System of the Future*, Wampfler AG, Weil am Rhein-Maerkt, Germany, 1998.

202. *The LIM Elevator Drive*, Elevator World, 1991, No 3, pp. 34–41.

203. *Thin Nonoriented Electric Steels*, Cogent Power Ltd., Newport, UK, 2005, www.cogent-power.com

484   References

204. Tomczuk, B., *Three-Dimensional Leakage Reactance Calculation and Magnetic Field Analysis for Unbounded Problems*, IEEE Trans. on MAG, Vol. 28, 1992, No. 4, pp. 1935-1940.
205. Tomczuk, B., *Analysis of 3-D Magnetic Fields in High Leakage Reactance Transformers*, IEEE Trans. on MAG, vol. 30, 1994, No. 5, pp. 2734-2738.
206. Tomczuk, B. and Babczyk, K., *Calculation of the Self- and Mutual Inductances and 3-D Magnetic Fields of Chokes with Air Gaps in the Core*, Electrical Engineering (Archiv. für Elektrotechnik), Springer-Verlag, Berlin, Vol. 83, 2001, pp. 41-46.
207. Tomczuk, B. and Sobol, M., *Time Analysis of an Oscillating Motor*, XVII Int. Symp. on Electromagn. Phenomena in Nonlinear Circuit EPNC'02, Leuven, Belgium, 2002, pp. 23-26.
208. Tomczuk, B. and Sobol, M., *Analysis of Tubular Linear Reluctance Motor (TLRM) under Various Voltage Supplying*, Int. Conf. on Electr. Machines ICEM'04, Cracow, Poland, 2004, pp. 331-333.
209. Tomczuk, B. and Sobol, M., *Field Analysis of the Magnetic Systems for Tubular Linear Reluctance Motors*, IEEE Trans. on MAG, Vol. 41, 2005, No. 4, pp. 1300-1305.
210. Tomczuk, B. and Sobol, M., *A Field-Network Model of a Linear Oscillating Motor (LOM) and its Dynamics Characteristics*, IEEE Trans. on MAG, Vol. 41, 2005, No. 8, pp. 2362-2367.
211. Tomczuk, B. and Waindok, A., *Magnetic Field Calculations of a Permanent Magnet Tubular Linear Motor (PMTLM)*, Conf. on Computer Appl. in Electr. Eng., Poznan, Poland, 2005, pp. 89-90.
212. Tomczuk, B., *Numerical Methods of Analysis of Electromagnetic Fields of Transformer Systems* (in Polish), OWPO, Opole, Poland, 2007.
213. Tomczuk, B., Schröder, G., and Waindok, A., *Finite Element Analysis of the Magnetic Field and Electromechanical Parameters Calculation for a Slotted Permanent Magnet Tubular Linear Motor*, IEEE Trans. on MAG, Vol. 43, 2007, No. 7, pp. 3229-3236.
214. Tomczuk, B., Zakrzewski, K., Waindok, A., *Field Analysis in Permanent Magnet Tubular Linear Motor (PMTLM) under Variable Scaled Geometries*, Electromotion, Vol. 14, No. 1, 2007, pp. 19-25.
215. Tomczuk, B. and Waindok, A., *Integral Parameters of the Magnetic Field in the Permanent Magnet Linear Motor*, Intelligent Computer Techniques in Applied Electromagnetics, series Studies in Computational Intelligence, Springer Verlag, Heidelberg, Germany, Vol. 119, 2008, pp. 277-281.
216. Tomczuk, B. and Waindok, A., *Tubular Linear Actuator as a Part of Mechatronic System, Solid State Phenomena*, Trans. Tech. Publications, Switzerland, Vol. 147-149, 2009, pp. 173-178.
217. Tomczuk, B. and Waindok, A., *Linear Motors in Mechatronics — Achievements and Open Problems*, in Transfer of Innovation to the Interdisciplinary Teaching of Mechatronics for the Advanced Technology Needs, OWPO, Opole, Poland, 2009, pp. 343-360.
218. Tomczuk, B., Zimon, J. and Waindok, A., *Field-Circuit Method for the Non-Steady State Analysis in the Active Magnetic Bearings*, 17th Int. Conf. COMPUMAG'09, Florianopolis, Brazil, 2009, pp. 183-184.
219. Tomczuk, B., Zimon, J. and Waindok, A., *Effects of the Core Materials on Magnetic Bearing Parameters*, 17th Int. Conf. COMPUMAG'09, Florianopolis, Brazil, 2009, pp. 39-40.

220. Tomczuk, B. and Waindok, A., *A Coupled Field-Circuit Model of a 5-Phase Permanent Magnet Tubular Linear Motor*, 21st Symp. on Electromagnetic Phenomena in Nonlinear Circuits EPNC'10, Dortmund, Germany, 2010, pp. 139–140.

221. *Transrapid Maglev System*, ed. Heinrich, K. and Kretzschmar, R., Hestra-Verlag, Darmstadt, Germany, 1989.

222. *Trilogy PM Linear Motors*, Trilogy Systems Corp., Webster, TX, USA, 1999, ww.trilogysystems.com

223. Trumper,D.L., Kim, W.J., Williams, M.E., *Design and Analysis Framework for Linear Permanent-Magnet Machines*, IEEE Trans. on IA, Vol. 32, 1996, No. 2, pp. 371–379.

224. Tsuchishima, H., Mizutani, T., Okai, T., Nakauchi, M., Terai, M., Inadama, S. and Asahara, T., *Characteristics of Superconducting Magnets and Cryogenic System on Yamanashi Test Line*, 15th Int. Conf. on Magnetically Levitated Systems and Linear Drives Maglev'98, Mt. Fuji, Yamanashi, Japan, 1998, pp. 237–243.

225. Turowski, J., Turowski, M. and Kopec, M., *Method of Fast Analysis of 3D Leakage Fields in Large Three-Phase Transformers*, Compel, James & James Scie Publishers, Vol. 9, 1990, London, UK, pp. 107–116.

226. Turowski J., *Technical Electrodynamics* (in Polish), 2nd edition, WNT, Warsaw, Poland, 1993.

227. Utsumi, T. and Yamaguchi, I., *Thrust Characteristics of a Rectangular Core LIM*, 2nd Int. Symp. on Linear Drives for Ind. Appl. LDIA'98, Tokyo, 1998, pp. 248–251.

228. Vaez-Zadeh, S. and Isfahani, H., *Multiobjective Design Optimization of Air-Core Linear Permanent-Magnet Synchronous Motors for Improved Thrust and Low Magnet Consumption*, IEEE Trans. on MAG, Vol. 42, 2006, No. 3, pp. 446–452.

229. *Vector Fields Software*, Cobham Technical Services, Kidlington, Oxfordshire, UK, www.cobham.com/about-cobham/avionics-and-surveillance/ about-us/technical-services/kidlington.aspx

230. Waindok A., *Computer Simulation and Measurement Verification of the Permanent Magnet Tubular Linear Motor (PMTLM) Characteristics*, (in Polish), Ph.D., Opole University of Technology, Opole, Poland, 2008.

231. Waindok, A. and Mazur, G., *A Mathematical and Physical Models of the Three-Stage Reluctance Accelerator*, 2nd Int. Students' Conf. on Electrodynamic and Mechatronics, Gora sw. Anny, Poland, 2009, pp. 29–30.

232. Wang, R. and Gieras, J.F., *Performance Calculations for a PM Hybrid Linear Stepping Motor by the Finite Element and Reluctance Network Approach*, 2nd Int. Symp. on Linear Drives for Ind. Appl. LDIA'98, Tokyo, 1998, pp. 400–403.

233. Wang, R. and Gieras, J.F., *Analysis of Characteristics of a Permanent Magnet Hybrid Linear Stepping Motor*, Int. Conf. on Electr. Machines ICEM'98, Istanbul, Turkey, 1998, pp. 835–838.

234. Wegerer, K., Ellman, S., Becker, P. and Hahn, W., *Requirements, Design and Characteristics of the Maglev Vehicle Transrapid 08*, 15th Int. Conf. on Magnetically Levitated Systems and Linear Drives Maglev'98, Mt. Fuji, Yamanashi, Japan, 1998, pp. 202–208.

235. Węgliński, B., *Soft Magnetic Powder Composites — Dielectromagnetics and Magnetodielectrics*, Reviews on Powder Metallurgy and Physical Ceramics, vol. 4, 1990, No. 2, pp. 79–153.

236. *What Every Engineer Should Know About Finite Element Analysis*, ed. Brauer, J.R., Marcel Dekker, New York, 1988.

237. Wiescholek, U., *High-Speed Magnetic Levitation System Transrapid*, 14th Int. Conf. on Magnetically Levitated Systems Maglev'95, Bremen, Germany, 1995, pp. 17–23.

238. White, D.C. and Woodson, H.H., *Electromechanical Energy Conversion*, John Wiley & Sons, New York, 1959.

239. *Yamanashi Maglev Test Line—Guide of Electric Facilities*, Central Japan Railway Company, Tokyo, 1992.

240. Yamada, H., *Handbook of Linear Motor Applications* (in Japanese), Kogyo Chosaki Publishing Co., Tokyo, Japan, 1986.

241. Yamazaki, M., Gotou, Y., Aoki, S., Hashimoto, S. and Sogabe, M., *Guideways and Structures on the Yamanashi Maglev Test Line and their Dynamic Response Characteristics*, 15th Int. Conf. on Magnetically Levitated Systems and Linear Drives Maglev'98, Mt. Fuji, Yamanashi, Japan, 1998, pp. 178–183.

242. Yoon S.B., Jung I.S., Kim K.C. and Hyun D.S., *Dynamic Analysis of a Reciprocating Linear Actuator for Gas Compression Using Finite Element Method*, IEEE Trans. on MAG, Vol. 33, 1997, No. 5, pp. 4113–4115.

243. Yoshida, K., Takaki, T. and Muta, H., *System Dynamics Simulation of Controlled PM LSM Maglev Vehicles*, 10th Int. Conf. on Maglev and Linear Motors, MAGLEV'88, Hamburg, Germany, 198, pp. 269–278.

244. Yoshida, K., Muta, H. and Teshima, N., *Underwater Linear Motor Car*, Int. Journal of Appl. Electromagnetics in Materials, Vol. 2, 1991, pp. 275–280

245. Yoshida, K., Liming, S., Takami. H. and Sonoda, A., *Repulsive Mode Levitation and Propulsion Experiments of an Underwater Travelling LSM Vehicle ME02*, 2nd Int. Symp. on Linear Drives for Ind. Appl. LDIA'98, Tokyo, 1998, pp. 347–349.

246. Yoshida, K., Takami. H., Kong, X. and Sonoda, A., *Mass Reduction and Propulsion Control of PM LSM Test Vehicle for Container Transportation*, IEEE Int. Electr. Machines and Drives Conf. IEMDC'99, Seattle, WA, USA, 1999, pp. 72–74.

247. Yoshioka, H., Suzuki, E., Seino, H., Azakami, M., Oshima, H. and Nakanishi, T., *Results of Running Tests and Characteristics of the Dynamics of the MLX01 Yamanashi Maglev Test Line Vehicles*, 15th Int. Conf. on Magnetically Levitated Systems and Linear Drives Maglev'98, Mt. Fuji, Yamanashi, Japan, 1998, pp. 225–230.

248. Zajac, P., *Demystifying Linear Motor Integration*, Motion System Design, January 2011, pp. 24–26.

249. Zakrzewski, K., *Physical Modelling of Leakage Field and Stray Losses in Steel Constructional Parts of Electrotechnical Devices*, Archiv. für Elektrotechnik, Springer-Verlag, Berlin, Vol. 69, 1986, pp. 129–135.

250. Zakrzewski, K. and Tomczuk, B., Comparison of Finite Difference Method (FDM) and Integral Equations Method (BIM) in Calculation of Piecewise Nonhomogeneous Magnetic Fields, in *Electromagnetic Fields in Electrical Engineering*, ed. Turowski, J. and Zakrzewski, K., James & James Science Publishers, London, 1990, pp. 139–142.

251. Zhu, Z.Q., Pang, Y., Howe, D., Iwasaki, S., Deodhar, R. and Pride, A., *Analysis of Electromagnetic Performance of of Flux-Switching Permanent Magnet Machines by Non-Linear Adaptive Lumped Parameter Magnetic Circuit Model*, IEEE Trans. on MAG, Vol. 41, 2005, No. 11, pp. 4277–4287.

252. Zhu, Z.Q., Chen, Z.,Howe, D., and Iwasaki, S., *Electromagnetic Modeling of a Novel Linear Oscillating Actuator*, IEEE Trans on MAG, Vol. 44, No 11, 2008, pp. 3855–3858.

253. Zienkiewicz, O.C. and Morgan, K., *Finite Elements and Approximation*, John Wiley & Sons, New York, 1983.

254. Zolghadri, M.R., Diello, D. and Roye, D., *Direct Torque Control System for Synchronous Motor*, 7th European Conf. on Power Electronics and Applications EPE'97, Trondheim, Norway, 1997, Vol. 3, pp. 694–699.

Patents

[P1] US31128. Otis, G. E. (1861). Improvement in hoisting apparatus.

[P2] US3884154. Marten F. (1975). Propulsion arrangement equipment with linear motor.

[P3] US4016441. Herr J.A., Jaffe W., Dob A.M., Francis M., Adams K.D. and Herron W.L. (1977). Linear motor.

[P4] US4163911. Simes J.G. and Gillott D.H. (1979). Permanent magnet translational motor for respirators.

[P5] US4217507. Jaffe W. and Peterson W.R. (1980). Linear motor.

[P6] US4220899. Von der Heide J. (1980). Polyphase linear motor.

[P7] US4235253. Rinde J.E. and Perry R.M. (1980). Linear motion, electromagnetic force motor.

[P8] US4259653. McGonigal J.J. (1981). Electromagnetic reciprocating linear actuator with permanent magnet armature.

[P9] US4315197. Studer P.A. (1982). Linear magnetic motor/generator.

[P10] US4344022. Von Der Heide J. (1982). Linear motor.

[P11] US4369383. Langley L.W. (1983). Linear DC permanent magnet motor.

[P12] US4370577. Wakabayashi N., Fukumoto T. and Ueda N. (1983). Linear motor.

[P13] US4387935. Studer P.A. (1983). Linear magnetic bearing.

[P14] US4402386. Ficheux R. and Pavoz M. (1983). Self-powered elevator using a linear electric motor as counterweight.

[P15] US4535260. Pritchard R.J. and Lindenmucht K.(1984). Magnetic linear motor.

[P16] US4461984. Whitaker G.C. and Safford J.H. (1984). Linear motor shutting system.

[P17] US4504750. Okedera H., Wakabayashi N., Yamada K. and Sugizaki Y. (1985). Linear motor.

[P18] US4510421. Schwarzler P. (1985). Linear magnets.

[P19] US4546277. Carbonneau J.T. and Stahel P. (1985). Linear motor.

[P20] US4581553. Moczala H. (1986). Brushless d.c. motor, especially linear motor, having an increased force-to-velocity ratio.

[P21] US4583027. Parker R.J. and Cornell A.W. (1986). Moving magnets linear motor.

[P22] US4595870. Chitayat A. (1986). Linear motor.

[P23] US4613962. Inoue Y. and Mizunoe K. (1986). Tracking device with linear motor.

[P24] US4631430. Aubrecht R.A. (1986). Linear force motor.

[P25] US4638193. Jones P.J. (1987). Linear impulse motor.

[P26] US4641065. Shibuki O., Matsuyama N., Nagasawa Y., Kawai K., Sakagami S. and Onoyoma T. (1987). Moving coil type linear motor.

[P27] US4689529. Higuichi T. (1987). Linear stepping motor.

[P28] US4678951. Nikaido A. (1987). Linear motor.

[P29] US4703297. Nagasaka N. (1987). Permanent magnet type linear electromagnetic actuator.

[P30] US4749921. Chitayat A. (1988). Linear motor with non-magnetic armature.

[P31] US4761573. Chitayat A. (1988). Linear motor.

[P32] US4761574. Nakagawa H. (1988). Linear pulse motor.

[P33] US4789815. Kobayashj F. Sakai K. and Yamagashi J. (1988). Linear motor.

[P34] US4793263. Pritchard R.J.and Lindenmucht K.(1988). Integrated linear synchronous unipolar motor with controlled permanent magnet bias.

[P35] US4794866. Brandis C., Schulze-Buxloh H. and Pirags S. (1989). Linear motor driven railway car.

[P36] US4819564. Brandis C., Schulze-Buxloh H. and Pirags S. (1989). Linear motor driven conveying installation and braking device therefore.

[P37] US4825111. Hommes W.J. and Keegan Jr. J. (1989). Linear motor propulsion system.

[P38] US4839543. Beakley B.E. and Flanders T.E. (1989). Linear motor.

[P39] US4858452. Ibrahim F.K. (1989). Non-commutated linear motor.

[P40] US4860183. Meada T., Ohta H. and Inazumi H. (1989). Planar linear pulse motor.

[P41] US4862809. Guadagno J.R. (1989). Supports for railway linear synchronous motor.

[P42] US4937481. Vitale N. (1990). Permanent magnet linear electromagnetic machine.

[P43] US4945268. Nihei H., Miyashita K. and Asano H. (1990). Permanent magnet type linear pulse motor.

[P44] US4965864. Roth P.E. and Roth B.A. (1990). Linear motor.

[P45] US4983868.Spiesser G. (1991). Disengageable linear stepper motor.

[P46] US4985651. Chitayat A. (1991). Linear motor with magnetic bearing preload.

[P47] US5017819. Patt P.J. and Stolfi F.R. (1991). Linear magnetic spring and spring/motor combination.

[P48] US5072144. Saito J., Matsumoto Y., Suzuki Y., Okamura Y., Hori H. and Shiraiwa N. (1991). Moving-coil linear motor.

[P49] US5075583. Sakagami S., Onayama T., Nagasawa Y. and Andou T. (1991). Brushless DC linear motor.

[P50] US5086881. Gagnon E., Jamminet J. and Olsen E. (1991). Elevator driven by a flat linear motor.

[P51] US5087844. Takedomi S. and Umehara H. (1992). Linear motor.

[P52] US5126604. Manning M.J.N. (1992). Linear motor conveyance system.

[P53] US5134324. Sakagami S. and Nagasawa Y. (1992). Moving magnets type linear motor for automatic door.

[P54] US5141082. Toshiaki I., Ikejima H., Sugita K., Sakabe S., Sugimoto H., Maehara T. and Kisimoto T. (1992). Linear motor elevator system.

[P55] US5149996. Preston M.A. and Stingle F.W. (1992). Magnetic gain adjustment for axially magnetized linear force motor with outwardly surfaced armature.

[P56] US5172803. Levin H.U. (1992). Conveyor belt with built-in magnetic-motor linear motor.

[P57] US5175455. Penicaut A.M.R. (1992). Permanent magnet linear door motor.

[P58] US5179305. Van Engelen G. (1993). Linear motor within a positioning device.

[P59] US5183980. Okuma S., Furuhashi T., Ikejima H. and Ishi T. (1993). Linear motor elevator device with a null-flux position adjustment.

[P60] US5201641. Richer S. (1993). Electrically driven diaphragm suction or pressure pump.

[P61] US5216723. Froeschle T.A. and Carreras R.F. (1993). Permanent magnet transducer.

[P62] US5235144. Matsui N., Kangawa U. and Shimazu T. (1993). Linear motor driven elevator.

[P63] Re34674. Beakley B.E. and Flanders T.E. (1994). Linear motor (reissue).

[P64] US5300737. Nakanishi Y. (1994). Tubular linear motor driven elevator.

[P65] US5341053. Yamazaki K., Noda K. and Nakamura E. (1994). Linear motor with permanent magnets.

[P66] US5352946. Hoffman B.D., Pollack S.H., Smit P. and Woolley J. (1994). Linear motor suspension system.

[P67] US5379971. Kim S.G., Franklin D.K. and Conner M.P. (1995). Emergency power system for door.

[P68] US5431109. Berdut E. (1995). Levitation and linear propulsion system using ceramic permanent magnets and interleaved malleable steel.

[P69] US5434459. (1995). Pulsed power linear actuator and method of increasing actuator stroke force.

[P70] US5486844. Preston, M. and King, R.(1996). Noise-canceling quadrature magnetic position, speed and direct sensor.

[P71] US5497038. Sink J.D. (1996). Linear motor propulsion drive coil.

[P72] US5602431. Satomi H. and Iwasa T. (1997). Linear motor.

[P73] US5621259. Multon B.F.A., Lucidarme H.L. and Prevond L.P.A. (1997). Linear motor.

[P74] US5644176. Katagiri S. (1997). Linear direct current motor.

[P75] US5696411. Takei Y. (1997). Linear direct current motor.

[P76] US5703418. Assa S. (1997). DC cooled linear motor.

[P77] US5717261. Tozoni O.V. (1998). Linear synchronous motor with screening permanent magnet rotor with extendable poles.

[P78] US5719451. Cook S.J. and Clark R.E. (1998). Linear magnetic actuator.

[P79] US5751075. Kwon B.H. and Lee H.K. (1998). Magnet assembly for linear motor.

[P80] US5757091. Sagobe M. and Higashi S. (1998). Permanent magnet field pole for linear motor.

[P81] US5808379. Zhao W. (1998). Bi-directional linear drive motor.

[P82] US5808381. Aoyama H. and Shimizu Y. (1998). Linear motor.

[P83] US5838079. Morohashi N. and Kato S. (1998). Synchronous linear motor using permanent magnet.

[P84] US5910691. Wavre N. (1999). Permanent-magnet linear synchronous motor.

[P85] US5920164. Moritz F.G. and Mosciatti R. (1999). Brushless linear motor.

[P86] US5936319. Chitayat A. (1999). Wireless permanent magnet linear motor with magnetically controlled armature switching and magnetic encoder.

[P87] US5949036. Kowalczyk T.M. and Piech Z. (1999). Double linear motor and elevator door using the same.

[P88] US5952742. Stoiber D. and Rosner P. (1999). Synchronous linear motor with improved means for positioning and fastening permanent magnets.

[P89] US6043572. Nagai S., Sasaki T. and Kamata S. (2000). Linear motor, stage device, and exposing device.

[P90] US6104108. Hazelton A.J. and Gery J.M. (2000). Wedge magnet array linear motor.

[P91] US6184597. Yamamoto H., Shibuya K. and Hamaoka K. (2001). Linear motor and linear compressor.

[P92] US6230760. Di Natale G. and Vollenweider B. (2001). Linear motor weft presenting apparatus.

[P93] US6252315. Heo J.T. (2001). Magnet fixing structure for linear motor.

[P94] US6365993. Calhoon G.A.,Bottegal P.T., Rinefierd R.J., Marchand F.J. and Slepina R.M (2002). Round linear actuator utilizing flat permanent magnets.

[P95] US6407471. Miyamoto Y., Hisatsune M., Meakawa K., Doi T. and Tanabe M. (2002). Linear motor.

[P96] US6417584. Chitayat A. (2002). Magnet configuration for linear motor.

[P97] US6429611. Li H. (2002). Rotary and linear motor.

[P98] US6441515. Shimura Y. (2002). Linear motor.

[P99] US6445093. Binnard M. (2002). Planar motor with linear coil arrays.

[P100] US6476524. Miyamoto Y., Meakawa K., Doi T. and Hisatsune M. (2002). Linear motor.

[P101] US6483208. Hooley A. (2002). Linear motor.

[P102] US6491002. Adams J. (2002). Intermittent linear motor.

[P103] US6495934. Hayashi Y. and Tanaka K. (2002). Method of manufacturing linear motor, stage apparatus equipped with linear motor and exposure apparatus.

[P104] US6495935. Mishler M.(2002). Linear motor driven unit.

[P105] US6501357. Petro J. (2002). Permanent magnet actuator mechanism.

[P106] US6538349. Lee J., Im T. and Boldea I. (2003). Linear reciprocating flux reversal permanent magnetic machine.

[P107] US6541880. Okada T., Joong K.H., Tahara K., Maki K., Yamamoto K., Takahashi M. Miyata K. and Takahata R. (2003). Linear motor.

[P108] US6548919. Maki K., Joong K.H., Katayama H. and Miyata K. (2003). Linear motor.

[P109] US6570273. Hazelton A.J. (2003). Electric linear motor.

[P110] US6614137. Joong K.H., Maki k., Iwaji Y., Miyazaki T. and Hanyu T. (2003). Linear motor, driving and control system thereof and manufacturing method thereof.

[P111] US6633217. Post R.F. (2003). Inductrack magnet configuration.

[P112] US6653753. Kawano S., Honda Y. and Murakami H. (2006). Linear motor.

[P113] US6657326. Yamamaoto H. and Shibuya K. (2003). Efficient cylindrical linear motor.

[P114] US6657327. Tajima S. (2003). Linear direct current motor.

[P115] US6661130. Yamazaki T., Ohnishi K., Miyagawa N. and Inoue T. (2003). Linear motor.

[P116] US6664664. Botors S.J. and Novotnak R.T. (2003). Printed circuit linear motor.

[P117] US6664665. Hsiao S.S. (2003). Coreless type linear motor.

[P118] US6664880. Post R.F. (2003). Inductrack magnet configuration.

[P119] US6705408. Kim J.H. and Shinohara S. (2004). Power tool with linear motor.

[P120] US6713899. Greubel K. and Knauff A. (2004). Linear synchronous motor.

[P121] US6713922. Piech Z. and Wagner P.D. (2004). Integrally skewed permanent magnet for use in electric machines.

[P122] US6724104. Katsuki M., Kawatsu K. and Gakuhari M. (12004). Linear motor drive apparatus.

[P123] US6731029. Shikayama T., Yoshida S., Irie N. and Yu J. (2004). Canned linear motor.

[P124] US6747376. Hashimoto A., Kimura Y., Nakahara Y., Watarai A. and Nishitani S. (2004). Linear motor.

[P125] US6750570. Grehant B. (2004). Flux switching linear motor.

[P126] US6750571. Tominaga R., Tamai M. and Tanaga K. (2004). Magnetically shielded linear motors and stage apparatus comprising same.

[P127] US6753626. Hwang J.H., Kim D.H. and Park S.S. (2004). Linear motor having an integrally formed air bearing.

[P128] US6753627. Kim H. and Maki K. (2004). Linear motor.

[P129] US6756705. Pulford R. (2004). Linear stepper motor.

[P130] US6787945. Miyata K. (2004). Linear motor.

[P131] US6800968. Shikayama T., Irie N. and Miyamoto Y. (2004). Linear motor.

[P132] US6809434. Dunkan G.D. and Boud J.H. (2004). Linear motor.

[P133] US6822349. Lunz E. and Durschmied F. (2004). Linear motor.

[P134] US6825583. Joung J.H. and Park J.W. (2004). Linear motor including cooling system.

[P135] US6831379. Ohto M., Tanabe M., Miyamoto Y. and Inokuchi H. (2004). Permanent magnet synchronous linear motor.

[P136] US6833638. Kang H. and Kim J. (2004). Integrated system for non-contact power feed device and permanent magnet-excited transverse flux linear motor.

[P137] US6844651. Swift G.L., Sommerhalter F.A. and Syskowski R.A. (2005). Encapsulated armature assembly and method of encapsulating an armature assembly.

[P138] US6849969. Kang H. and Jeon J. (2005). Transverse flux motor with permanent magnet excitation.

[P139] US6849970. Watanabe K. (2005). Linear motor.

[P140] US6856049. Hirata A. (2005). Voice coil linear actuator, apparatus using the actuator, and method for manufacturing the actuator.

[P141] US6867512. Delaire G. and Gagnon F. (2005). Actuator having permanent magnet.

[P142] US6873066. Yammaoto T. and Maeda T. (2005). Linear motor.

[P143] US6879064. Kobayashi M., Kita I., Morita I., Inagaki K. and Inoue A. (2005). Linear motor and linear-motor based compressor.

[P144] US6879066. Hashimoto A., Kimura Y., Nakahara Y., Watarai A. and Nishitani S. (2005). Linear motor.

[P145] US6888269. Arndt A., Canders W.R., Mascher K., Schuler K., Wisken H., May H. and Weh H. (2005). Magnetic linear drive.

[P146] US6919654. Harned T.J. and Huard S.R. (2005)Linear motor with magnet rail support, end effect cogging reduction, and segmented armature.

[P147] US6922025. Smith J.F. (2005). Zero ripple motor system.

[P148] US6930411. Berghouse M., Jelusie M., Kaiser M., Lambrecht M. amd Wehner E. (2005). Linear motor.

[P149] US6930412. Chang E.S., Lin C.Y., Hsu C.M. and Kung H.J. (2005). Closed box structure of the horizontal linear motor machine tool.

[P150] US6930413. Marzano (2005). Linear synchronous motor with multiple time constant circuits, a secondary synchronous stator member and improved method for mounting permanent magnets.

[P151] US6977450. Asou T. and Aida T. (2005). Linear motor and linear guiding apparatus.

[P152] US6977451. Onishi Y. (2005). Ironless AC linear motor.

[P153] US6983701. Thornton R.D. and Clark T.M. (2006). Suspending, guiding and propelling vehicles using magnetic forces.

[P154] US7015613. Lilie D.E.B. and Vollrath D. (2006). Linear electric motor.

[P155] US7042119. Miyashita M. and Horikoshi A. (2006). Linear motor.

[P156] US7071584. Kawano S., Honda Y. and Murakami H. (2006). Linear motor.

[P157] US7091679. Schroeder R., Gronewold R., Kohler C., Stevens R. and Michalske G. (2006). Magnetic trust motor.

[P158] US7109610. Tamai H. (2006). Integrated wireless linear motor.

[P159] US7154198. Kawai Y. (2006). Linear motor.

[P160] US7166938. Kang H. and Kim J. (2007). Horizontal and vertical transportation system using permanent magnet excited transverse flux linear motors.

[P161] US7170202. Watarai A., Naka K., Itou K., Inoue M. and Nakamoto M. (2007). Linear motor.

[P162] US7176590. Fujimoto Y. (2007). Spiral linear motor.

[P163] US7199492. Hashimoto A., Kimura Y., Yamashiro S., Nakahara Y., Watarai A., and Michio N. (2007). Armature of linear motor.

[P164] US7201096. Cetinkun S., Egelja A.M. and Sorokine M.A. (2007). Linear motor having a magnetically biased neutral position.

[P165] US7220090. Wakazono Y., Imanishi K. and Achiva F. (2007). Linear motor operated machine tool.

[P166] US7230355. Lin H. and Heilig J.A. (2007). Linear hybrid brushless servo motor.

[P167] US7235936. Oba T. and Takahashi A. (2007). Linear vibration motor.

[P168] US7242117. Sugita S. and Misawa Y. (2007). Linear motor.

[P169] US7291941. Miyamoto Y. and Yamada T. (2007). Tandem arrangement linear motor.

[P170] US7312540. Miyamoto Y. and Inoue T. (2007). Linear motor armature and linear motor.

[P171] US7317266. Beakley B., MArsh J.J, Shaul R.T. and Juhasz B.L. (2008). Anti-cogging method and apparatus for linear motor.

[P172] US7336007. Chitayat A. and Faizullabhoy M. (2008). System and method to control a rotary-linear actuator.

[P173] US7339289. Wang J., Zhu Y., Cao J., Yin W. and Duan G. (2008). Synchronous permanent magnet planar motor.

[P174] US7339290. Sugita S. and Misawa Y. (2008). Linear motor.

[P175] US7378763. Jack A., Pennander L, and Dickenson G. (2008). Linear motor.

[P176] US7385317. Sugita S. and Tang Y. (2008). Linear motor not requiring yoke.

[P177] US7429808. Lehr H., Shrader S, and Walter S. (2008). Gliding field linear motor.

[P178] US7449802. Sasaki Y., Nakao T. and Shiratori K. (2008). Vibration wave linear motor.

[P179] US7456526. Teramachi A., Aso T., Tanaka Y., Kaneshige H. and Xu Y. (2008). Linear motor actuator.

[P180] US7456528. Thirunarayan-Kumar N. and Smith J. (2008). High performance linear motor and magnet assembly therefore.

[P181] US7471018. Nozawa H. and Narita T. (2008). Linear motor and manufacturing method of linear motor.

[P182] US7474019. Kang D.H., Chang J.H., Kim J.W. and Chung S.U. (2009). Permanent magnet excited transverse flux linear motor with normal force compensation structure.

[P183] US7482902. Kampf M. and Protze C. (2009). Linear magnetic drive.

[P184] US7514824. Kasahara S. and Ohno M. (2009). Sliding system with onboard linear motor.

[P185] US7531923. Shikayama T. and Irie N. (2009). Coreless linear motor.

[P186] US7538456. Miyamoto Y., Yamada T. and Koba T. (2009). Moving magnet type linear actuator.

[P187] US7573162. Onishi Y. (2009). Linear motor.

[P188] US7573170. Petro J.P. and Wasson K.G. (2009). Motor modules for linear and rotary motors.

[P189] US7576452. Shikayama T. and Sadakane K. (2009). Coreless linear motor and canned linear motor.

[P190] US7582991. Sugita S. and Misawa Y. (2009). Linear motor.

[P191] US7586217. Smith J.F. and Thirunarayan K.N. (2009) High performance motor and magnet assembly thereof.

[P192] US7595571. Thirunarayan K.N. and Smith J.F. (2009) High performance motor and magnet assembly thereof.

[P193] US7604520. Kotlyar O. (2009). Electrical linear motor for marine propulsion.

[P194] US7619377. Yamada K., Maemura A., Morimoto S., Shikayama T., Tanaka Y. and Takaki M. (2009). Linear motor system.

[P195] US7622832. Moriyama T. (2009). Linear motor and linear moving stage device.

[P196] US7626308. Kang D.H., Chang J.H. and Kim J.W. (2009). Permanent magnet transverse flux motor with outer rotor.

[P197] US7659641. Miyamoto Y. and Koba T. (2010). Moving magnet type linear slider.

[P198] US7675202. Huang B. (2010). Isotropic ring magnet linear voice coil motor.

[P199] US7696651. Miyamoto Y. (2010). Linear motor.

[P200] US7701093. Tang Y. and Sugita S. (2010). Linear motor.

[P201] US7732951. Mukaide N. (2010). Moving-magnet type linear motor.

[P202] US7745963. Jenny A. (2010). Linear motor with integrated guidance.

[P203] US7786631. Sasaki T., Watanabe K. and Ishikawa K. (2010). Linear motor.

[P204] US7812482. Aso T., Miyamoto T. and Yamanaka S. (2010). Rod-type linear motor.

[P205] US7825548. Maemura A., Kawazoe Y., Suzuki T. and Fukuma Y. (2010). Cylindrical linear motor.

[P206] US7825549. Wang X. (2010). Linear motor with reduced cogging.

[P207] US7834488. Finkbeiner M., Guckel J. and Thomas F. (2010). Electric linear drive unit.

[P208] US7834489. Matscheko G. and Jajtie Z. (2010). Synchronous linear motor.

[P209] US7839030. Tang Y. and Sugita S. (2010). Linear motor.

[P210] US7847442. Rohner R., Hitz M. and Ausderau D. (2010). Linear motor.

[P211] DE2002081. Kudermann K. (1971). Eletrischer Antrieb für Lastenfoerderer.

[P212] JP2002199679. Fukishima A. (2002). Inductor type electric machine having magnet equipped armature.

[P213] FR2840124. Cavarec P.E. and Gergaud O. (2003). Moteur lineaire cylindrique.

[P214] PL206292. Tomczuk, B. and Waindok, A. (2006). Multiphase tubular linear motor (in Polish).

[P215] GB2450465. Iwasaki S., Deodhar R.J. and Zhu Z.Q. (2008). Switching flux permanent magnet electrical machine.

Index